环境公共治理丛书

本书为中欧环境治理项目成果（项目编号：EuropeAid/132-005/L/ACT/CN）

本项目成果不代表欧盟观点

环境保护公众参与的国际经验

林卡　吕浩然　主编

中国环境出版社·北京

图书在版编目（CIP）数据

环境保护公众参与的国际经验 / 林卡，吕浩然主编.
—北京：中国环境出版社，2015.12
（环境公共治理丛书）
ISBN 978-7-5111-2452-4

Ⅰ.①环… Ⅱ.①林… ②吕… Ⅲ.①环境保护—研究 Ⅳ.①X

中国版本图书馆CIP数据核字（2015）第149487号

出 版 人	王新程
责任编辑	周 煜　宋慧敏
责任校对	尹 芳
封面设计	彭 杉

出版发行　中国环境出版社
　　　　　（100062　北京市东城区广渠门内大街 16 号）
　　　　　网　　址：http://www.cesp.com.cn
　　　　　电子邮箱：bjgl@cesp.com.cn
　　　　　联系电话：010-67112765（编辑管理部）
　　　　　　　　　　010-67112738（管理图书出版中心）
　　　　　发行热线：010-67125803，010-67113405（传真）

印　　刷	北京中科印刷有限公司
经　　销	各地新华书店
版　　次	2015 年 12 月第 1 版
印　　次	2015 年 12 月第 1 次印刷
开　　本	787×960　1/16
印　　张	23.75
字　　数	380 千字
定　　价	78.00 元

【版权所有。未经许可请勿翻印、转载，侵权必究。】
如有缺页、破损、倒装等印装质量问题，请寄回本社更换

《环境公共治理丛书》

指导委员会

主　任：环境保护部政策法规司副司长　别涛
　　　　环境保护部环境与经济政策研究中心副主任、中欧环境治理项目主任　原庆丹
　　　　浙江省环境保护厅副厅长　卢春中
副主任：中欧环境治理项目欧方主任　Dimitri de Boer
委　员：商务部国际经贸关系司一等商务秘书　陈红英
　　　　环境保护部政策法规司法规处副处长　李静云
　　　　环境保护部环境与经济政策研究中心战略室主任、中欧环境治理项目执行主任　俞海
　　　　浙江省环境保护厅政策法规处处长　叶俊
　　　　浙江省环境保护厅科技与合作处处长　李晓伟
　　　　浙江省环境宣传教育中心主任　潘林平

《环境公共治理丛书》编委会

主　编：虞　伟

编　委：（按拼音排序）

Hinrich Voss（英国）　　Laurent J.G. van der Maesen（荷兰）

Neil Munro（英国）　　钱水苗　　王崟屾　　朱狄敏　　朱海伦

总 序

生态文明建设是建设中国特色社会主义的重要内容,事关民生福祉和民族未来,党中央、国务院高度重视生态文明建设,先后出台了一系列重大决策部署,推动生态文明建设取得了重大进展和积极成效。但总体上看我国生态文明建设水平仍滞后于经济社会发展,资源约束趋紧,环境污染严重,生态系统退化,发展与人口资源环境之间的矛盾日益突出,已成为经济社会可持续发展的重大瓶颈制约。

当前,中国正面临环境与发展的新常态,即将到来的"十三五"是实现国家环境保护战略目标的关键时期,要破解发展中遇到的环境问题,既要我们腾笼换鸟,凤凰涅槃,推动经济绿色转型,更要强化国家环境治理体系顶层设计,给出中国环境治理体系和治理能力现代化的方向和路径。

中欧环境治理项目是中欧环境政策对话和欧盟中国战略文件确定的重点项目,项目由欧盟出资1500万欧元,支持中国政府在环境公共治理领域开展合作对话,环境保护部作为项目主管单位并授权环境保护部环境与经济政策研究中心承担项目管理与执行工作。项目聚焦环境信息公开、公众参与环境规划与决策、环境司法以及企业环境社会责任四个领域。旨在通过国家层面为中央政府直接提供政策支持,并在地方层面开展创新试点,从而推进中国环境治理体系和治理能力现代化。

作为中欧环境治理项目 15 个地方伙伴关系项目之一——嘉兴模式中的公众参与环境治理及其在浙江的可推广性项目(项目号:

EuropeAid/132-005/L/ACT/CN），由浙江省环境宣传教育中心作为项目主要承担单位，浙江大学、浙江工商大学、英国格拉斯哥大学、英国利兹大学、荷兰国际社会质量协会等作为合作伙伴。项目实施期为 2012 年 9 月至 2015 年 3 月，共计 30 个月。项目通过分析"嘉兴模式"价值、特征，建立一套公众参与环境决策过程的新机制，并将该机制在浙江省或者全国推广。在中外项目执行团队成员的共同努力下，通过中欧互访、调研、讲座、培训、研讨会等形式，加强了中欧双边的合作伙伴关系，并形成了一系列成果，推动了"嘉兴模式"的传播和推广，提交的政策建议报告也为环境保护部和浙江省地方立法作出了贡献。

项目结束之际，我们将研究成果整理汇编成 5 册，分别是：《中国环境保护公众参与：基于嘉兴模式的研究》《环境保护公众参与："嘉兴模式"调查报告》《嘉兴模式与浙江省公众参与环境保护的机制构建》《公众参与环境保护：实践探索和路径选择》《环境保护公众参与的国际经验》。这些成果通过中欧学者的合作研究，站在国际、国内两个视角下对"嘉兴模式"进行了全方位的解读，揭示了中国特色社会主义政治体制下，如何改变政府社会管理方式，实现各社会主体之间的利益协调，通过培育社会组织建立政府部门与民间力量的有效互动机制来化解环境利益的冲突，实现政府与公众良性互动的现实途径，这为研究当下中国地方环境协商民主进程和环境治理能力现代化建设提供了范例。

由于时间仓促和水平有限，不足之处敬请读者批评指正。

编　者

2015 年 11 月

前　言

随着经济的快速发展，环境问题正日益突出，有关环境参与的讨论逐渐流行起来，并成为公众关注的重点议题。人们开始不断批判以环境为代价追求经济成长而造成的恶果，把环境治理看成是与社会质量和生活质量紧密相关的问题。为此，中国政府强化了其在环境治理方面所作出的努力，并出台了一系列环境保护法律法规来应对由环境污染问题所带来的挑战。当然，环境问题不仅仅是政府治理的问题，更是社会管理的问题。这一问题的解决不仅仅要依靠政策制定和技术手段的采用，更需要公民的广泛参与。随着工业化的发展和城市化的推进，人们日益感受到生活环境与其切身利益息息相关，环境治理需要广大民众的介入和广泛参与。在此背景下，环境民主这一概念逐渐为民众所接受，并已投身到环境治理的事业中去探索各种参与途径，政府也在积极推进公众对环境治理的监督和社会创新活动。

要推进这一过程，我们需要强化公民环境参与意识和相关实践，也要积极学习欧美国家环境治理的经验。这些国家工业化进程历史悠久，环境污染问题出现得较早，从而针对环境污染所采取的社会行动发生得也较早。在其后推进现代化的进程中，这些国家也在环境治理方面积累了许多成功经验。学习这些经验对中国的环境治理的发展具有积极意义。中国工业化、城市化进程起步较晚但推进速度很快，是在一个压缩了的时空中追求现代化。这使得欧美国家在工业化早期中遇到的一些问题和后工业化时代中所出现的问题可能在中国会同时出现。这种特点意味着欧美国家在环境民主进程中的早期和近期所取得的不同经验，对于中国目前应对环境挑战都会具有借鉴作用。通过对欧美国家经验的学习和讨论，可以帮助我们寻找中国应对环境问题的解决途径。

同时，环境治理不仅是各国各自面临的问题，同时也是一个全球问题。环境治理是一项基于人类共同使命的全球化事业。在我们共同居住的地球村上，一地发生的生态灾难会影响到地球的其他角落。为此，强化国际交流，进行研究互动并把环境治理作为一个全球化议题来讨论势在必行。通过国际交流去寻找环境治理的有效方法以解决我们所面临的共同问题，是应对环境问题全球化的客观要求。为此目的，本书将围绕环境治理的公众参与问题，并对欧美各国所取得的经验展开讨论，也回顾这些国家在环境参与方面的一些理论、观点和政策。全书由以下论文所构成：

在《公众环境参与的欧洲经验》一文中，Guanli Zhang、Ka Lin 和 Neil Munro 界定了欧洲公众参与环境治理的理论、概念、途径和方法等概念，从而为全书的展开提供了理论基础。它也为本书界定公众参与所涉及的议题领域提供了一般框架和基本范围。在这一讨论中，作者涉及了有关欧洲环境民主议题讨论的核心议题，如参与途径、实践经验以及政策模式等，并强调影响这些进程的政府和社会的关系。

随后，Sander Chan《环境保护的自愿参与视角：来自欧盟的经验》的研究讨论了如何通过政府法规推进企业进行环境保护的自愿行动。该研究针对环境保护的自愿行动展开，从政策研究的视角讨论了欧盟的环境政策相关议题及如何通过各国政府立法和政策制定来推进公民的环境参与。Hinrich Voss 的《英国的环境公众参与概况》则力图展开个案研究。它以英国为例，讨论了环境问题中公众参与的核心问题，这些讨论具有较强的理论内涵，通过对个案的讨论来阐发公众参与的过程、关键问题和社会影响力。

Juha Kaskinen 教授对芬兰的环境规划和决策中公众参与的影响的研究也为我们提供了研究环境公众参与的范例。在此研究中作者描述了芬兰环境保护政策的制定及其机制，并倡导通过信息网络和以众包的形式来强化环境信息的披露及其环境信息的传播，并把众包这一形式作为公众参与环境的重要途径。由斯德哥尔摩国际水研究院的学者提供的研究论文《水资源管理中的公众参与——欧盟水框架的实践与经验》，关注水环境保护的公共参与问题。该研究以公众参与的各种途径和方法强调了利益相关者在公众参与的治理中所扮演的作用。该研究特别讨论了三个个案：一是瑞典等波罗的海国家的水资源保护个案，二是德国莱茵河的水资源保护，三是苏格兰的个案。这些研究强调了公众在水资源保护方面的作用及其影响。

Kai Wang 和 Laurent J.G. van der Maesen 关于海牙公众参与环境治理的经验和 Gentian Qejvanaj 与 Ka Lin 所提供的中意环境治理的比较研究则蕴含了国际比较的视野。Wang 和 Maesen 倡导基于社会主体的分析视角对环境公众参与展开研究，强调在此过程中分析政府、学界、社会网络和私营部门各方所具有的参与动机及其互动的结果。这一研究基于荷兰海牙的公众参与经验来展开，并在分析的过程中揭示公众参与的社会基础和政治驱动力。Gentian Qejvanaj 和 Ka Lin 的论文则讨论了欧盟环境决策和欧盟各国对该环境政策的执行状况。通过对意大利的中央、地方和欧盟这三者关系的分析，揭示了意大利作为欧盟国家在执行和推行环境保护政策方面所面临的问题和政策驱动力。通过与中国体制的比较，意大利体制所具有的这些特征能够更为清晰地体现出来，从而反映了欧盟各国在执行欧盟的法令中所面临的矛盾。

在欧盟国家之外，本书也涉及美国的环境治理和公众参与。Tseming Yang 的论文讨论了美国的环境信息披露制度及其对公众参与的激励作用研究。它倡导信息公开、环境教育，并注重利益相关者的利益协调等因素的重要性。由于信息披露是公众参与环境治理的前提条件之一，该研究说明了美国在信息披露方面所制定的规则对环境治理行动的影响。最后，Peter Herrmann《环境民主——新的挑战》一文从理论的视角把公众参与环境治理与社会质量的考察相结合，讨论了环境民主对于全球可持续性发展所具有的重要意义。该论文强调环境民主与全球问题的关联和经济发展与社会质量问题的关系。同时，该研究也反思了人类进步、人的权利（包括环境的权利）以及应对全球挑战的策略。这些讨论从宏观角度阐述了相关议题，强调了本书的主旨并进行了理论升华，为全书的讨论提供了很好的结论。

在本书中，我们可以看到许多研究都是通过欧洲学者和非欧洲学者或中国学者和外国学者合作研究的方式形成的。因而本书的成果不仅会让中国读者受益，也会给国际读者以启发。鉴于此，为便于中国学者和外国学者方便使用这些研究成果，本书用中英双语形式编写。我们期望该书作为欧洲学者和非欧洲学者互动的成果，能够对推进公众参与环境治理的理论作出一定的贡献，也为中国学者了解欧洲经验提供参考。

Preface

In recent years the Chinese government has paid increasing attention to environmental problems. Along with the fast pace of economic development, environmental problems have becoming more pressing by the day, and environmental participation has gradually become a popular topic for discussion including by the general public. People have begun to criticize the negative consequences of the pursuit of economic growth at the cost of the environment, and to regard environmental governance as a problem which is closely connected to social quality and the quality of life. To deal with this, the Chinese government has strengthened its efforts in environmental governance and a series of environmental protection laws and regulations have been passed in order to deal with the challenge posed by environmental pollution. Of course, the environmental problem is not only a governance problem, it is also a problem of social management. Solving this problem requires not only policy making and the application of technical solutions but also citizens' participation. In pace with industrial development and the progress of urbanization, people are increasingly feeling the close connection between the environment they live in and their own interests. Environmental governance requires the involvement and broad participation of the population. Against this background, the concept of environmental democracy has gradually been accepted by the population, it has been introduced into the conduct of environmental governance as a way of creating various paths for participation, and government has also enthusiastically promoted public supervision of environmental governance and activities relating to social innovation.

In order to drive this process forward, we need to strengthen citizens' awareness of environmental participation and related practices, and also to enthusiastically study the experience of European and North American countries in environmental governance. These countries have a long history of progress with industrialization, the problem of environmental pollution appeared relatively early, and therefore social actions directed against environmental pollution also occurred relatively early. In the subsequent process of modernization, these countries also accumulated a great deal of successful experience in environmental governance. Studying these countries' experience has great relevance to the development of China's environmental governance. The process of China's industrialization and urbanization began relatively late, but has progressed very fast, involving the pursuit of modernization within a compressed time frame. This means that certain problems encountered in the early stages of industrialization in European and North American countries and problems which appeared in the post-industrial period may in China appear simultaneously. This means that both early and recent diverse experiences in processes of environmental democracy in Europe and North America can provide lessons for China in meeting its present environmental challenges. Studying and discussing European and American experience can help us to find ways to solve environmental problems in China.

At the same time, environmental governance is not just a problem which each country faces individually, but rather it is a global problem. Environmental governance is a globalized enterprise, based on humanity's common purpose. In the global village which we inhabit together, an ecological disaster in one place can affect other parts of the globe. For this reason, it is imperative to strengthen international exchange, carry out research cooperation and to discuss environmental governance as part of a globalized agenda. With this recognition, this research will initiate a discussion of international experiences in environmental governance, and recall the theory, perspectives and policies of European and American experience of environmental participation. The research is composed of the following sections:

In "European Experience of Public Participation in Environmental Governance", Guanli Zhang, Ka Lin and Neil Munro outline the theory concepts, paths and methods of European public participation in environmental governance, and in this way provide a theoretical foundation for the book. They also define the general framework and scope of the areas involved in citizen participation. In this discussion, the authors touch on the core concepts involved in European discussions of environmental democracy, such as channels of participation, practical experience, policy models and so on, whilst emphasizing the government-society relationship which influences these processes.

Next, Sander Chan's "Voluntary Environmental Approaches – Experiences in the EU" discusses how to promote enterprises' voluntary environmental protection actions by means of governmental regulations. This research concerns voluntary environmental protection activities; it discusses the relevant European Union environmental policies from the point of view of policy research as well as how to promote citizens' environmental participation by means of legislation and policy-making in each member state. Hinrich Voss's "Environmental public Participation in the UK" presents a case study. Taking the UK as an example, it discusses core questions of public participation in environmental governance. His discussion has quite strong theoretical implications, and by means of individual examples illustrates processes of public participation, key problems and societal influence.

In addition, Juha Kaskinen has provided us with a fine example of how to study environmental public participation with a study on the influence of public participation on environmental regulations and policy-making in Finland. In this research, he describes Finland's environmental protection policy-making process and institutions and advocates the use of the internet and crowdsourcing to strengthen environmental information disclosure and dissemination, making crowdsourcing an important path of citizens' environmental participation. The paper "Public Participation in Water Management— EU Practices and Experiences" by scholars at the Stockholm International Water Institute focusses on public participation in protection of the aquatic environment. Through an exploration of the water governance agenda, this research emphasizes the roles played by stakeholders in participatory governance through diverse methods and channels of public participation. Moreover, this research explores three particular cases:

the first is the management of aquatic resources by Baltic countries like Sweden; and the second is the protection of aquatic resources in the Rhine River in Germany, and the third is the case of the North Sea and Scotland. This research emphasizes the role and influence of the public in protecting aquatic resources.

Comparative research by Kai Wang and Laurent J.G. van der Maesen on experience with public environmental participation in the Hague and Gentian Qejvanaj and Ka Lin's comparison of Chinese and Italian environmental governance bring an international comparative perspective. Wang and van der Maesen advocate an approach to analyze from the point of view of social subjects, emphasizing motivations for participation and the interactions of government, academia, social networks and the private sector. This research is based on the experience of the Hague in the Netherlands, and in the course of the analysis it uncovers the social basis and political drivers of environmental participation. Gentian Qejvanaj and Ka Lin's paper discusses European environmental policy and its implementation in different countries. Through an analysis of central government, local and European Union relations in Italy, it uncovers the challenges and policy drivers faced by Italy as a European Union member state in implementing and promoting environmental protection policy. Through a comparison with the Chinese system, the features of the Italian system are revealed more clearly, reflecting the contradictions faced by all countries in implementing European Union laws and regulations.

Apart from European Union countries, this book also touches on American and Thai experience with environmental governance and public participation, specifically through Tseming Yang's "The US System of Environmental Information Disclosure and its Role in Encouraging Public Participation". Yang's paper explains the benefits of information disclosure regulations for environmental governance. It advocates information openness, environmental education, as well as the importance of such factors as the harmonization of the interests of stakeholders. Since the availability of enviornmental information is one of the preconditions of public enviornmental participation, this paper discusses the inflence of establishing the rules about information discloure in US on people's actions of enviornmental protection. Finally, Peter Herrmann's "Environmental Democracy—New Challenges" investigates from a theoretical perspective public participation in environmental governance and social quality, and from the point of view of development studies discusses the importance of environmental democracy for global sustainable development. This research emphasises the linkages between environmental democracy and global issues and the relation of economic development to social quality. At the same time, this research reflects on human progress, human rights (including environmental rights) and strategies for dealing with global challenges. The discussion sets forth the relevant agenda from a macro perspective, emphasising the purport of this book and distilling relevant theory to provide a very satisfying conclusion.

In this research, we can note that a great deal of work has been done on a cooperative basis: European scholars with non-European scholars, scholars from developing countries with scholars from developed countries, and Chinese and foreign scholars have pushed the research forward through common efforts. As a result, the fruits of this research will not only benefit Chinese readers, but also give inspiration to international readers. With this in mind, this book has been edited in two languages so that Chinese scholars and foreign scholars can both use it. We hope that this research, as the fruit of joint efforts by both European and non-European scholars, can make a certain contribution to the theory of public participation in environmental governance, and can provide basic knowledge for Chinese scholars to understand European experience.

目 录

公众环境参与的欧洲经验 .... 张冠李　林卡　尼尔·曼路　1
European Experience of Public Participation in Environmental Governance
.... Guanli Zhang，Ka Lin，Neil Munro　25

环境保护的自愿参与视角：来自欧盟的经验 .... 陈文山　69
Voluntary Environmental Approaches—Experiences in the EU .... Sander Chan　89

英国的环境公众参与概况 .... 辛里奇·沃斯　123
Environmental Public Participation in the UK .... Hinrich Voss　141

实践中的环境民主——芬兰环境规划及决策中的公众参与
.... 尤哈·卡斯基宁　165
Environmental Democracy in Practice—Citizen Participation
　　in Finnish Environmental Planning and Decision-Making ...... Juha Kaskinen　172

水资源管理中的公众参与
　　——欧盟水框架的实践与经验 .... 斯德哥尔摩国际水研究院　183
Public Participation in Water Management—EU Practices and Experiences
.... Stockholm International Water Institute　210

海牙公众参与环境治理的"波德模式" .... 王恺　劳伦特·冯·米森　250
The "Polder Model" of Public Participation in Environmental Governance
　　in The Hague .... Kai Wang and Laurent J.G. van der Maesen　263

作为规范导引和作为政策现实的环境政策：中意环境治理的
　　比较研究...根亭　林卡　281
Environmental Policies as Normative Guide and as Realities of Policy
　　Implementation: A Comparative Study of Environmental Governance
　　in Italy and China ....Gentian Qejvanaj and Ka Lin　295

论美国的环境信息披露制度及其对公众参与的激励作用...................杨泽明　316
The US System of Environmental Information Disclosure and its Role in
　　Encouraging Public Participation..................................Tseming Yang　324

环境民主——新的挑战 ....彼得·赫尔曼　336
Environmental Democracy—New Challenges ....Peter Herrmann　344

后　　记..359
致　　谢..360

公众环境参与的欧洲经验

张冠李　林卡　尼尔·曼路*

导　言

从1973年罗马俱乐部强调环境危机的理念并率先倡导可持续性发展的理念以来，人们对于环境问题的讨论已经从社会研究的边缘走向中心。这一变化一方面是由于工业化进程加快、广大第三世界国家面临的经济发展压力加大，以及全球社会治理理念的深入等因素造成的；另一方面，它也是人类日益恶化的生存环境状况的真实反映。由此，1992年召开的联合国环境与发展大会通过了《21世纪议程》，并将其作为世界范围内可持续发展方面的行动纲领。该《议程》将改善自然环境列为21世纪实现可持续发展的关键目标，并要求各国政府制定相应的环境法规以缓解大气污染、水体污染、全球变暖、生态多样性危机，以及其他生态灾难所造成的社会压力及社会风险。同时，保护自然环境这一要求倡导全人类行动起来，把各种社会行动有机地融入全球治理中。在这一背景下，公众环境参与、环境信息公开等议题也已成为当前全球关注的焦点问题。

倡导环境治理的公众参与就要求把对环境问题的讨论从技术领域拓展至社会领域。公众参与环境治理有着多重意义：它既有助于加强对环境的监管、促进污染防治技术的发展，也可以有效地组织社会舆论并在某种程度上形成社会压力来推动政府与相关机构的行动。同时，公众参与环境治理本身也是一个不断强化自身环境观念、改造其行为模式并形成新生活习惯的过程。这一过程中，人们通过

---

*张冠李，牛津大学圣安东尼学院地理与环境专业博士生；林卡，浙江大学公共管理学院教授；尼尔·曼路，格拉斯哥大学中国政治学讲师。本文原文为英文，中译者为张育琴（浙江大学公共管理学院）。

调整自身行为来影响企业与社会组织的行为，达到重塑行为规范、形成环境友好为导向的社会文化的目的。因此，环境治理不应仅仅是一个技术与管理层面的问题，更是一个社会层面的问题。要实现良好的环境治理，公众参与是必不可少的，是我们解决环境问题的基本立足点。只有通过全社会的共同努力，我们才能够应对环境对人类生存与社会发展的挑战。

基于这种认识，我们要对于公众参与环境治理的途径、方式、成效等问题进行多方面的研究，并对二者的内在联系展开讨论。这些研究要包含从理论的角度界定环境治理公众参与的范畴与范式；探究公众参与环境治理的具体形式、实践经验及相关政策；并基于实证研究探索公众参与环境治理的成功经验。这些研究要能够对于公众参与环境治理的议题具有理论与实践的双重意义。当然，对环境问题的研究属于交叉学科，其研究议题所涉及的领域十分广泛，议题众多，因此对该议题的研究需要形成一定的聚焦点。

目前，人们对于环境问题的研究大多处在技术层面（特别是大气污染的测量、海水污染的监控、土壤污染的监测、生物多样性下降的风险评估、工业污染的监测及生产过程的环境效应等技术），也存在大量从管理学角度探究环境保护立法及其法律法规制定合理性的研究，但从社会维度展开对环境治理的议题研究仍有待进一步发展。从这一立足点出发，我们需要考察环境治理的参与进程、各类社会群体的互动、社会规范与公共政策行动的关系等问题。

采取社会视角来开展有关环境治理的研究就要关注公众参与议题。当前，公众参与环境治理已作为为各种国际组织所倡导的一般法则，"参与式治理"的理念也体现在联合国、世界银行等国际组织的政策纲领和发展研究报告中。由于不同国家与地区的工业化程度与市场化程度存在差异，其经济发展水平参差不齐，人们的环境意识与其相应的社会行动也不尽相同。公众的环境参与理念在各国所形成的影响力及其导致的政策进程会有差异。因此比较各国环境治理的实践（特别是公众参与的模式与程度），可以为我们理解环境的社会治理提供很好的经验基础。各有千秋。

根据一些比较研资料，较早实现工业化的国家对于其在工业化进程中所付出的环境代价的认识较为深刻，在环境保护领域所采取的行动也较为彻底。而经济发展相对滞后的国家，工业化对其所造成的环境影响较为明显而集中地呈现在人们面前，环境保护的任务十分迫切。然而，由于这些国家要在很短的时间内走完发达国家近百年的工业化历程，这一任务十分艰巨，而且其在发展进程中所面临的环境压力很大。因而，他们采取的应对环境问题的策略也有所不同。有的国家认同先发展后治理的思路，把环境治理的责任向后推延；也有的国家尽管工业化程度并不高（如许多南亚国家），但环境恶化所引起的问题（包括土壤贫瘠化、水资源的污染等）已直接威胁到了人民的生存，迫使其很早就采取了一定的措施改善环境。不过，无论是发达国家还是发展中国家，他们在应对环境问题的挑战时各自取得的经验，都可以相互交流。各国通过交流来构建政策的框架，并从中各取所需，寻找适合本土实际的对策。

  通过这些交流，欧洲国家所取得的经验对环境问题的全球性探索具有重要的意义。首先，这些国家的工业化进程较早，其环境问题已充分暴露，因此它们在环境治理的实践中积累了许多经验。以上世纪中叶的伦敦为例，其以大雾弥漫作为城市标签而被称为"雾都"；而今通过不断的环境治理行动（包括工厂搬迁、城市规划、建设环境友好型社会的相关运动等），其环境状况已经得到很大的改善。同样，德国的莱茵河也曾因为受污染严重被称为"黑水河"。经过近半个世纪的治理，莱茵河重新恢复良好的水质状况。种种历史表明，欧洲各国在工业化进程中都经历过以环境为代价追求发展从而引起严重社会问题的阶段。正是由于这种经历，欧洲各国的环境风险控制意识与环境保护意识也随之强化，以至于在上世纪70年代末，一些欧洲学者就强调经济增长与环境可持续发展对于人类发展的意义。而在一些发展中国家，这些意识直至70年代以后才逐渐被政策制定者所关注。因此，欧洲国家的发展历程以及由此所产生的环境保护意识、环境治理经验可以为许多后发国家借鉴，使我们在面对环境治理的问题上少走弯路。

  其次，公众参与环境治理与民主化进程息息相关。公众参与环境治理需要一些前提条件，包括赋权（界定公众参与环境治理的权责）、教育（普及相关环境知识）、信息透明化（开放获取环境状态、环境风险等信息的渠道以便民众评估环境、发表意见）和机制（通过机制保证公民的参与渠道，从而使得公众参与环境治理成为可能）。从这些方面来说，战后逐步完善的欧洲社会民主机制为发展公众环境参与和环境民主提供了政治基础和制度条件。这可以部分解释欧洲在公众参与环境治理中所具有的优势。在欧洲一些国家，人们可以通过参与环境政策的制定与实施来形成良好的环境意识,使之成为一种大众意识形态(如当前绿色主义理念在欧洲的广泛传播)。以此为基础,进而确立诸如环境权利等概念(在欧洲公民环境基本权利已得到法律保障)。上述民主制度与法制规范使得欧洲许多国家可以在环境问题上的公众参与实践方面走在了世界前列,成为世界其他国家与地区借鉴的蓝本。

最后,作为一个全球性的问题,环境治理与全球化的进程密切相关(Naughton,2010)。欧洲各国所进行的区域一体化是全球化的重要因素。欧洲对区域环境政策制定所做出的努力也是国际合作治理环境问题的一部分。从国际经济循环的角度来看,欧洲国家对环境问题的共同关注由两方面的因素所推进:第一,由于实现工业化后欧洲人民的生活水平得到普遍提升,因此对于环境标准有了更高的要求。这种高标准使得欧洲的环境保护事业能够走在世界前列。第二,由于欧洲的污染密集型产业和原材料加工逐渐向边缘国家转移,使得欧洲能够以较低的环境代价享有当前的社会发展与生活水平。当然,这种工业与技术的全球化转移不可避免地为许多快速工业化的发展中国家付出了环境与劳动力成本的代价。为此,发达国家对发展中国家的温室气体排放以及其他环境指标施加压力,并无形中提高了人类社会对环境保护的总体标准,由此形成了相应的政策倡导(EC,2000,2001,2003a,2003b)。欧洲各国也充分认识到其权责对等的重要性,并通过与发展中国家的积极合作,在技术、资金领域对其环境治理予以支持。欧盟各国成为推进发展中国家环境治理的主要力量。

由此,在讨论公众参与环境治理这一议题时,我们有必要回顾欧洲各国在政策及实践领域中的经验教训,并通过对于欧洲经验的学习来阐释欧盟在环境社会治理上所取得的成功,达成具有借鉴意义的政策评估。在全球环境保护的公众参与事业中,这些努力是基于各国对于全球的环境保护重要性的认识所决定的。本研究将从价值导向、法律法规、政策维度与社会实践四个方面阐释欧洲经验,并对此做出相应的分析。在展开讨论前,我们首先回顾公众参与环境治理的相关概念及基本理论。

公众参与环境治理的相关概念与理论

公众参与环境治理的理论基于对环境权利、环境民主与参与模式三个概念的界定。在欧洲，这些概念为欧洲环境法令与环保项目所反映，并进行发展和创新。欧洲国家先进的环境治理理念为其公众参与环境治理的制度的确立打下了坚实的理论基础。这些概念与理论在世界其他国家与地区的环境治理实践中也起到了一定的指导作用。

1. 环境权利

1972年，联合国人类环境会议在斯德哥尔摩召开，并通过了《人类环境宣言》。该《宣言》认为，人类享有自由、平等和充足生活条件的基本权利，也应该享有尊严地生活在这样环境中的权利。此后，在维也纳召开的欧洲环境部长会议进而制定了《欧洲自然资源人权草案》，把环境权利确认为基本人权的一部分。在随后40多年的环境治理实践中，环境权利这一概念也得到了不断的充实与发展。

欧洲语境下的基本环境权利包含环境知情权与环境参与权等方面。《奥胡斯公约》指出：对公民基本环境权利的保障是实现环境正义的基础（de Abreu Ferreira，2010）。公民可以通过以下三个途径享受环境正义的权利：环境信息的披露的保证、公众参与环境政策的制定以及公众质疑环境政策的渠道。在此，实现环境正义的三个途径恰是对公民环境知情权与参与权的概括（Mares，2011）。

环境知情权是指公民可以自由获取其居住所在的社区、工作单位以及更广泛生活区域内的环境信息的权利。《奥胡斯公约》第三条对"环境信息"做出界定：它不仅包括环境的组成实体（空气、水、土壤、地貌、资源以及生物多样性等），还包括影响环境的要素（能源、噪声、辐射、人类活动、环境协议、政策、法律、环境开发项目等一切改变环境的因素）。除此之外，环境信息还指向前述环境实体、环境要素与人类健康、安全的关系。《奥胡斯公约》责成拥有相关信息的公共机构以多种渠道对公众进行信息披露，以便公民了解环境问题及风险评估情况（Aarhus Convention，1998）。此外，依法保障公民的环境知情权还有利于公民个体与社会力量及时发现问题，并将经验观察、地方环境知识与环境评估中的科学数据相结合，从而更为精准地定位问题，实现环境风险的早发现、早控制（Ayers，2011；Beckmann，2009；Hakala，2012；van Laerhoven，2014）。

作为公民基本环境权利的另一个组成部分，环境参与权在《奥胡斯公约》第一条中被定义为"公民及公民组织在环境问题中参与决策、寻求正义的权利"（Aarhus Convention，1998；另见 de Abreu Ferreira，2010）。这一概念源于欧洲国家鼓励社会参与立法法制的传统，并为近年来的学术讨论所推进。这些有关欧洲公众环境参与权的讨论主要集中在公民及公民社会组织参与环境治理进程的法理依据（cf. Glasbergen，2007；Ruggie，2004；Bäckstrand，2006；Biermann and Pattberg，2008；Khagram and Ali，2008；Andonova，2010）和参与机制的设计（cf. Harrison，2002；Edwards，2011；Mares，2011；Mehnen，2013；Roggero，2013）。这些理论探讨推动了欧洲各国通过立法与改革，保障其环境基本权利的发展，也成为各国环境治理体制设计的重要参考。

2. 环境民主

环境民主是描述公众环境参与进程的核心概念。由于环境问题覆盖面广、涉及人数众多，它较之于一般的社会问题包含更为多元化的利益诉求。在欧洲，其传统的自由民主模式鼓励利益集团间竞争，但这一模式对公民福利与利益的表达能力有限，无法充分地表达环境问题中利益相关者的诉求。因此，近年来，欧洲国家常常把"民主程序"引入环境治理的诉求（Mason，1999；Zeiter，2000）。依据环境治理公众参与的形式，欧洲环境民主可以被分为三类：参与式民主、商议式民主与交往式民主。在上述不同的民主理论框架下，学者对环境民主的概念与运作机制可以做出不同的解读。

首先，参与式民主倡导在政策制定过程中纳入公民的直接参与。相对于传统的代议制民主体制，这种民主体制更为激进，被称为"强民主"（Barber，1984，1988）。在参与式民主的模式中，（Fischer，1993）将环境民主定义为"通过将民众纳入决策从而解决环境问题的过程（Fischer，1993）"。倡导"增加参与"而非"减少参与"，强调"普遍性参与"而非"代表性参与"。因此，在解决环境问题时需要通过最广泛的公众参与使得其作为一个整体性问题得以解决。引入公民直接参与环境民主形式有助于定位风险，可以有效提升治理中对于环境问题的反馈效率。同时，这也被认为是促进个体层面环境正义的基本手段，并有助于同时实现公众教育与社会组织化两个目标。一方面通过社会参与培养公民环境意识，另一方面通过社会参与建立环境行动网群（Zeiter，2000）。

与激进的参与式民主不同，商议式民主强调参与者通过商议产生相互影响，在共同认知的基础上以公开批评或说服的方式促成符合公共利益的政策（Dryzek，1990；Zeiter，2000）。不少学者认为，环境问题可以通过商议的方式解决。由此，Dryzek（1992）给出了商议式环境民主的定义：在平等与相互尊重的基础上，通过长时期渐进地解决环境问题的模式。商议式环境民主被认为是一种理想化的环境民主类型。在此过程中，政府鼓励言论沟通并接纳多方努力，以协同的方式追求公共环境利益。不过，成功的商议一方面需要作为发言者的民主程序参与者有高度的逻辑性，另一方面也需要通过制度保证发言者的地位平等。据此，一切有悖平等原则的障碍均被排除，所有言论都能被充分聆听与尊重。

交往式民主肯定商议民主中"商讨"的重要性，但质疑商议民主的核心假设，即"商讨是一种普适的、价值中立的、去文化的、去民主性的民主过程。"Young认为，商议民主的不平等来自文化规范：特定的语言及言语方式包含着被主流文化规制所认可的权力，因而造就了商议参与者的天然不平等（Young，1990，1997）。例如，白人中产男性因其言语优势在商议民主中获得权力。由此，交往式的环境民主鼓励非商议性行为，以此使主流之外的参与者充分表达其环境领域的利益诉求。

在欧洲公民的环境参与实践中，参与式民主被认为是最为符合当代欧洲政治土壤与治理文化的模式（Robinson，1993；Dryzek，1992）。有学者指出，通过商议能够实现民主需满足一系列条件，包括社会平等、公民较高的教育水平、文化的同质性、一定的社会财富积累、现代的社会与文化规范以及政府权力的高度分化等，这恰好是欧洲国家发展状况的描述（Offe，1997）。基于实证案例的研究表明，"商议"以及商议式民主模式在传统社会以及发展中国家也具有较强的适应性（Gupte，2007；Omanga，2014）。

显而易见，作为公众参与环境治理的环境民主，无论处以何种模式中，其核心始终在于强调把多元参与主体纳入环境治理，并鼓励公民与公民社会发挥其独特的作用。可以说，围绕环境民主的争鸣恰是对公众参与环境治理的论证，这些讨论也丰富了欧洲公众环境参与研究的理论架构。

3. 参与模式

当前，欧洲的公众环境参与已形成了多途径、多主体、多形式的参与模式。这些模式内嵌于各国的政治制度与政治文化之中，并在实践中取得了良好的成效。

在治理途径方面，欧洲国家推崇政府、市场与公民社会的合作。欧洲传统的治理理论假设政府、市场与公民社会均有着明晰的行为边界，由于其各自独特的运作逻辑而无法实现"跨域"的管理（Glasbergen，2011）。Arnstein 的参与阶梯理论（Ladder of Partnership）则颠覆了对治理这一概念制度化、合法化、正式化的旧的解读。基于对公共领域与私人领域日渐融合的现状，Arnstein 认为，公共事务治理的合作途径正显示出强大的生命力。依据阶梯理论，随着各种治理参与主体的增入与合作方法的调整，治理合作的输出形式将由最初始的建立互信，经由确认共同利益、固化规则体制，逐级发展成为重塑市场与政治体制（Arnstein，1969）。这一合作式治理演化阶梯的解释力在欧洲国家的环境治理实践中屡屡得到验证（cf. Glasbergen，2011；Rydin，1999；van der Haijden and ten Heuvelhof，2012）。

就公民社会中的治理主体而言，除公民个体外，社区与绿色组织也作为重要的行为体参与并影响环境政策的制定与实施。社区作为环境治理的当地参与的主体力量，为环境政策的制定与实施提供了微观视角。基于社区日常的、非正式的信息获取途径，社区的环境治理工作往往能够较好地兼顾弱势群体与高危群体的环境权益保障（Hakala，2012）。如 van Laerhoven（2014）所指出，社区层次的环境政策制定能够显著提升环境公共服务的质量。这不仅因为社区自治组织对当地的环境需求更为敏感，还因为社区往往作为特定人群参与公共事务的场所，有时社区甚至是这些人群获取公共服务的唯一渠道。在欧洲社会社区力量作为环境治理不可或缺的主体，有着重要地位。另一方面，非政府组织也因其灵活机动与专业性强的特点在欧洲环境治理的实践中担负了执行监督、理念传播、专业知识提供以及矛盾调解等重要职能，并为传统的由政府主导的公共治理提供了良好的补充。

同时，欧洲公众参与环境治理的形式灵活多样。当前，欧洲公民社会通过圆桌会议、知识培训、行动动员会等多种方式，组织社会力量并发挥其特长，在环境治理的议题确立、规范制定、政策执行等环节起到了重要作用。此外，欧洲公民社会力量正着力完善社会技术监控网络体系的建设。当前的许多民众团体介入水文、$PM_{2.5}$的监测等事例均为这一努力的典型案例。这种公众参与的形式将技术与人力两方面的要素相结合，有着十分广泛的覆盖面。另需补充的是，欧洲的公众环境参与还力图促进环境治理的法制化进程，在荷兰、德国等国，市民陪审团开始介入环境污染案件的审理（Barber and Barlett，2006）。这是社会治理在司法层面的体现。

上述多途径、多主体、多形式的公众环境参与模式的实践已在欧洲环境治理的实践中形成积极成果。世界其他国家与地区践行环境的社会治理时，可以借鉴欧洲的这一参与模式。

价值导向的经验

以上述相关概念为框架，我们可以针对欧洲各国公众参与环境治理的经验展开进一步探讨。首先，在理念方面，欧洲自20世纪70年代蓬勃发展起来的绿色主义为其环境参与提供了价值导向。一些由绿色主义所延伸出的一系列理论在欧洲社会的背景中产生并成长。它们对于欧洲的环境参与实践活动具有双重意义：一是作为意识形态的重要导引；二是作为政策制定的价值基础。

要回顾欧洲的环境治理与公众参与的思想发展历程，我们首先要回到可持续发展的概念中来。在此，罗马俱乐部发布的《增长的极限》报告指出：地球上的土地资源、不可再生资源、污染承载能力都存在着极限。我们如果继续无限制地追求增长就可能很快达到环境资源的极限，使得工业生产能力和人口都将发生不可控制的衰退。因此，罗马俱乐部提出了"零增长"概念。此后，1992年召开的联合国环境和发展大会确认了以"可持续发展"作为21世纪的全人类发展观。它要求将经济发展建立在社会公正和环境、生态可持续的前提下，既满足当代人的需要，又不对后代人构成威胁。这一理念既协调了发展与环境的关系，又兼顾了代际公平，很快便得到广泛响应。

在实践中，欧洲社会是践行可持续发展观的先锋。1995年欧盟环保咨询论坛进而提出了"可持续发展的12项原则"作为欧盟成员国发展规划制定的总纲领。

根据这些政策导向，欧洲国家纷纷成立相关机构，包括德国的可持续发展委员会（REN）与法国的可持续发展部际委员会（CIDD），来统合各部门行动，实现可持续发展目标。欧洲人希望自己能在可持续发展方面成为世界的"领跑者"，因为他们环境保护意识强烈且拥有成熟的环保技术及成功经验。

其次是城市生态系统的建设。城市生态系统的良好运行对欧洲高度的城市化水平有着至关重要的作用，因此建设生态城市（eco-city）已成为 21 世纪城市规划领域发展的核心目标。"生态城市"概念由联合国教科文组织（UNESCO）在其"人与生物圈"计划的研究过程中提出，原指趋向尽可能降低对于能源、水或是食物等必需品的需求量，也尽可能降低废热、二氧化碳、甲烷与废水的排放的低碳城市。随后，在欧洲国家建设生态城市的实践进一步将生态城市的标准扩展至社会生态、自然生态与经济生态三个维度。当前，欧洲国家已引入城市生态足迹作为生态城市的评价体系（Holden，2004；Pickett et al.，2001）。该体系通过 9 类资源指标核算单位面积土地能够支持的人口数量。这种系统化的评价思路与欧洲国家所普遍采用的系统化治理思路相契合（Domptail et al.，2013）。

最后是参与性的环境治理。为了实现可持续发展与生态城市建设的目标，欧洲国家鼓励社会力量参与到环境治理中来。这种参与式的环境治理也内嵌于欧洲传统的政治文化与社会价值之中。由于当地居民有可能是全球环境变化的直接受害者，因此，许多全球性环境问题亟需地方层面的治理（Ayers，2011）。当前欧洲国家鼓励加强在区域内的环境治理，由于分享共同环境体验的社区居民往往具有相似的行动基础与利益诉求，因此通过统合社区居民的环境行动，能够使社区内自发地形成组织、生成共同行动的规则体系从而分享资源（Balsinger，2012；Conca，2012）。这种基于地方组织的参与式环境治理模式也有效地克服了技术官僚体制对于经验的漠视。当前，参与式的环境治理不仅作为一种治理形式在欧洲国家得以普遍使用，更已成为一种价值导向影响着欧洲国家，乃至世界其他国家与地区环境治理模式的选择。

法规制度的经验

在环境保护的立法层面上，欧洲国家亦走在世界前列。当前，欧洲已形成了覆盖大气、水、能源效率、生物多样性等诸多领域的环境指标体系，为其环境保护实践的行动准则与法律依据。"欧洲标准"在世界其他国家与地区的环境治理中业已成为最严格的环境标准的代名词。另一方面，在欧洲国家的环境立法的程序中，来自各方的意见也被广泛采纳。良好的制度设计保证了欧洲公民在立法立项中的参与，这体现了确保公民环境权的精神。通过对一系列国家环境保护立法的实证考察不难看出，欧洲的环境立法进程走在世界前列，其通过法律形式保证环境民主与环境善治的经验值得他国借鉴。

1. 环境标准的制定

在环境保护的标准制定上，由欧盟牵头实施的包括大气、水、能源效率等诸多领域的环境评价与标准体系，在指导成员国环境治理实践的同时，也规范着它们对于环境资源利用的行为。首先，在大气环境的标准体系方面，《欧洲空气质量准则》的诞生引领了全球对于空气质量评价的潮流。早在1972年，世界卫生组织和联合国环境规划署在其联合倡导的环境健康基准计划中发布了多项空气污染物的环境健康基准文件。与此遥相呼应的是，欧洲地区工业化国家开始了其环境政策制定的进程。

以此为契机，很多欧洲国家制定典型的城市空气污染物的空气质量标准或目标。这一系列标准进而于1987年被欧共体委员会统一整合为《欧洲空气质量准则》。首次发布的《欧洲空气质量准则》规定了12种有机空气污染物和16种无机空气污染物的指导值，其中包括各国当前制定环境空气质量标准普遍规定的一般污染物项目：一氧化碳（CO）、铅（Pb）、二氧化氮（NO_2）、臭氧（O_3）、二氧化硫（SO_2）和颗粒物。可以说，《欧洲空气质量准则》是世界其他国家与地区空气质量标准体系的范本。2005年，世界卫生组织欧洲事务办公室发布了《空气质量准则》（2005年全球升级版），规定了适用于全球范围的颗粒物、臭氧、二氧化氮和二氧化硫的指导值，这也标志着《空气质量准则》开始由欧洲走向世界。

欧洲在水环境的标准方面也发布了一系列的规定。早在1975年欧洲共同体委员会就发布了第一条有关饮用水水源地的第75P1440PEEC号指令。迄今为止，欧共体及欧盟发布的关于水体水质维护、水污染控制的相关指令共计20余条（Hu，2005）。由于水域环境的跨界特性，欧洲关于水环境的评价与标准体系均着眼于流域的整体管理，强调各国合作。2000 年 12 月欧洲议会和欧盟执委会共同颁布的建立共同水体政策行动框架的指令（2000P60PEC）对成员国在江、河、湖泊、海洋和地下水等水体的流域管理计划提出指示，要求进行跨行政区流域的协调管理，最终保证区域的水环境持续改善。这种在标准上实现统一跨区合作的水体治理模式为其他国家与地区的管理实践提供了良好的范本。

欧洲国家对于能源效率的共同关注源于其经济一体化的进程。当前，欧盟的政策导向和市场发展都驱使其能源消费结构向着新型、清洁和可再生的方向转变。为配合能源结构的转型，欧盟出台了一系列能源效率的标准，这一标准对能源消费的热耗散以及消费过程中产生的温室气体、有毒废物进行了规定。欧盟委员会于 2005 年 6 月颁布的《关于能源效率的绿皮书》明确了以提高能源效率为基础减少能源消费，从而达到减少二氧化碳排放目的的发展方略，并规定到 2020 年将能耗降低至当前水平的 80%。在微观层面，欧洲国家出台了具体的行业标准，有效限制了高能耗产业的生存空间，并通过市场渠道调整了消费者的能源消费习惯。以机动车排放标准为例，欧共体与欧盟每四年更新一次汽车能耗与废气排放标准。自 1992 年实行欧洲一号标准起，迄今欧洲国家已更新了四次标准，这一系列汽车能耗与废气排放标准体系不仅在欧盟成员国内强制实施，一直以来也为广大发展中国家所沿用，这为汽车制造行业的能效升级提供了有效的外部压力。

除了具体的环境保护指标之外，欧洲国家还将宏观层面的可持续发展指数纳入其发展评价体系。这些指标的设立着眼于环境与发展的关系，用以评估社会发展的可持续性。欧盟采用 Dashboard of Sustainability 面板工具并选取了一系列混合指标，对自然资源、人力资本、社会资本、实体资本的数量与质量进行综合考量。使用可持续发展指数宏观评价环境与资本的关系有助于在政策制定中获取全面的视角，使国家科学地制定中长期发展方略（Stiglitz et al.，2009）。当前，欧洲国家普遍把可持续发展指数的制定与实施通过立法程序内化于其政治过程中。相较于联合国人类发展指数与世界经济论坛所发布的环境可持续性指数，欧盟采取的 Dashboard of Sustainability 指标体系具有高度制度化的特点，因此对欧洲国家环境政策的制定和实施具有一定的指导意义（Garnåsjordet，2012）。

2. 环境立法立项中的公众参与

欧洲各国不仅在环境保护标准的制定上处于世界领先之列，还一直致力于推动公民在环境立法中的参与。在 1998 年的《奥胡斯公约》中，联合国欧洲经济委员会将公民的"基本环境权利"定义为公民不可剥夺的环境知情权与参与权，并确认其作为人权不可分割的一部分（UNECE，1998）。欧洲公民的环境参与权利当前在欧盟委员会、欧洲人权法院与欧盟成员国家三个层面实现法律保障，并且三个层次的机构对于环境参与权的解读趋于一致（de Abreu Ferreira，2010）。可以说，欧洲是全球立法保障公民环境权利的先行者。

对公民环境知情权与参与权的保障在欧洲国家环境政策的制定与重大规划项目的立项中多有体现。以英国为例，1990 年颁布的《城镇规划法》提出"公众审查规划会"程序规定各郡在制定发展规划时，环境部大臣应将其公布周知，对此有异议的公民可以直接在会上向环境部大臣提出意见。这一会议允许一切感兴趣的公民参加。法令要求在为每一个发展项目颁发许可证前，都应召开公开的地方调查会，以募集来自民众的意见（Town and Country Planning Act，1990）。

相应地，在法国 1998 年颁布的《环境法典》中，也有鼓励公众参与的原则。该法专门设立了"信息与民众参与"一编，分为对治理规划的公众参与、环境影响评价的公正参与、有关对环境造成不利影响项目的公众调查和获取信息的其他渠道四章，具体细致地规定了公众参与环保的目的、范围、权利与程序，并补充了增加环境决策透明度和有组织的咨询等内容（Zhu，2007）。

欧洲的法律法规为公民在环境参与中的权利提供了机制化的保障，在长期的立法与执法实践中，欧洲国家为环境治理塑造了良好的民主氛围，这有利于治理实践的效果达成与效率提升。依法保障公民的环境参与权是当前世界其他国家与地区法制建设进程的重要组成部分，欧洲国家在法律法规层面制定的先进经验为其提供了较好的范本。

政策维度的经验

欧洲国家在环境治理的政策实践方面也走在世界的前列。这些国家从各自的国情出发，形成了一系列环境治理模式。正如 Ayers（2011）指出，公民作为经济发展对环境带来负面影响的受害者，有充分的理由参与其生存环境的管理与监督活动。这一理念成为各国环境治理政策制定的立足点。不难发现，在欧洲国家的环境政策模式中，各国都直接或间接地强调环境治理的多元主体，通过动员个人、社区、社会企业等力量，充分发挥公民在环境治理实践中的重要作用。这些模式为其他国家与地区的环境政策制定与实施提供了宝贵的经验。在此，我们可以对一些欧洲国家环境治理政策模式的经典案例研究进行一些回顾。

1. 社会企业参与地方治理：奥斯塔德模式（The Orestad Model）

当今的城市在其发展过程中均面临着竞争压力与可持续性的双重挑战：前者要求城市加快经济建设的步伐，吸引资本流入，进而打造区域经济中心的地位；后者则要求城市协调经济发展与居民的生活质量的关系。哥本哈根奥斯塔德新城的经验表明，基于国家指令的传统治理模式与依托国家拨款的传统财政模式无法对这两个挑战做出同时的回应。为了实现城市竞争力与可持续性的协同发展，奥斯塔德新城引入新型治理工具，即混合组织以重构城市治理中的权力组织形式（Andersen，2003；Majoor and Jorgensen，2007）。

1995 年，奥斯塔德发展公司（Orestad Development Corporation）建立，以社会企业的形式介入奥斯塔德新城的治理，并在某种程度上取代了传统地方政府管理与规划的职能。社会企业的融资具有独立性，能够通过售卖、租赁、抵押区域内的土地获得区域治理的资金。奥斯塔德新城财政的独立性确保了其自主治理成为可能。在奥斯塔德发展公司的主导下，该城完成了新型公共交通体系的建设。这种多枢纽、多形式的公共交通体系在规划过程中广泛吸收公民意见，将公共交通与土地利用有机结合，不仅提高了交通效率，也有效解决了公民所关注的空气污染、噪声、安全以及温室气体排放等问题。虽然奥斯塔德模式自问世以来不断受到来自政策界的质疑（指责其在总体发展纲领的制定方面具有局限性，且对于市场波动的抵抗力不足），但这种将社会企业引入地方治理的创新在实践中已显示出其高效、亲民的优势：社会企业能够更直接地接触民众，聆听居民的要求并以此为出发点制定区域发展方略（Book et al.，2010）。

奥斯塔德模式在丹麦的成功实践也源自于国家对社会企业较高的接纳程度。

在奥斯塔德模式中，虽然社会企业的引入挑战了国家对于土地、公共交通等资源的控制能力，然而丹麦国家强力的政府的传统政治文化使得政府具备足够信心与容忍度，这在无形中对社会企业的行为也起到了规范作用（Book et al.，2010）。在奥斯塔德模式中，营造国家政府、社会企业与公民良好互动氛围的制度架构是地方可持续性治理的实践典范。

2. 政治转型国家的自然资源管理：捷克案例

20 世纪末，一系列中欧和东欧国家经历了社会制度的转型，其环境治理模式及环境政策也有着较大的调整和转变。以捷克为代表，Earnhart 指出这一时期中、东欧转型国家的环境治理实践结合了其政治体制、社会架构的特点。因此，讨论这一时期东欧国家的一些成果经验对这些国家转型和政策调整具有指导的价值（Earnhart，1996，2001）。

经验表明，中欧与东欧的前社会主义阵营国家因环境保护立法的缺乏及环保执行力度弱而导致了大规模环境退化。在转型过程中，捷克对其政府的组织架构进行了调整，直接导致冷战时期集中的政府权力出现了部门性分化。1990 年 1 月，捷克设立了共和国与联邦两个层次的环境保护部门，使得环境领域的行政执法脱离了经济部门的直接控制，获得了一定的独立性（Bowman and Hunter，1992）。另一方面，一些学者如 Wolchik（1991）、Andrews（1993）、Slocock（1996）、Earnhart（2001）等指出，共产主义时期捷克严格的社会控制使得公众参与环境的机会趋近于零。其国内没有独立的环境社会组织，也没有以环境保护为导向的社会集体行动。转型后的捷克来自公民社会的环保声音获得释放渠道，公民可以通过投诉、公告、调查与上访的形式表达其在环境领域的诉求，这一改变导致了对于环境问题的反馈案件数量显著增加。与此同时，产生于共产主义时期环境行动网络的捷克环境非政府组织也逐渐成为了重要的院外政治力量。

然而，在环境治理的实践中，捷克的民主化进程并未有效地将公民纳入环境决策中来。这与中、东欧前共产主义国家的传统政治文化息息相关（Weiland，2010）。虽然有意识地设立了公民意见表达的渠道，但捷克的环境管理机构仍存在曲解民意的状况，拒绝承认公民意见的合法性与正当性。另一方面，捷克存在环境管理机构与经济机构协调不当的现象，仍延续经济导向的处罚形式而拒绝使用资源导向的处罚形式,这使得纠错式的环保执法无法得以施行。捷克环境治理在政策维度中的经验与教训为我们提出了警示:环境治理模式的调整需要与政治体制和社会架构的改革并行,在民主化环境治理的制度设计中切实地融入公众参与必不可少。

3. 青少年的环境社会动员:芬兰与意大利的政策经验

研究显示,在不同年龄段的人群中,青少年对于环境问题的关注度最高(Rohrschneider,1991;Halla et al.,2013)。培养青少年的环境意识,不仅能够在短期有着良好实际收效(因为青少年本身就是一支重要的社会力量),而且还能收获代际效果,从而长期保持环境治理成果。因此,发挥青少年在社会动员中的作用是吸收公众参与环境治理实践的重要议题。在这一方面,意大利与芬兰的经验值得借鉴。

基于意大利的实证案例表明,青少年拥有自身的社会交往网络,并藉由这一网络实现相互的意见影响与观念塑造。环境保护是青少年交往网络中的重要议题。意大利政府着意使青少年的社会交往网络与更大范围的社区接轨,因此设立了少年议会这一机构,并对这一制度给予立法与资金上的支持,以鼓励青少年表达意见(Alparone and Rissotto,2001;Tonucci and Rissotto,2001)。与意大利相似的是,芬兰政府也就发挥环境动员中的青少年力量进行了一系列的政策实验与实践。当前,芬兰半数以上的城市均设立了少年委员会(Children's Council),他们通过选举产生"议员",出席部分成人议会的会议。以赫尔辛基市为代表的部分城市还引入 Norwegian Porsgrunn 模型,给予少年议会财政拨款,意在培养青少年政治决策、政策执行以及资金使用等全方位能力。基于少年议会的制度设计,两国均着力引导青少年在环境治理中的参与一系列环保项目,诸如由芬兰 Ristinummi 地区的"儿童环境署"(11-13 year-old eco-agent)所推出的公共绿地设计计划、由一群 6 岁儿童提出的芬兰 Pihlajisto 地区建筑群内人工运河与丛林的设计等。这些项目都获得了巨大的成功(Horelli,2001)。

然而,如 Horelli(1998,2001)指出,意大利与芬兰当前均存在将儿童的环境议案现实化的障碍。如何实现青少年环境社会动员能力与当前成人主导的政治体系的统合,仍然是欧洲国家乃至其他试图发挥青少年在环境保护领域力量的国家所面临的重要挑战。

4. 国家公园管理：英国与德国的政策模型

对于国家公园的治理涉及生态环境、自然资源管理两大要素。如何实现国家公园区域内资源的可持续开发、生态系统维持与当地居民需求的平衡，是当前各国环境治理中不可回避的重要议题。欧洲国家在长期的自然公园管理实践中总结出一系列相关经验，其核心在于将多层次的治理主体纳入国家公园的管理中来，以建立充分吸收公众参与的环境资源管理架构。

基于德国 Lauenberg Lakes 自然公园的治理实践案例，Mehnen 总结了当前德国在国家公园环境治理的三大参与主体及其各自角色。在机构层面，德国并行设置了正式与非正式的管理机关，旨在吸纳多方意见，制定管理规则；在个人层面，德国鼓励多种形式的利益表达，并为个体的利益相关者设立了诸如交互式协商、妥协性协议、投票（voting）等多种参与政策制定与治理实践的途径；在社区层面，德国尝试建立"意见库"，以了解公民关注的核心问题，进而为政策制定打下基础。这种三层次的管理体系是德国自然保护区治理的政策创新（Mehnen，2013）。

无独有偶，英国政府所推行的国家公园治理 PROGRESS 计划也体现为一个多元主体的参与式治理框架。在 PROGRESS 计划中，取向各异的利益相关者均被赋予意见表达的机会（Edwards，2011）。这种参与式的管理体系具有巨大的优势：一方面，广泛接纳利益相关者的环境治理参与应和了平等、互信的管理哲学；另一方面，它也有助于政府对区域内复杂情况的宏观把握；与此同时，这一举措统合了涉及环境治理的科学知识与地方经验，使得政策更符合实际并易于执行（Reed，2008）。PROGRESS 计划自 2007 年问世以来，已被 New Forest 国家公园等诸多英国国家级自然保护区所践行，且效果良好。

然而，需要承认的是，国家公园环境治理中多元主体参与式的管理模式仅为鼓励利益相关者参与环境资源管理的政策制定与制度设计，却无法弥合其本质上的意见差异。由此，联合国环境规划署（UNEP，2007）建议通过长期的社会学习（social learning）过程来建立共识，进而统合利益相关者相互矛盾的利益诉求。这将成为欧洲国家自然保护区环境管理的下一个政策目标。

社会实践经验

纵观欧洲公民社会在环境治理中的参与，不难发现其参与实体多元化与网络化的特点。首先，公众参与环境治理的渠道是多种多样的：除了以个体与家庭为基本单位的行动之外，公民也通过社区、环境非政府组织等机构有组织地介入环境的管理与治理。其次，随着通信技术的发展，当前公民社会的环境治理参与模式呈现网络化的新特点。借助网络技术与数字技术，行动者能够实现高效的交流，并及时地共享、发布相关环境信息。在这一背景下，欧洲公众参与环境治理的领域由传统的污染监控与治理扩展到社会规划领域的政策制定。相应地，欧洲国家政府也积极调整其角色，通过环境决策中的去集权化鼓励更多的公众参与，进而提升政府信度。源自社会实践视角的观察使我们更为直观地了解欧洲公民以及公民社会参与环境治理的现状。欧洲高度发达的公众参与环境治理实践为世界其他地区与国家提供了宝贵的经验。

1. 多元化的参与实体

欧洲国家成熟的公民社会使得在环境治理中的公众参与实体相对多元化。首先，在行动者个体层面上，欧洲公民普遍较高的环境意识是其自发环保行动的原动力。Ronis 与 Aberg 基于德国与瑞典实证研究提出的模型将潜意识行动与理性选择列为环境机会结构中行动者个体的两条并行的行动生成路径（Ronis, 1989; Aberg, 1996）。"习惯"作为一种非逻辑性的行动，使欧洲公民的环境治理参与并不局限于技能、知识与资金等要素，而呈现为一种普遍的社会文化。

其次，家庭也是欧洲公民社会参与环境治理进程中的重要行动单位。当前，欧洲国家的环境保护项目有很大部分以家庭行动为基础，这类自下而上的参与式环保项目有着广泛的覆盖面，并且在运作过程中收效甚佳。与此同时，家庭也是公民环保意识形成的基本单元。Bubolz 与 Sontag 提出的家庭环保路径模型检视了家庭成员在意见塑造过程中的相互影响，并指出家庭在公民个体环保习惯养成、环境信息获取以及行动资源获取中的重要作用（Bubolz and Sontag, 1993）。

再次，在欧洲兴起的区域治理理论强调了社区在环境治理中的重要性。由于环境问题的广泛地理分布，不同时间、空间维度的环境问题往往呈现出一定的特异性，这种特异性根植于问题发生区域的社会背景上（Ayers，2011；Squires and Renn，2011）。引入社区参与环境治理，能够为决策过程提供基于经验性观察的"本土知识"。这有效地克服了技术官僚制度下环境政策因"一刀切"而导致的执行困难与效果不佳。欧洲社区参与环境治理的成功与欧洲国家商谈式民主的政治架构是密不可分的，后者为社区在环境保护中的参与提供了制度保障。

最后，环境治理的非政府组织是欧洲公民社会在环境保护领域内最为活跃的一支力量。在一系列欧洲国家的环境治理项目中，非政府组织担负了执行监督、理念传播、专业知识提供以及矛盾调解等重要职能。此外，欧洲区域一体化所衍生出的特殊政治架构更赋予非政府组织广阔的活动空间。如 Beckmann 针对欧盟农业环境项目的研究所指出，非政府组织能够有效协调国家与欧盟行动，使得项目在国家层面得到更好的执行（Beckmann，2009）。

然而，当前欧洲非政府组织对于环境治理的参与也存在一定的局限性：由于欧盟和部分欧洲国家的环境决策多受政治精英非正式交往方式的影响，非政府组织难以在政策制定中发挥其影响力。欧盟环境决策的信息流向为自上而下的模式，虽然当前环境非政府组织的数量大大增加，但因其在信息获取上存在天然劣势，无法在环境决策中输入自身力量（Hurtak，2007；Polini-Staudinger，2005；Hallstorm，2004）。多个欧洲国家在意识到环境非政府组织在环境政策执行领域的重要角色以及环境政策制定领域的巨大潜力，都对其政治结构进行了渐进式的改革，以期更好地发挥非政府组织在环境治理中的作用（Hurtak，2007）。

2. 网络化的行动方式

当前，欧洲国家的公民环保行动日趋网络化。互联网作为公民社会力量组织与行动的重要依托，不仅能够为行动者提供即时交流的平台，也能够成为其信息发布的重要媒介。一方面，互联网的使用大大增加了环境行动网络的行动效率，促进公民社会力量间的信息交流与资源共享。这种网络化的环境行动具有低组织成本的特征，这对于公益性质的小规模环保组织而言具有重大的意义。另一方面，依托技术网络的环境保护行动能够很好地扩大影响力，接触更广泛的人群，由此传播相关的信息与理念。

将互联网引入公众环境参与也利于保障公民的环境基本权利。以罗马尼亚为例,当前在罗马尼亚全境推行的 E-Environment 项目,作为 E-Romania 国家战略的重要组成部分,将国内环境数据指标全部接入互联网,以便社会组织及个人查阅。与此同时,E-Romania 网络平台也作为公众反馈环境问题、参与环境监控的重要渠道(Mares, 2011)。这一网络平台同时实现了公民的环境知情权与环境参与权,由此被认为是一则重要的制度创新。

3. 广泛化的参与领域

就公民社会参与环境治理所涉及的领域而言,当前欧洲公民社会力量的关注问题日趋广泛,已从对污染治理的传统关注拓展到自然资源管理与社会规划等领域。首先,欧洲国家的公民社会一直以来均在环境污染的监控、预防与治理扮演着重要角色。在污染治理政策的制定层面上,欧洲国家的公民社会组织往往能够通过多种渠道"发声",以积极的院外活动施加政治压力,实现其环保诉求的表达。在针对污染治理的政策的执行过程中,公民社会良好地发挥了其监督者与协助者的作用:一方面,环保组织作为环境污染事件中直接利益相关者的第三方力量,能够公正地评估环境政策的执行,因此,引入作为监督者的环保组织被认为是增加执政方公信力的重要手段(Poloni-Staudinger, 2005);另一方面,公民社会组织作为政府服务的提供者,能够有效发挥其灵活性与专业性的特点,及时发现问题并组织对策。概言之,欧洲公民社会力量在对污染事件的长期关注与治理参与下,业已形成了成熟的模式与经验,能够积极动员、快速反应,并在治理实践中"及他人之不及",填补政府力量、市场力量力不能及的领域。

其次,欧洲公民社会对于自然资源管理问题的重视日益增加。公民社会在自然资源管理中的重要作用业已得到充分的理论论证(cf. Rauschmayer et al., 2009; Dietz and Stern, 2008; Stoll-Kleemann and Welp, 2008)。面对资源开发的高度不确定性、错综复杂的利益结构以及自然资源管理制度的普遍缺失,欧洲国家亟需自然资源管理的模式创新。引入公众力量参与决策对解决资源管理方面的困境有着良好的收效(Dreyer, 2011)。正如 Reed 所指出,公民社会在自然资源管理实践中践行分权、平等、互信的哲学,提倡符合决策背景的决策方式,并自觉地整合科学知识与本土知识(Reed, 2008)。欧洲多国自然资源管理所采用的"分析-

商议程序"均体现了对于公民社会力量及其倡导的"社会学习"过程之重要性的认识（Squires，2011），这种依托民间力量的自然资源管理模式当前亦开始见于发展中国家，与政治改革并进，渐进式地植入其社会土壤之中（Swatuk，2005；Tsang，2010）。

除此之外，欧洲公民社会的环境治理参与还涉及社会规划领域：在国家公园开发、生物多样性保护等问题上，公民社会力量具有得天独厚的优势。一方面，基于丰富的经验与本土知识，公民社会组织往往得以敏锐地甄别问题并提出解决方案；另一方面，对于专业技术的掌握使得公民社会组织常常领跑政府决策，将科技与社会规划有机结合。当前广泛见于西欧国家公园开发的仿真模型技术即是一典型案例：在英国的PROGRESS计划中，学者与环境非政府组织共同推广地理信息系统在国家公园管理中的应用，以此为利益相关者直观地呈现国家公园开发的设计细节与整体前景。通过该技术，公众能够更为直接地获取信息，由此更好地参与相关政策的制定（Edwards，2011）。然而，诚如Harrison所指出，在社会规划领域的公民环境治理参与仍以开放的政府决策模式为前提条件，仅当政府推崇商议式民主并鼓励公众在政策制定中的参与时，科技力量方能作为公众获取信息的途径，用以促进环境治理中的民主化（Harrison，2002）。在审视欧洲公民社会的参与式环境治理时，我们必须意识到将欧洲经验置于本土政府决策文化框架下所存在的局限性。

4. 转型中的政府-社会关系

欧洲公民社会在环境治理实践中的长期活跃进一步塑造了该议题下政府与社会的关系。基于对公民社会力量在环境治理领域的重要作用的充分认识，欧洲国家的政府不约而同地进行了角色的相应调整，通过去集权化进程下放权力，鼓励公民、社区以及社会组织在环境问题上的参与式治理，并尝试在企业内引入志愿性创制，以此协同市场力量在环境治理中共同发挥作用。欧洲国家政府在环境治理中的一系列实践，应和了源自公民社会对于公共治理的强烈参与意愿，在环境民主化的进程中，欧洲国家政府的信度亦得到了很好地加强。这种政府-社会关系的转型在欧洲公共治理的实践中取得了共赢的结果，很大程度上解释了当前欧洲国家环境治理理念、效果均领先于全球其他国家、地区的现状。

为了推动环境民主、实现环境的参与式治理，欧洲国家均做出了制度层面的调整。这种渐进式的改革以去集权化为核心，通过弱化中央权力、简化行政层级、引入专业机构、开放意见渠道更好地在环境政策的制定、环境治理的实践中整合社会力量并融入公众参与（Holzhauer，1997）。去集权化是公共治理参与机制的必要条件，它解决了决策过程中的信息不公开与本地参与缺失的弊端。如 Akbulut（2012）的参与机制树中所显示，去集权化有助于利益相关方掌握信息并使得决策程序充分照顾地方权力结构与传统制度安排，从制度层面为公众的治理参与开放渠道。这种去集权化的制度调整不仅广泛见于欧洲传统的民主政治体内（cf. Polini-Staudinger，2005；Hurtak，2007；Holzhauer，1997），也内在于新近实现民主化的东欧国家所推行的政治改革进程之中（Adaman and Arsel，2005；Akbulut，2012）。

除了自身制度的创新与角色转型，欧洲国家政府还通过一系列机制动员市场力量以实现环境领域的善治，其核心在于引入议价者（bargainers）各方之间的合作，把原本外在于企业的环境资源内化成为企业产权所属，由此促进企业自发的环境保护行为（Barron，1998；Cavaliere，1998，2000）。志愿性创制是政府向企业"放权"并重新定义环境产权的一项重要实践，当前在英国得到广泛的应用。在这一政策下，企业与政府在市场框架内被平等地赋予权利，由此通过市场机制开展平等互惠的合作。在环境治理的政策制定与执行中，志愿性创制充分地调动了中小型企业的积极性，这不仅使得环境决策的主体更为多元、模式更为灵活、执行更具效力，也有效改善了政府的形象、促进了政府与企业间的互动（Peters，2004）。

欧洲国家政府向社会、企业的放权在实现环境民主与环境善治的同时，也增强了政府信度。作为评价现代民主政府政策与政治决策的重要指标（Bovens，2005；Mulgan，2003），政治信度标志着决策型政府的政策制定过程与服务型政府的服务行为表现的优劣。频频出现的信任危机使得欧洲国家对其政府信度的建设做出思考。学者认为，立法与执法过程中的责任界定不明、专家话语与技术统治论之风盛行以及公众在公共治理过程中的边缘化是导致政府信度缺失的主要原因（Joss，2010）。调整政府在公共治理决策过程中的主导地位，接纳社会力量与市场力量对于治理的参与，进而营造民主的决策氛围，有利于政府信度的构建。不难看出，欧洲国家政府在环境治理中整合公民社会的参与，不仅提高了决策效率与执行效果，也巩固了作为其执政合法性来源的政府信度，这种执政经验值得世界其他国家地区学习借鉴。

评估讨论

审视欧洲国家在公众环境参与的立法与政策实践经验，不难发现，鼓励环境治理中的公众参与恰恰顺应了欧洲政治发展的潮流与社会文化的演进。一方面，欧洲民主政治的土壤为公民环境权利以及环境民主等概念与理论的生发提供了养分；另一方面，欧洲公民浓厚的政治参与意识培育了强大的公民社会力量，其对于环境治理的介入恰恰体现了当前欧洲社会主流的参与式政治文化。

与此同时，欧洲国家公众环境参与数十年的实践经验证明，这种充分接纳多元社会力量的治理模式在其政治、经济与社会架构内有着良好的收效。公民社会力量不仅能够良好地反映边缘群体的利益诉求，也为环境问题的界定提供了宝贵的地方性知识。被赋予权力的社会政治力量对环境参与有着高度的热情，通过介入环境政策制定与执行过程，它们能够在如何解决自己身边的问题中发表意见，这无形中提升了政府的执政效率。

欧洲在环境治理的公共参与实践中领跑全球其他国家与地区。然而，欧洲公众环境参与的成功经验对于他国的适应性仍然是一个有待讨论的问题。一般而言，国家的经济发展水平、政治文化与社会传统，均对移植"欧洲经验"的成功率存在一定影响：一国的经济发展水平决定了其在政策目标上对环境与发展的权衡；一国的政治集中程度决定了其对于社会力量参政议政的容忍程度；而一国的社会文化传统代表着主流的社会价值，进而决定公民及公民社会力量参与社会治理的意愿。当今世界，国家的发展仍存在显著的不均衡，各国工业化进程与民主化步伐有着较大差异，在环境领域下的公众参与式治理虽然具有种种优势，但若将欧洲经验生搬硬套，难免出现南橘北枳的情况。

对于中国而言，环境与发展并非位于政策光谱的两端而不可统合，相反地，节能环保与提升能源效率成为中国经济发展的目标与国家安全的组成部分（Tsang，2010）。另一方面，中国屡屡出现的环境危机也使得环境议题一跃成为社会关注的焦点。当前，虽然中国的中央政府已经在政策文件中表明其厉行环保的态度，但中央层面的指令在地方层面并不能得到良好执行。原因是这些指令无法有效调解地方层面短期经济发展与长期环境效应的冲突。由此，我们可以得出，通过公民社会弥合中央与地方政治力量在环境问题上的差异是解决中国环境治理困境的一剂良方。相较欧洲社会，中国尚缺乏参与环境治理的成形社会力量。因此，如何在中国本土培育起一支专业性强、积极性高、参政意识浓厚的公民社会力量并在相应的制度保障前提下调动其参与环境治理，是当前执政者需要思考的重要问题。

European Experience of Public Participation in Environmental Governance

Guanli Zhang, Ka Lin, Neil Munro*

Introduction

Since 1973, when the Club of Rome took up the concept of environmental crisis and began to promote the model of sustainable development, attention to environmental problems has moved from the margins of society to the centre. This change has been driven by the progress of industrialization, the pressure of economic development in the broader Third World, and the advancement of globalized concepts of social governance. At the same time, it has been a real and objective reflection of the daily deterioration of the environment's ability to sustain human existence. In view of this, Agenda 21, adopted by the United Nations Conference on Environmental and Development in 1992, has become a guiding document for sustainable development plans worldwide. Agenda 21 makes improving the environment a key objective for sustainable development plans in the twenty-first century. It requires every country to establish relevant environmental regulations to ameliorate atmospheric pollution, water pollution, global warming, various ecological crises and social pressures created by other ecological disasters, as well as to guard against other social risks. At the same time, achieving improvements in the natural environment requires humanity to act,

---

* Guanli Zhang: Doctoral researcher in Geography and the Environment, St Antony's College, University of Oxford; Ka Lin: Professor in College of Public Administration, Zhejiang University; Neil Munro: Lecturer in Chinese Politics, University of Glasgow.

organically introducing social actions into global governance from the bottom up. Environmental governance is the joint outcome of national legislation, executive actions and social actions by citizens. Against this background, an agenda focusing on citizens' environmental participation and the openness of environmental information has become a focus of global attention.

The combination of environmental governance and public participation broadens environmental problems from the technical to the social domain. The significance of citizens' participation in environmental governance is multifaceted: it not only helps strengthen environmental supervision, and promote progress with the technical aspects of pollution prevention, at the same time, it can also effectively organize public opinion and create social pressure. This kind of pressure has helped stimulate governments and other organizations to take a positive attitude to environmental governance. At the same time, citizens' participation in environmental governance continuously strengthens their own environmental awareness, changes behavior patterns, and forms new life habits. In this process, by changing their own behavior, people influence enterprises and social organizations to change their behavior patterns, and by shaping standards of behavior, they foster an environmentally friendly culture in society. For this reason, environmental governance is a problem not merely on a technical or managerial level, but also on a social level. In order to realize good environmental governance, public participation is an indispensable link. It is a basic standpoint for the solution of environmental problems. Only through society's common efforts can we meet environmental challenges to human existence and social development.

Against this backdrop, researchers have begun to address various problems related to the inherent relationship between environmental governance and public participation, as well as the means, forms and effectiveness of participation. Some of this research attempts to distinguish categories of environmental governance and public participation from a theoretical viewpoint. Other research focuses on specific forms and practical experiences of public participation as well as the corresponding policies. Some empirical research, based on comparative methods, explores successful experiences of public participation, and seeks to understand its inherent motives and rationality. These various research efforts have promoted public participation in environmental governance from different viewpoints and have proved valuable, both theoretically and in terms of practice. However, because the scope of this research is extremely broad, and there are many agendas, and because environmental problems touch on many disciplines, it is necessary to choose a specific focal point in environmental research.

At present, research efforts on many environmental problems require a certain level of technical expertise, especially as concerns the control and measurement of pollution in the atmosphere, seas and soil, estimating biodiversity loss, monitoring of industrial pollution and the environmental effects of production processes. Contemporary research devotes much attention to the discussion of the relevant techniques. At the same time, there are great many articles which approach the problem from the point of view of management theory, explorations of the formulation of environmental laws and regulations, discussions of the rationality their establishment and of loopholes in environmental legislation. From this standpoint we can discuss progress with the forms of participation in environmental governance, the interaction of diverse social subjects, the relationship of public policy implementation to social norms, and so on.

In order to promote research on environmental governance from a social viewpoint, we can use the comparative method. At present, citizen participation in environmental governance is treated as a general principle, promoted by all kinds of international organizations. The concepts of "participation and governance" are embodied in policies, development reports and various social action program of the United Nations, World Bank and other international organizations. However, the influence of these concepts and the type of policy processes they lead to in different countries vary considerably. Comparing different countries' experience in environmental governance, especially their patterns and degrees of citizen participation, can provide us with a good experiential basis for understanding environmental social governance. Because of differences between countries and regions in levels of industrialization and urbanization, and differences in levels of economic development, people's environmental awareness and corresponding social behaviour likewise show very large differences. Experiential observation shows that countries which began their industrialization early have a relatively profound recognition of the environmental price paid for industrialization and therefore in the area of environmental protection take relatively comprehensive actions. And countries which lag behind in development face even greater environmental pressures: because these countries need to finish in a very short time the process of industrialization which developed countries took close to one hundred years to complete, the environmental price of industrialization can express itself in a particularly overt form.

On the other hand, the tactics adopted by each country to deal with environmental challenges are not completely identical. Some countries accept the idea of development first and clean-up later, transferring the responsibility for environmental governance to their successors. In other countries, such as many in South Asia, the problems brought by environmental deterioration, including soil degradation, water resource pollution and so on, already directly threaten people's existence, even though the level of industrialization is not very high. This compels them to take prompt action with decisive measures to improve the environment. At present, each country, whether developing or developed, has its own unique experience, and each can learn from the others, thereby making the environmental agenda a global one. Every country can construct through exchange a "policy toolbox" and use it to seek solutions appropriate to actual local conditions.

In terms of furthering discussion of environmental problems as a global issue, the experience of European countries is highly significant. First, these countries industrialized relatively early, their environmental problems have been fully exposed, and they have gradually accumulated a lot of experience in environmental governance.

To take Britain as an example, in the middle of last century London was known for its great fogs and was known as the "Foggy Capital". And now, thanks to continual environmental governance actions (including factory relocations, urban planning, and the development of movements for building a more environmentally friendly society), the situation is much improved. At the same time, Germany's Rhine formerly was subject to serious pollution and was called the "Black Water River". After about half a century of environmental governance, the waters of the Rhine have recovered their original good quality. European countries pursued development at the cost of the environment in the course of industrialization, which caused serious social problems.Thus, environmental risk control and environmental protection consciousness came into being in European countries. At the end of the 1970s, some European scholars were already pointing to the significance of zero economic growth and sustainable development of the environment for human development. In many developing countries, it would take up to 30 years before these ideas were gradually accepted by policy-makers. Therefore, many late developing countries can borrow from European countries' developmental history and the environmental awareness and environmental governance experience it generated. This has shortened our road to recognition of environmental problems, and inspired us to learn from that experience.

Moreover, citizen participation in environmental governance is closely bound up with the process of democratization. Social governance of the environment requires certain preconditions including delegation of authority (that is, the delineation of powers and responsibilities for citizen participation in environmental governance), education (the spread of environmental awareness), growth in information transparency (opening channels of information about the state of the environment and environmental risks, enabling the public to evaluate the environment and express opinions), mechanisms (through which channels of public participation are guaranteed and public participation in environmental governance is made possible). The gradual improvement of post-War European social democracy provided the political foundation and systemic conditions for citizen participation in environmental governance and for environmental democracy, partially explaining European success in this area. Through the establishment and implementation of policies for participation in environmental governance, residents were able to acquire environmental awareness, and gradually transform it into a feature of popular culture (as shown in the wide dissemination in Europe of the concepts of green ideology), and on this basis established such concepts as environmental rights. The basic environmental rights of citizens of European countries received legal guarantees. The democratic system and rule of law norms described above put Europe in the front rank in terms of its practice of public participation in environmental governance, and much of this experience can provide lessons for other countries and regions of the world.

Third, since the environment is a global problem (Naughton, 2010), its governance is closely related to the progress of globalization. The territorial blending and integration of all European countries is an important component of globalization. The effort put into the establishment of European regional environmental policy is the model for international cooperation in the governance of transboundary environmental problems. In terms of international economic cycles, there were two reasons driving European countries to strengthen their common attention to environmental problems. First, because European people's living standards had generally risen after the completion of industrialization, their requirements in terms of environmental standards had also risen. This was one reason why they led the world in environmental protection. Second, because many of Europe's pollution intensive industries and much of its raw material processing had moved to peripheral countries, Europe was able to afford its current level of development and living standard at a relatively low environmental cost. However, it should be noted that this kind of global transfer of industry and technology, as well as bringing fast implementation of industrialization to developing countries, also positioned them downstream of the developed countries, for which the former paid a relatively high price in terms of environmental and labour costs. At present, in international trade developed countries exert pressure on developing countries over greenhouse gas emissions and other environmental indicators. This has lifted the overall environmental standards of human society and been accompanied by relevant policy measures. In this process, the countries of the EU have become the main force pushing forward (cf. EC, 2000, 2001, 2003a, 2003b). European countries have recognized the connection between power and responsibility, and through positive cooperation, have supported environmental governance in developing countries in technical and financial areas.

Therefore, in discussing public environmental participation, we should recall the experience and lessons of policies and practice in European countries, strive to explain through the study of European experience the EU's successes in environmental social governance, and to provide policy assessments with useful lessons. This article will examine and analyze European experience from the point of view of value directions, legal regulations, policy dimensions, and social practice.

Before beginning this discussion, it is necessary to recall the concept and basic theory of public participation in environmental governance.

The concept and theory of public environmental participation

Public participation in the process of environmental governance touches on a broad spectrum of social activities. In discussing this process, we have to draw on the three concepts of environmental interests, environmental democracy and participation, as well as their associated theories. European countries' advanced model of environmental governance laid a solid theoretical foundation for institutionalizing public environmental participation, many of the relevant concepts being formulated in European environmental regulations and environmental protection program. These concepts and theories have helped to guide environmental governance practice in other countries and regions in a positive direction.

1. Environmental interests

The United Nations Conference on the Human Environment, held in Stockholm in 1972, adopted a "Declaration on the Human Environment". This declaration expressed the conviction that people have basic rights to freedom, equality and adequate living conditions in an environment that permits them to live a life of dignity and wellbeing.

Soon after that, the "Draft Declaration on Human Rights to Natural Resources in Europe", which was adopted by the Council of Environment Ministers meeting in Vienna, confirmed environmental rights as part of fundamental human rights. Since then, over the course of more than forty years of environmental governance, the concept of environmental rights has acquired continual substantiation and development.

At that time, in the European context fundamental environmental rights included citizens' inalienable rights to know environmental information and to environmental participation (de Abreu Ferreira, 2010). According to the Aarhus convention, guaranteeing citizens' fundamental environmental rights is the foundation stone of environmental justice. Citizens' rights to environmental justice are realized in three ways, namely guaranteeing the disclosure of environmental information in accordance with regulations; public participation in environmental policy formulation; and channels for questioning environmental policy(Mares, 2011).Evidently, the three paths to securing environmental justice are a good summary of the rights to public environmental information and to participation.

The right to environmental information is the right of people to freely obtain environmental information for the area where they live and work as well as for their wider living space. Article three of the Aarhus Convention defines environmental information: it covers not only the media comprising the environment (air, water, soil, terrain, natural resources and biodiversity etc.) but also factors affecting the environment (energy, noise, radiation, human activity, environmental agreements, policies, laws, program for developing the environment etc.). In addition, environmental information includes information on the relationship of the above mentioned media and environmental factors to human health and security (Aarhus Convention, 1998). The Aarhus Convention charges public bodies which hold information to use diverse channels to disseminate it to the public in order to enable citizens to have a general understanding of their environment and to be able to assess the risks facing them. Guaranteeing people's legal right to environmental information helps individual citizens and social groups to see in a timely way the relationship between local environmental knowledge based on experience or observation and scientific data gathered for environmental assessments, which facilitates more precise formulation of questions as well as early detection and control of environmental risks (Ayers, 2011; Beckmann, 2009; Hakala, 2012; van Laerhoven, 2014).

Another important element in the basic environmental rights of citizens, the right to environmental participation is defined in Article 1 of the Aarhus Convention as follows: "Citizens and civic groups have the right to participate in policy making on environmental issues and to seek justice"(Aarhus Convention, 1998; see also de Abreu Ferreira, 2010). This concept can be traced back to European countries' tradition of encouraging social participation in law-making, and its recent development cannot be separated from scholarly advances in theory. Theoretical discussions of citizen environmental participation in Europe focus on the legal bases of citizens' and civic groups' participation in environmental governance (cf. Glasbergen, 2007; Ruggie, 2004; Bäckstrand, 2006; Biermann and Pattberg, 2008; Khagram and Ali, 2008; Andonova, 2010) and on the design of participatory institutions (cf. Harrison, 2002; Edwards, 2011; Mares, 2011; Mehnen, 2013; Roggero, 2013). A series of theoretical investigations on citizen environmental participation and rights has greatly advanced European countries' progress through institutional reforms and legislation. At the same time, it has become an important source of reference for the design of institutions of environmental governance in other countries and regions.

2. Environmental democracy

Environmental democracy describes the core concept of public environmental participation. Because environmental problems affect the broad mass of people, they encompass an even greater plurality of interests than other social problems. On the other hand, even though Europe's traditional model of liberal democracy encourages interest groups to complete, interest groups have very limited ability to express the public welfare and public interests. Because the division of traditional interest groups cannot fully express the aspirations of affected parties in regard to environmental problems, European countries commonly seek to involve "democratic procedures" in environmental governance (Mason, 1999; Zeiter, 2000) .According to the form of democracy practiced by citizens in their environmental participation, European environmental democracy can be divided into three types: participatory democracy, deliberative democracy and associative democracy. In the framework of the democratic theory described above, scholars have arrived at different understandings of the concept of environmental democracy and its operational mechanisms.

First, participatory democracy advocates direct participation by citizens in the political process of policy-making. Compared to the traditional system of representative democracy, this kind of democracy is more radical and has been called "strong democracy"(Barber, 1984, 1988). In the framework of participatory democracy, Fischer(1993)defined environmental democracy as "involving the people in policy making and thereby determining the process for resolving environmental problems". Just as Fischer described, environmental democracy advocates "increasing participation" and "universal participation" instead of "restricting participation" and "representative participation". This requires solving environmental problems as general problems by means of the widest possible citizen participation. The form of environmental democracy which involves citizens in direct participation helps to identify risks, and can effectively raise the effectiveness of feedback about environmental risks. At the same time, it has been recognized as one way of promoting environmental justice at individual level, and at the same time promoting the two goals of public education and social organization. Through social participation, participation on the one hand cultivates environmental awareness, and on the other hand, establishes networks for environmental community action (Zeiter, 2000).

Unlike radical participatory democracy, deliberative democracy emphasizes the mutual influence of participants through deliberation. On the basis of common understanding and by means of open criticism and persuasion, it aims at facilitating cooperation and contributing to the formulation of a policy which meets the public interest(Dryzek, 1990; Zeiter, 2000). Quite a few scholars believe that environmental problems can be solved through deliberation. Dryzek (1992) defined deliberative environmental democracy as a model for progressively solving environmental problems over a relatively long period of time by means of the pursuit of the environmental public interest through free flowing discussion, admitting multilateral efforts, on the basis of equality and mutual respect. Deliberative environmental democracy has been considered as an ideal type of environmental democracy. Successful deliberation on the one hand requires participants in the democratic process who take on the role of spokespersons to have a high degree of logicality; and on the other hand it requires systemic guarantees of the equality of spokespersons: all hindrances which are contrary to the principle of equality must be eliminated so that all speeches are equally listened to and respected.

Associative democracy recognizes the importance of discussion as in deliberative democracy, but puts in doubt the core hypothesis of deliberative democracy, namely that "discussion is a universal, value neutral, de-cultured, de-democratized democratic process". Young (1990, 1997) considers that the lack of equality in deliberative democracy comes from cultural norms: special language and ways of speaking carry power, are recognized by the rules of the mainstream culture, and therefore create a natural inequality amongst participants in the discussion. A white, middle class male receives power in deliberative democracy because of the way he talks. Consequently, associative democracy encourages non-deliberative behavior, in order to allow participants from outside the mainstream to fully express their interests in the area of the environment.

In the practice of environmental participation by European citizens, participatory democracy has been considered the most suitable for the soil of contemporary politics in Europe and its culture of governance (Robinson, 1993; Dryzek, 1992). Although some scholars point out that the realization of democracy through deliberation requires fulfillment of a number of conditions, including social equality, a relatively high educational level amongst citizens, cultural homogeneity, a certain accumulation of social wealth, modern social and cultural norms as well as a high degree of political differentiation, which is precisely a description of European countries' general state of development (Offe, 1997), nevertheless empirical case study research shows that "deliberation" and the model of deliberative democracy are relatively suitable for traditional societies and developing countries (Gupte, 2007; Omanga, 2014).

It is not difficult to notice that environmental democracy as a process of public environmental participation, regardless of what form of external expression it takes, has as its core from beginning to end the involvement of plural subjects in environmental governance and the encouragement of citizens and civil society to fulfil their specific functions. One can say that the debates surrounding environmental democracy are proof of the progress of citizens' environmental participation, and have enriched its theoretical framework.

3. Forms of participation

The by now already formed European model of public environmental participation is characterized by multiple paths, multiple subjects and multiple forms. This model is embedded in each country's political system and political culture, and has proved itself in practice to be very effective.

As far as governance paths are concerned, European countries hold in esteem cooperation amongst the government, markets and civil society. Europe's traditional governance model presupposes that government, markets and civil society all have distinct areas of competence, and because each has its particular logic of operation, there is no way to effectively implement "cross-sector" management (Glasbergen, 2011). Arnstein's theory of the ladder of participation overturned the old understanding of the concept of governing as institutionalization, legitimation and formalization. Based on the hypothesis that the public and private sectors are gradually blending together, Arnstein proposed that cooperation in public affairs shows great vitality. According to the ladder theory, following the entry of increasing numbers of subjects participating in pluralized governance and the adjustment of methods of cooperation amongst them, cooperative governance will gradually develop from the initial establishment of mutual trust, and through the recognition of common interests and the consolidation of systems of rules, become a system for remoulding markets and the political system (Arnstein, 1969). The explanatory power of this evolutionary ladder of cooperative governance has been repeatedly demonstrated in practice in European countries' environmental governance (cf.Glasbergen, 2011; Rydin, 1999; van der Haijden and ten Heuvelhof, 2012).

As far as governance subjects from civil society are concerned, apart from individual citizens, communities and green groups are also important actors, participating in and influencing the formulation and implementation of environmental policy. As local participants in environmental governance, communities provide a micro-level view of environmental policy formulation and implementation. Based on communities' daily use of informal channels for obtaining information, environmental governance which incorporates community groups is always relatively good at guaranteeing the interests of underprivileged and high-risk groups (Hakala, 2012). As van Laerhoven (2014) points out, policy formulation at local level can noticeably raise the quality of environmental public service provision. This is not only because self-governing community groups are more sensitive to local environmental needs, but also because the community is often the only channel through which certain groups of people participate in public affairs and obtain public services. Therefore, community groups have become as an indispensable subject of environmental governance, and in Europe the importance of their position is almost unshakeable. On the other hand, non-governmental organizations, because of their flexibility and high levels of professionalization, fulfill the important functions of supervising implementation, advocacy, providing expert advice and resolving conflicts, and in this way effectively complement the traditional government led form of public administration.

Third, European public environmental participation reveals great vitality and variety. Contemporary European civil society, through such forms as round tables, knowledge exchange events, action meetings, and so on, reveals its strengths as a form of self-organization, fulfilling important functions in agenda setting, establishing norms, and policy implementation for environmental governance. Moreover, European civil society groups have put great effort into constructing social technologies and networks for effective monitoring. The typical example is citizens groups' ongoing involvement in water governance and monitoring of $PM_{2.5}$ air pollution standards. This type of participation effectively combines the advantages of technology with those of popular involvement, and has wide applicability. It should be added that European popular environmental participation also aims to promote the legalization of the process of environmental governance. In Holland and Germany, citizens' juries have begun to play a role in trials of environmental pollution cases (Barber and Barlett, 2006), which is an expression of social governance on the level of the judiciary.

The model of popular environmental participation described above, featuring multiple channels, multiple subjects, and multiple forms, has shown itself to be rather effective, and has brought considerable achievements to the practice of European environmental governance. For other countries and regions embarking on social governance of the environment, the European model of participation is worthy of study and emulation.

Experience with shaping values

Within the above conceptual framework, we can further investigate European experience with popular environmental participation. First, in terms of principles, the flourishing of green ideology since the 1970s has helped shaped the values of environmental participation. The set of theories described below all emerged in European societies as extensions of green ideology. They have two important implications for European environmental participation and practice: first, they are important guides to ideology; second, they provide foundations for policy-making.

Recollecting the history of European thinking on public participation in environmental governance, we need to return first to the concept of sustainable development. The Club of Rome's "Limits to Growth" report pointed out that the earth's soil resources, non-renewable resources, and carrying capacity for pollution all have limits, and if we continue to pursue growth without limits, we will very quickly reach those limits, causing industrial production capacity and population to fall into an uncontrollable decline. Based on the Club of Rome's "zero growth" concept, the United Nations Conference on Environment and Development held in 1992, confirmed sustainable development as a development concept for all of humanity in the 21st century, requiring economic development to be based on social justice and the sustainability of the environment and ecology, that is, to satisfy the needs of the present generation without threatening the ability of future generations to meet their needs. This development concept regulated the relationship between the environment and development, at the same time taking into account generational fairness, and very quickly obtained wide support.

In this regard, Europe is in the vanguard of sustainable development. The "12 principles of sustainable development" adopted by the European Environmental Consultative Forum in 1995 constitute the guiding principles of development planning for European Union member states. In order to coordinate the actions of different ministries and realize sustainable development goals, European countries each established corresponding structures such as Germany's Sustainable Development Commission (REN) and France's Inter-ministerial Committee for Sustainable Development (CIDD). Europeans hope to become global "front runners" in sustainable development thanks to widespread environmental awareness, mature environmental protection technologies and successful environmental protection experiences.

Europe's high level of urbanization makes the healthy functioning of urban ecosystems particularly important. The building of eco-cities has already become the core goal of development in 21st century urban planning. The United Nations Education, Science and Culture Organization (UNESCO) proposed in the course of developing its "Humans and the Biosphere" plan the development of low-carbon cities, reducing as far as possible usage of essential resources such as energy, water and food, and at the same time reducing emissions of waste heat, carbon dioxide, methane, and waste water. The practice of establishing eco-cities' in European countries will lead to the further extension of eco-city standards to the three dimensions of social ecology, natural ecology and economic ecology. At present, European countries are using urban ecological footprints as a system of evaluation for eco-cities (Holden, 2004; Pickett et al., 2001). This system uses nine categories of resource indicators to calculate the number of people who can be supported per unit area. This kind of systematic evaluation fully corresponds to the thinking behind European countries' commonly applied systems of governance (Domptail et al., 2013).

In order to realize the goals of sustainable development and the construction of eco-cities, European countries encourage social groups to participate in environmental governance. This kind of participatory environmental governance is embedded in European's traditional political culture and social values. Because local residents are the direct victims of global environmental problems, many such problems urgently require solutions at the local level (Ayers, 2011). Because community residents who share a common environmental experience tend to have similar behavioral patterns and interests, the encouragement of regional environmental governance by contemporary European countries has enabled communities, through integrated community environmental actions, to spontaneously form groups, to build systems for delivering common action plans and share resources (Balsinger, 2012; Conca, 2012). This kind of environmental governance model based on local group participation has effectively overcome the tendency of technologically minded bureaucrats to neglect local knowledge and knowledge based on experience. To date, participatory environmental governance has not only been widely applied in European countries, but, more than that, has become a kind of value orientation, influencing the choices of European countries and even other countries and regions of the world in their choice of models for environmental governance.

Legal and statutory experience

In terms of environmental protection legislation, European countries are also world leaders. European countries have by now already formed comprehensive systems of environmental indicators covering air, water, energy efficiency, biodiversity and other areas, providing standards for action and legal bases for environmental protection in practice. "European standards" have already become a byword in environmental governance for the strictest standards. Moreover, in the course of passing European environmental legislation, opinions from widely different quarters have already been adopted. Good system design has guaranteed European citizens' participation in the process of passing laws and establishing programs, embodying the spirit of environmental citizenship. Through empirical investigation of environmental protection legislation in a series of countries, it is evident that Europe leads the world in progressive environmental legislation, and its experiences in environmental democracy backed by law and statute and in environmental good governance are worthy of other countries' study and emulation.

1. Establishing environmental standards

In establishing environmental standards, the system of environmental evaluation and standards, whose implementation is overseen by the European Union and which cover the atmosphere, water, energy efficiency and many other areas, in addition to guiding the practice of environmental governance in member states, also regulates their behavior in the utilization of environmental resources. First in terms of the system of standards for the atmosphere, the emergence of European air quality standards led to a global tide of air quality evaluation. As early as 1972, the World Health Organization and the United Nations Environment Program, under the Environmental Health Criteria Program which they jointly promoted, published numerous documents on basic environmental health standards for air pollutants. Coordinated with this was the beginning of the establishment of environmental policies in industrialized countries of Europe.

At this point, many European countries established air quality standards and targets for the most common urban air pollutants. In 1987, the Council of the European Communities integrated these under "European Air Quality Standards". The first set of European air quality standards established guidance values for 12 organic air pollutants and 16 inorganic air pollutants, including program for ordinary pollutants for which general regulations on environmental air quality had been previously established, such as carbon monoxide, lead, nitrogen dioxide, ozone, sulphur dioxide and particulate matter. One can say that European air quality standards have become the model for systems of air quality standards in other countries and regions around the world. In 2005 the European affairs office of the World Health Organization published "Air Quality Standards" (updated global edition of 2005), establishing indicative values applicable on a global basis for particulate matter, ozone, nitrogen dioxide and sulphur dioxide, thus establishing that the "Air Quality Standards" had begun to move beyond Europe into the wider world.

The European system of water quality standards is often promulgated in the form of directives. As early as 1975, the Council of the European Communities published directive 75P1440P/EEC on drinking water headwaters. So far there have been more than 20 directives on the protection of water quality in water bodies and control of water pollutants (Hu, 2005). Because of the transboundary nature of water basins, European systems of environmental evaluation and standards take into account integrated river basin management, and emphasize cross-national cooperation. In December 2000, in order to ultimately guarantee the sustainable improvement of the regional aquatic environment, the European Parliament and the European Council jointly promulgated a directive (2000P60P/EC) on the establishment of a common framework providing for management of water basins, including rivers, lakes, seas and underground water bodies in member states, and requiring them to adopt integrated plans for water basins crossing administrative boundaries. This model of governance of water bodies on the basis of the realization of integrated cross-boundary cooperation under common standards provided an excellent model for management practices in other countries and regions.

European countries' common focus on energy efficiency stems from their common processes of economic integration. At present, European policy and the development of markets both encourage change in energy consumption structures towards new, clean and renewable energy sources. In order to encourage the transformation of energy structures, the European Union has issued a series of energy efficiency standards, regulating heat loss as well as the production of greenhouse gases and harmful pollutants in energy consumption. In June 2005 the European Council's "Green Paper on Energy Efficiency" confirmed the strategic goal of reducing energy consumption by raising energy efficiency, and thereby reducing carbon dioxide emissions, and set the goal of reducing energy consumption from current levels by 20%. On a micro level, European countries have issued concrete professional standards, effectively limiting the space in which high energy consumption industries may thrive, and through market channels has adjusted consumers' energy consumption habits. To take the example of motor vehicle emissions standards, the European Communities and European Union have revised cars' fuel efficiency and emissions standards. Since 1992 when Europe implemented Standard One, European countries have so far revised the standard four times. This series of system of standards for vehicle fuel efficiency and emissions has not only been implemented on a compulsory basis in European member states, but also has continuously been applied in the wider world of developing countries and has exerted a positive external pressure on the car manufacturing industry to raise energy efficiency.

Apart from concrete environmental protection indicators, European countries have also incorporated sustainable development indices (SDIs) into their systems for evaluating development at the macro level. The aim is to take into account the relationship between the environment and development, and to conduct an integrated assessment of the sustainability of social development. The Dashboard of Sustainability used by the European Union has adopted a series of combined indicators to carry out integrated measurement of quantitative and qualitative aspects of natural resources, human capital, social capital and physical capital (Stiglitz et al., 2009). Use of sustainable development indicators to assess the relationship between capital and the environment helps to attain a complete overview in policy making so that countries can establish beneficial medium-to-long-term development strategies (Stiglitz et al., 2009). Through legislative procedures, European countries are internalizing the establishment and implementation of sustainable development indicators as part of their political processes (Garnåsjordet, 2012). Compared to the United Nations Human Development Index (HDI) and the Environmental Sustainability Index (ESI) adopted by the World Economic Forum, the European Union's Dashboard of Sustainability is characterized by a high degree of institutionalization, which makes it more useful as a guide to European countries' environmental policy-making and implementation.

2. Public participation in environmental legislation and environmental programmes

European countries not only occupy a leading position in the world in terms of setting environmental standards, but also have continuously emphasized the promotion of public participation in environmental legislation as part of an effort to protect and guarantee citizen' environmental rights. In the 1998 Aarhus Convention, the UN Economic Commission for Europe defined citizens' "fundamental environmental rights" as citizens' inalienable rights to environmental information and to participation, and confirmed them as inseparable elements of human rights (UNECE, 1998). European citizens' rights to environmental participation currently enjoy legal guarantees at three levels, the European Commission, the European Court of Human Rights and European Union member states, and there is a tendency towards harmonization of understandings of rights to environmental participation at the three levels (de Abreu Ferreira, 2010). We can say that Europe is a global pioneer in legislation to guarantee citizens' environmental rights.

The guarantees of citizens' rights to know and to participate in European countries' environmental legislation and in planning decisions about important projects are manifested in many different ways. To take Britain as an example, the Town and Country Planning Act (1990) provides for a procedure of "public approval of planning", requiring that whenever a county makes development plans, the Minister of the Environment must publicize them widely so that citizens who object to them can make direct representations to the Minister at the planning meeting. Every interested citizen is permitted to attend the planning meeting. Statutes require that before any major project receives planning permission, an open local consultation meeting must be held in order to solicit opinions from the people (Town and Country Planning Act, 1990).

Similarly, in France's "Environment Code" it is not difficult to notice the principle of encouraging public participation. This law has a special section dealing with "information and public participation," including four articles dealing respectively with public participation in the governance of planning, formal participation in environmental impact assessments, public investigations of relevant projects having harmful effects, and provision of information through alternative channels. These articles specify in detail the purposes, scope, rights and procedures of public participation in environmental protection, and also providing additional content on increasing the transparency of environmental policy-making and organized consultation (Zhu, 2007).

European laws and statutes provide institutionalized guarantees of citizens' rights to environmental participation, and practice in legislation and implementation over a long period of time shows that European countries have molded a democratic atmosphere which has been beneficial for environmental governance. It has also facilitated the practical achievement of good governance and the increase of its effectiveness. Guaranteeing citizens' environmental participation according to law is an important component of the process of building legislation in other countries and regions of the world, and European countries' advanced experience on the level of law and statutes has provided quite a good model.

Experience with policy dimensionality

European countries are also world leaders in practical policies for environmental governance. Emerging from the specific situations of particular countries, they have formed a series of models of environmental governance. As Ayers (2011) pointed out, as the victims of the negative effects of economic development on the environment, citizens have every reason to participate in the management and supervision of the environment in which they live. This concept has become the starting point for the establishment of each country's policies for environmental governance. It is evident that in these European models of environmental policy, each country emphasizes, directly or indirectly, citizens' important role in environmental governance through mobilization of individuals, community groups or social enterprises. These models constitute valuable experience for policy making and implementation in other countries and regions. Below are a few typical examples of environmental governance policy models in European countries.

1. Participation in local governance by social enterprises: the Orestad model

In the course of development, contemporary cities all face the double challenge of competitiveness and sustainability. The former requires cities to speed up the pace of economic development, attract capital inflows and improve their positions as regional economic centers; the latter on the other hand requires cities to compromise between economic development and residents' quality of life. The experience of Orestad in Copenhagen shows that the traditional governance model based on directives and the traditional public finance model based on grants are not able to effectively deal with both of these challenges simultaneously. In order to achieve the coordinated development of competitiveness and sustainability, the new town of Orestad introduced a new governance instrument, the hybrid organization, aimed at reconstructing the organization of power in local governance (Anderson, 2003; Major and Jorgensen, 2007).

The Orestad Development Corporation (ODC), founded in 1995, became involved in the governance of Orestad new town as a social enterprise, replacing the traditional management and planning functions of local government. The social enterprise has financial independence and is able to raise funds for governance and development by selling, leasing or mortgaging land in the area. The financial independence of Orestad new town made possible its autonomy in governance. Under the leadership of the Orestad Development Corporation, the new town completed construction of a new public transport system. The wide solicitation of citizens' opinions during the planning of this new public transport system, which occupies a pivotal position, and the organic combination of public transport and land use, not only increased the efficiency of transport but also effectively resolved citizens' concerns about air pollution, noise, safety, greenhouse gas emissions and so on (Book et al., 2010). Although since its initiation the Orestad model has continuously been called into question by policy circles, who find fault with its limitations of in terms of deciding on overall development policy and its insufficient ability to withstand market turbulence, the innovative involvement of social enterprises in local governance has proven its effectiveness and sensitivity to the people's demands. Social enterprises can more directly enter into contact with the people, listen to citizens' demands and on this basis determine development plans for the area.

The successful experience of the Orestad model in Denmark is also down to the high degree of receptiveness of the state to social enterprises. In the Orestad model, the entry of social enterprises challenged the state's ability to control such resources as land and public transport, but the traditional political culture of strong government in this peninsular country of Scandinavia endowed the government with sufficient confidence and tolerance, which also helped to regulate social enterprises' behavior (Book et al., 2010). The Orestad model's institutional structure, which fostered an atmosphere of effective cooperation amongst the government, social enterprises and citizens, is a practical example of sustainable local governance.

2. Natural resource management in a country in political transition: the Czech case

At the end of the twentieth century, Central and East European countries experienced transformations of their social systems, and in the process their models of environmental governance and environmental policy also passed through large adjustments and transformations. Using the Czech Republic as an example, Earnhart (1996, 2001) summarized the experience of environmental governance in Central and East European transition countries during this period, and, integrating the political system with social structure, discusses the lessons that can be learned from this experience.

Empirical investigation have shown that in countries of the former socialist camp in Central and Eastern Europe loopholes in environmental protection legislation and weak capacity for implementation led to widespread environmental deterioration. In January 1990 Czechoslovakia established environmental protection ministries at federal and republic level, freeing administration and implementation in the area of the environment from the direct control of economic ministries and giving them a certain amount of independence (Bowman and Hunter, 1992). On the other hand, as pointed out by such scholars as Wolchik(1991), Andrews(1993), Slocock(1996)and Earnhart (2001), strict social control in Czechoslovakia during the Communist period restricted opportunities for citizens' environmental participation to almost zero, and the country had no independent environmental social groups, let alone a collective social movement aimed at environmental protection. After transition in Czechoslovakia channels for civic groups to voice environmental concerns were opened, and citizens were able to express their environmental demands in the form of complaints, announcements, instigations and petitions, which led to a noticeable increase in the amount of feedback on environmental issues. At the same time Czechoslovak non-governmental organizations, born out of a network of environmental activist groups in the communist period, gradually became an important political lobby.

However, in the practice of environmental governance, the democratization of Czechoslovakia did not lead to the effective involvement of citizens in environmental policy-making. This outcome was closely related to traditional political culture of former Communist countries in Central and Eastern Europe(Weiland, 2010). Although in ideological terms channels of expressing citizens' opinions were established, environmental management structures in Czechoslovakia were still in the habit of distorting public opinion and refusing to recognize the legitimacy of citizens' opinions. On the other hand the phenomenon of an improper relationship between economic structures and environmental management structures and the continuing reliance on monetary-based penalties and the refusal to adopt resource-based penalties made it impossible to implement corrective environmental policies. In terms of policy dimensions, the experience and lessons of Czechoslovak environmental governance provide a warning to us: the adjustment of the model of environmental governance should be carried out in conjunction with reforms to the political institutions and social structures, and genuine citizen participation in the design of institutions of democratized environmental governance cannot be neglected.

3. Social mobilization of youth for the environment: the policy experience of Finland and Italy

Research has shown that amongst different age groups younger citizens pay relatively more attention to environmental questions (Rohrschneider, 1991; Halla et al., 2013). Encouraging younger citizens' environmental awareness can be effective not only in terms of intergenerational transmission, ensuring the long-term maintenance of successes in environmental governance, but also can have a very good effect in the short term, because younger citizens are themselves an important social force. For this reason, how to mobilize younger citizens to socially engage with environmental issues has become an important component of policies and practices designed to attract citizens to participate in environmental governance. In this respect the experiences of Italy and Finland are worthy of study and emulation.

Empirical case studies from Italy show that younger citizens have their own social networks and it is possible to mould viewpoints and influence relevant opinions by tapping into these networks. Environmental protection is an important topic in younger citizens' social networks. Taking great pains to ensure that younger citizens' social networks were joined up with those of the larger community and in order to encourage younger citizens to express their opinions, the Italian government established a Children's Council and gave this institution legislative and financial support (Alparone and Rissotto, 2001; Tonucci and Rissotto, 2001). Similarly, the Finnish government instituted a series of experiments and practices aimed at effective environmental mobilization of younger citizens. At present, more than half of Finnish towns have youth parliaments which elect their own deputies, some of whom have spoken at adult parliaments. Some cities, of which Helsinki is an example, adopted the Norwegian Porsgrunn model, allocating government grants to youth parliaments, aimed at fostering younger citizens' all-round abilities to make and implement policies and deploy funding (Horelli, 2001). Based on plans drawn up by the youth parliaments, both countries strove to encourage youth participation in environmental governance, and some projects were tremendously successful, such as the public green space plan proposed by an "11-13 year old eco-agent" in Finland's Ristinummi district, or the design for an artificial canal and forest in a building complex in Finland's Pihlajisto district (Horelli, 2001).

However, as Horelli (1998, 2001) points out, there are barriers to the realization of proposals made by the Italian and Finnish youth parliaments. How to reconcile the social mobilization of youth for the environment with the political system led by adults is a serious challenge faced by European countries as well as other countries in the world which are attempting to mobilize the energy of youth in the area of environmental protection.

4. Management of national parks: the British and German policy models

Management of the ecological environment and natural resources are two key factors in the governance of national parks. How to balance the sustainable use of the national park's natural resources and the preservation of its ecosystems with the needs of the local residents is an important and unavoidable topic in every country's environmental governance. To summarize a great deal of relevant experience, in the practice of long term management of national parks in European countries, the key is to involve multiple layers of governance subjects in order to establish an environmental management framework in which public participation takes a full role.

Based on the case of management of the Lauenberg Lakes natural park in Germany, Mehnen (2013) identified three major participatory subjects in the management of contemporary German national parks and their respective roles. At the level of structures, in order to draw on multiple opinions in the establishment of management principles, Germany deploys both formal and informal management organs. At the individual level, Germany encourages multiple forms of interest articulation, establishing for example such channels for participation in policy making and governance practice as "mutual adjustment", "negotiated agreements" and voting. At the community level, in order to understand the key problems which the public focusses on and provide a stronger basis for policy-making, the government has attempted to establish "opinion pools". These three levels of the management system constitute Germany's policy innovations in the governance of protected natural areas.

Similarly, the PROGRESS plan, promoted by the British government for the governance of national parks also constitutes a framework for participatory governance by multiple subjects. In the PROGRESS plan stakeholders with different interests are given opportunities to express their opinions (Edwards, 2011). This participatory governance system has a huge advantage: on the one hand, participation in environmental governance by a broad array of stakeholders complies with a management philosophy based on equality and mutual trust; on the other hand, it helps the government to grasp the complex situation in the area at a macro level; at the same time, this system combines scientific knowledge about environmental governance with local experience, and helps to make policy more realistic and easier to implement (Reed, 2008). Since its introduction in 2007, the PROGRESS plan has been used in numerous national level nature protection areas in Britain, including the New Forest National Park, and has produced very good results.

However, the multiple-subject, participatory model for management of national parks is only a means of encouraging stakeholders to participate in policy making and institutional design, and doesn't provide a way of bridging substantive differences of opinion. Therefore UNEP (2007) proposes that common understandings should be established through social learning in order to reconcile contradictions between the interests of different stakeholders. This will be the next policy goal of European countries' management of natural protected areas.

Experiences with social practice

Taking an overview of participation in environmental governance by European civil societies, the pluralization of participating groups and their networked nature are quite evident. First, the channels for citizen participation in environmental governance are multiple; apart from actions for which the basic unit is the individual and households, citizens also enter into environmental management and governance in an organized way through communities, environmental NGOs and so on. Moreover, with the development of communication technologies, the participation of civil society in environmental governance reveals a tendency to become networked. By using the internet and digital technology actors can exchange information effectively and share what is relevant in a timely way. Against this background, European citizens' participation in environmental governance has broadened from the traditional pollution control and regulation to policy making on social norms. Correspondingly, European governments have positively adjusted their role, increasing government accountability by decentralizing power in the area of environmental policy making by encouraging public participation. Surveys from the point of view of social practice allow us to more directly understand the participation of European citizens and civil society in environmental governance. The practices of Europe's highly developed civil societies have provided valuable experience for other countries and regions of the world.

1. Multiple participating entities

European countries' mature civil societies ensure that the number of participating entities in environmental governance is relatively large. Ronis' (1989) and Aberg's (1996) empirical research in Germany and Sweden identified subconscious action and rational choice as two parallel routes for actors in the environmental opportunity structure. Habit as an illogical course of action ensures that European citizens' participation in environmental governance is not merely restricted to such factors as technical ability, knowledge and finance but also shows itself to be a relatively widespread social culture.

Moreover, households are an important unit of action in the process of participation in environmental governance by European civil societies. At present, a large proportion of environmental protection projects are based on household actions, and these kinds of projects based on bottom-up actions have relatively wide coverage and their implementation is relatively effective. At the same time, the family is the basic unit for the formation of citizens' environmental consciousness. Bubolz and Sontag's (1993) model of the "family ecological approach" examined the mutual influence which family members have in molding opinions, and confirmed the importance of the family in the fostering of environmental protection habits, the acquisition of environmental information and the development of capacity for action by individual citizens.

Third, the theory of regional governance which arose in Europe emphasizes the importance of communities in environmental governance. Because of the wide geographic spread of environmental problems, environmental problems with different time and spatial dimensionality always reveal a certain specificity which is rooted in the social background in which they arose (Ayers, 2011; Squires and Renn, 2011). This has effectively overcome the problem of environmental policies being treated as uniform by technically minded bureaucracies, leading to difficulties with implementation and low effectiveness. Successful participation by European communities in environmental governance and the political structures of European deliberative democracy are inseparable. The latter provides the systemic guarantee of communities' participation in environmental governance.

Fourth, environmental NGOs are one of the liveliest forces in the area of environmental protection in European civil societies. In environmental governance projects in a series of European countries, non-governmental organizations have taken on such important functions as overseeing implementation, disseminating ideas, providing expert knowledge and resolving disputes. Moreover, the special governance structures derived from European integration have given non-governmental organizations wide freedom of action. As Beckmann (2009) points out in his study of EU agri-environmental schemes (AES's), non-governmental organizations have been able to effectively coordinate national and EU-level actions, making implementation of the schemes more effective at national level.

However, participation by European non-governmental organizations in environmental governance has certain limitations: because environmental policies in the EU and some European countries experience relatively strong influence by the informal patterns of interaction of political notables, it is hard for non-governmental organizations to exert effective influence over policy-making. Information flows in EU environmental policy making follow a top-down model, and even though the number of environmental NGOs has substantially increased, because of their inferior position in information acquisition, it is impossible for them to take part directly in policy-making (Hurtak, 2007; Polini-Staudinger, 2005; Hallstorm, 2004). Considering the important role of non-governmental organizations in environmental policy-implementation and their huge potential in environmental policy-making, many European countries have progressively reformed their political structures in order to allow non-governmental organizations to more effectively fulfill their functions in environmental governance (Hurtak, 2007).

2. Networked forms of action

At present citizens' environmental actions in European countries are becoming increasingly networked. As an important aid to organization and action by civil society forces, the internet can become not only a platform for timely exchange between actors, but also can become an important medium for information dissemination. The use of the internet has greatly increased the effectiveness of environmental action networks and promoted information exchange and resource sharing amongst civil society groups. Moreover, this kind of networked environmental action has noticeably reduced organization costs, which has enormous significance for small public interest groups focused on the environment. Environmental protection actions which use digital technology and the internet can greatly expand their influence by contacting a large number of people in order to disseminate information and ideas.

Using the internet to involve people in environmental participation also helps to guarantee citizens' basic environmental rights. To take Romania as an example, the E-Environment program, implemented nationwide in Romania, as an important part of the E-Romania national strategy, will put all indicators from domestic environmental databases online for social groups and individuals to consult. At the same time, the E-Romania platform also serves as an important feedback mechanism for the public on environmental issues and a channel for participation in environmental supervision (Mares, 2011). This web platform has at the same time realized citizens' rights to know environmental information and their right to environmental participation, which makes it an important institutional innovation.

3. The broader domain of participation

As far as the domains affected by public participation in environmental governance are concerned, the problems which contemporary European civil society groups focus on grow broader by the day, from the traditional focus on pollution control to the areas of natural resource management and social norms. First of all, civil society groups in European countries have long played an important role in environmental pollution control regulation and governance. In terms of policy-making on the governance of pollution, civil society groups in European countries have long been able to "speak out" through different channels, to exert political pressure through proactive lobbying and to express their demands in relation to environmental protection. As far as implementation of pollution control policy is concerned, civil society has effectively fulfilled its role of supervision and assistance. On the one hand, civil society groups are third parties, external to the stakeholders whose interests are directly involved in environmental pollution, and are able to objectively assess implementation of environmental policy, with the result that involving environmental protection groups as monitors has become an important technique for increasing public trust in policy implementation (Poloni-Staudinger, 2005). On the other hand, as providers of public services, civil society groups are able to effectively deploy their flexibility and specialization, spotting problems quickly and organizing solutions. In summary, in circumstances of longstanding focus on and participation in governance of pollution incidents, European civil society groups have already formed a mature model, able to mobilize effectively, respond rapidly and in governance "reach the places that others can't reach", filling in gaps left by government and market forces.

Moreover, the emphasis given to natural resource management questions by European civil society grows by the day. The important role of civil society in natural resource management has already been the subject of full theoretical demonstrations(cf. Rauschmayer et al., 2009; Dietz and Stern, 2008, Stoll-Kleemann and Welp, 2008). Faced by the high degree of uncertainty in natural resource development, intricately complex structures of interests and common deficiencies in natural resource management systems, European countries were in dire need of an innovative model of natural resource management. Involving the public in finding policy solutions to difficulties in natural resource management has had beneficial effects (Dreyer, 2011). As Reed (2008) points out, in the practice of natural resource management, the role civil society accords with a philosophy of empowerment, equality and mutual trust, encourages policy-making methods which suit the policy-making background, and naturally integrates scientific knowledge with local knowledge. The "analytical-deliberative processes" adopted by European countries in natural resource management all embodied the recognition of the importance of civil society groups and the processes of "social study" which they advocated (Squires, 2011). This model of natural resource management reliant on social groups is starting to be seen in developing countries, moving in step with political reforms and gradually becoming implanted in their social soil (Swatuk, 2005; Tsang, 2010).

Apart from this, environmental governance participation by European civil societies also touches upon the area of social norms: on the questions of opening national parks and protecting biodiversity, civil society groups enjoy exceptional advantages. On the one hand, based on rich experience and local knowledge, civil society groups can often find solutions to sensitive questions through careful examination and screening; on the other hand, familiarity with specialized technology allows civil society organizations to set the pace for government in policy-making, organically combining technology with social norms. The simulation model for developing national parks, widely seen in Western Europe these days, is a typical example: in Britain's PROGRESS plan, scholars and environmental NGOs jointly promoted the use of Geographic Information Systems (GIS) in management of national parks, and in this way provided detailed visual information and an integrated forward view on the development of the park to stakeholders. Through this technology, the public was able to obtain information more directly and thus better participate in the formulation of the relevant policies (Edwards, 2011). However, as Harrison (2002) points out, an open model of policy making remains a precondition for public participation in environmental governance in the area of social norms. It is only when government holds deliberative democracy in esteem and encourages the public to participate in policy making that technological strengths can serve as a means for the public to obtain information and thus promote democratization of environmental policy. When examining the participation of European civil society in environmental governance, we must take into consideration the ways in which the local culture of policy-making may limit the transferability of European experience.

4. Government-society relations in transformation

The lively nature of the participation of European civil society in the practice of environmental governance over a long period of time helped to mould government-society relations in this area. Based on a full recognition of the important role of civil society groups in the area of environmental governance, European governments tacitly adjusted their roles, delegating power through a process of decentralization, encouraging participatory forms of environmental governance by the public, communities and social organizations, and attempting to encourage voluntary initiatives within enterprises so that market forces could collectively play a positive role. The various experiences of European governments in environmental governance responded to a strong desire to participate in governance on the part of civil society groups, and in the process of environmental democratization, the accountability of European governments was also greatly strengthened. This kind of transformation of government-society relations in the practice of European public governance produced mutually beneficial results, and to a large degree explains why European countries, both conceptually and in terms of results, occupy a leading position in the area of environmental governance compared to other countries and regions of the world.

In order to promote environmental democracy and to realize participatory environmental governance, European governments all introduced institutional adjustments. The core of this kind of progressive reform was decentralization of power. By weakening central power, simplifying the layers of administration, introducing specialized agencies, and opening up channels for expression of opinion, these changes aimed to integrate social forces and to introduce public participation into environmental policy making and the practice of environmental governance (Holzhauer, 1997). Decentralization was a necessary condition for participatory mechanisms of public governance. It resolved the problems of lack of openness with information in policy-making and lack of local participation. As shown by Akbulut's (2012) "participatory mechanism tree", decentralization helped stakeholders to grasp relevant information and ensured that policy-making procedures took into account the local power structure and traditional institutional arrangements, opening up the channels for public participation in governance at a systemic level. This kind of decentralizing institutional reform was widely seen not only in Europe's traditional democratic political institutions (cf. Poloni-Staudinger, 2005; Hurtak, 2007; Holzhauer, 1997), but was also inherent to the process of political reforms in recently democratized East European countries (Adaman and Arsel, 2005; Akbulut, 2012).

Apart from institutional innovation and transformation of roles, European governments also achieved good environmental governance by means of mechanisms to mobilize market forces. Their essence was to introduction cooperation amongst "bargainers" and to make natural resources, which had originally been considered external to the capital of the enterprise, into the enterprise's property, so as to encourage them to protect the environment spontaneously (Barron, 1998; Cavaliere, 1998, 2000). An important practice in government delegation of power to enterprises and reassigning environmental property rights was voluntary innovation, which was widely applied in Britain. Under this policy, enterprises and government were given equal powers in the market, which was meant to open up equal and mutually beneficial cooperation through the market mechanism. In environmental policy making and implementation, voluntary innovation fully mobilized the energies of small and medium enterprises, which not only increased the number of subjects of environmental policy-making, made the model livelier and increased the effectiveness of implementation, but also improved the image of government and promoted joint action between government and enterprises (Peters, 2004).

At the same time as realizing environmental democracy and environmental good governance, the delegation of power by European governments to society and to enterprises also increased government accountability. As an important hallmark of modern democratic governance (see Bovens, 2005; Mulgan, 2003), the degree of political trust indicates the levels of advantage or disadvantage for an authoritarian state that oriented their behaviors to the decision-making process, and for a service-oriented state that act to provide services. Frequently occurring crises of trust made European governments examine their political accountability. Scholars believe that the main reasons for the lack of political accountability were a lack of clarity in the division of responsibilities for legislation and implementation of legislation, the proliferation of expert language and technocratic rule, and the marginalization of the public in processes of public governance (Joss, 2010). The adjustment of the government' leading position in public policy-making, the involvement of social groups and market forces in governance, and the fostering a democratic policy-making atmosphere are all beneficial to the building of political accountability. It is not hard to see that integrated participation of civil society in environmental governance in European countries not only raised the efficiency of policy-making and the effectiveness of implementation, but also consolidated political accountability which is the source of political legitimacy. This kind of governing experience is worthy of study and emulation by other countries and regions.

Critical assessment

Through an investigation of European countries' policy practice and legislation on public participation in environmental governance, it becomes evident that the encouragement of public participation in environmental governance closely followed tendencies in European political development and the evolution of social culture. On the one hand, the soil of European democratic politics nourished such concepts and theories as citizens' environmental rights and environmental democracy. On the other hand, the highly developed consciousness of political participation amongst European citizens fostered strong social forces, and their involvement in environmental governance fully embodied the participatory political culture of contemporary European social democracies.

At the same time several decades of practical experience with public environmental participation in European countries shows that this governance model involving multiple social forces in governance has proved its effectiveness in political, economic and social terms. Civil society groups can not only effectively reflect the interests of marginal groups, but can also provide valuable local knowledge to the definition of environmental problems. Empowered social and political forces are highly enthusiastic about environmental participation, and through their involvement in environmental policy-making and implementation, they have been able to express opinions on how to solve their own problems, which has generally raised the effectiveness of government.

European countries are ahead of other countries and regions of the world in the practice of public participation in environmental governance. However, the question of how far successful European experience of public environmental participation is applicable to other countries is still a question worthy of some discussion. Generally speaking, a country's level of economic development, political culture and social traditions will exert some influence on the success rate of transferring the "European experience". A country's level of economic development will determine the balance between the environment and development in its policy goals; a country's degree of political centralization will determine its ability to tolerate participation by social forces in political debates; and a country's social and political traditions represent its mainstream social values which determine the level of desire for participation in social governance by citizens and civil society groups. In the contemporary world, countries' development continues to be uneven. There are relatively large differences in countries' progress with industrialization and their pace of democratization. Although participatory governance in the area of the environment has numerous advantages, copying the European experience wholesale in a different context would be a mistake.

As far as China is concerned, environment and development are not at two irreconcilable ends of the spectrum. On the contrary, conserving the environment and increasing energy efficiency have become integral parts of China's economic development goals and national security (Tsang, 2010). On the other hand, China' repeated environmental crises have caused the environment to leap to the top of the public agenda. Of course the Chinese central government has already in its policy documents expressed its determination to strictly enforce regulations, but central directives cannot effectively resolve the contradiction between short term economic development and long term environmental effectiveness at local level. Central government policies are by no means effectively enforced at local level. For civil society to close the gap between central and local political forces on environmental issues would be an important step towards resolving China's environmental governance difficulties. Like East European societies, China still lacks a fully formed process for social forces to participate in environmental governance. How to cultivate on China's soil civil society groups with a high degree of expertise, enthusiasm, and political engagement, in order to fully mobilize citizens' environmental participation, and how to provide the appropriate institutional guarantees for this, are important topics which current incumbents need to consider.

References

[1] Aberg H, Dahlman S, Shanahan H, Saljo R. 1996. Towards sound environmental behaviour: exploring household participation in waste management. Journal of Consumer Policy 19. pp. 45-67.

[2] Adaman F, Arsel M. 2012. Political economy of the environment in Turkey. In Heper M, Sayarı S. (eds). Handbook of Modern Turkey. London: Routledge. pp. 317-327.

[3] Akbulut B, Soylu C. 2012. An inquiry into power and participatory natural resource management. Cambridge Journal of Economics 36. pp. 1143-1162.

[4] Andersen J. 2003. Gambling politics or successful entrepreneurialism? The Orestad in Copenhagen. In Moulart F, Swyngedouw E, Rodriquez A. (eds). The Globalised City: Urban Redevelopment and Social Polarisation in the European City. Oxford: Oxford University Press. pp. 91-106.

[5] Andonova LB. 2010. Public-private partnerships for the earth: politics and patterns of hybrid authority in the multilateral system. Global Environmental Politics 10. pp. 25-53.

[6] Andrews R. 1993. Environmental policy in the Czech and Slovak Republic. In Vari A, Tamas P. (eds). Environment and Democratic Transition: Policy and Politics in Central and Eastern Europe. Boston: Kluwer Academic Publishers. pp. 5-48.

[7] Arnstein S. 1969. A ladder of citizen participation. Journal of the American Institute of Planners 35 (4). pp. 216-224.

[8] Ayers J. 2011. Resolving the adaptation paradox: exploring the potential for deliberative

adaptation policy-making in Bangladesh. Global Environmental Politics 11 (1). pp. 62-88.

[9] Bäckstrand K. 2006. Multi-Stakeholder Partnerships for Sustainable Development. European Environment 16. pp. 290-306.

[10] Balsinger J, Van Deveer SD. 2012. Navigating Regional Environmental Governance. Global Environmental Politics 12 (3). pp. 1-18.

[11] Barber B. 1984. Strong Democracy. Berkeley: University of California.

[12] Barber B. 1988. The Struggle for Democracy. Boston: Little Brown.

[13] Barron WF. 1998. Unpriced economic incentives for environmental management. Working Paper(Hong Kong: Centre of Urban Planning and Environmental Management, The University of Hong Kong).

[14] Beckmann V, Eggers J, Mettepenningen E. 2009. Deciding how to decide on agri-environmental schemes: the political economy of subsidiarity, decentralisation and participation in the European Union. Journal of Environmental Planning and Management 52 (5). pp. 689-716.

[15] Biermann F, et al. 2007. Multi-stakeholder partnerships for sustainable development: does the promise hold? In Glasbergen P, Biermann F, Mol APJ. (eds). Partnerships, Governance and Sustainable Development: Reflections on Theory and Practice. Cheltenham: Edward Elgar. pp. 239-260.

[16] Book K, Eskilsson L, Khan J. 2010. Governing the balance between sustainability and competitiveness in urban planning: the case of the Orestad Model. Environmental Policy and Governance 20. pp. 382-396.

[17] Bovens M. 2005. Public accountability. In Ferlie E, Lynne L, Pollitt C (eds). The Oxford Handbook of Public Management. Oxford: Oxford University Press. pp. 182–208.

[18] Bowman M, Hunter D. 1992. Environmental reforms in post-communist Central Europe: from high hopes to hard reality. Michigan Journal of International Law 13. pp. 921-980.

[19] Bubolz MM, Sontag MS. 1993. Human ecology theory. In Boss P, Doherty W, LaRossa R, Schumm W, Steinmetz S (eds). Sourcebook of Family Theories and Methods: A Contextual Approach. New York: Plenum. pp. 419-448.

[20] Cavaliere A. 1998. Voluntary Agreements, Over-compliance and Environmental Reputation

Milan: Fondazione Eni Enrico Mattei (FEEM).

[21] Cavaliere A. 2000. Over-compliance and voluntary agreements: a note about environmental reputation. Environmental and Resource Economics 17 (2). pp. 195-202.

[22] Conca K. 2012. The Rise of the Region in Global Environmental Politics. Global Environmental Politics 12 (3). pp. 127-133.

[23] de Abreu Ferreira. 2010. Fundamental environmental rights in eu law: an analysis of the right of access to environmental information. In Pavlich D. (ed). Managing Environmental Justice. Amsterdam: Rodopi. pp. 125-142.

[24] Dietz T, Stern PC.(eds). 2008. Public Participation in Environmental Assessment and Decision Making. Washington, DC.: The National Academic Press.

[25] Dreyer M, Renn O. 2011. Participatory approaches to modelling for improved learning and decision-making in natural resource governance: an editorial. Environmental Policy and Governance 21. pp. 379-385.

[26] Dryzek JS. 1990. Discursive Democracy. Cambridge: Cambridge University Press.

[27] Dryzek JS. 1992. Ecology and discursive democracy: beyond liberal capitalism and the administrative state. Capitalism, Nature, Socialism 3 (2). pp. 18-42.

[28] Earnhart D. 1996. Environmental penalties against enterprises and employees: labour contracts and const-shifting in the Czech Republic. Comparative Economic Studies 38 (4). pp. 1-34.

[29] Earnhart D. 2001. Environmental prontection efforts under communism and democracy: public involvement and political influence. Comparative Economic Studies 43 (2). pp. 35-58.

[30] Edwards VM, Smith S. 2011. Lessons from the application of decision-support tools in participatory management of the New Forest National Park, UK. Environmental Policy and Governance 21. pp. 417-432

[31] European Communities (EC). 2000. Directive 2000/60/EC of the European Parliament and of the Council of 23 October 2000 establishing a framework for Community action in the field of water policy. Official Journal of the European Communities.

[32] Fischer F, Forester J. (eds). 1993. The Argumentative Turn in Policy Analysis and Planning. Durham: Duke University Press.

[33] Garnåsjordet PA. et al. 2012. Sustainable development indicators: from statistics to policy.

Environmental Policy and Governance 22. pp. 322-336.

[34] Glasbergen P. 2007. Setting the Scene: the partnership paradigm in the making. In Glasbergen P, Biermann F, Mol APJ (eds.). Partnerships, Governance and Sustainable Development: Reflections on Theory and Practice. Cheltenham: Edward Elgar. pp. 1-25.

[35] Gupte M, Bartlett RV. 2007. Necessary preconditions for deliberative environmental democracy? Challenging the modernity bias of current theory. Global Environmental Politics 7 (3). pp. 94-106.

[36] Hakala E. 2012. Cooperation for the enhancement of environmental citizenship in the context of securitization: the case of an OSCE project in Serbia. Journal of Civil Society 8 (4). pp. 385-399.

[37] Hallstorm LK. 2004. Eurocratising enlargement? EU elites and NGO participation in European environmental policy. Environmental Politics 13 (1). pp. 175-193.

[38] Harrison C, Haklay M. 2002. The potential of public participation geographic information systems in UK environmental planning: appraisals by active publics. Journal of Environmental Planning and Management 45 (6). pp. 841-863.

[39] Holden E. 2004. Ecological footprints and sustainable urban form. Journal of Housing and the Built Environment 19. pp. 91-109.

[40] Holzhauer A. 1997. Environmentalism and Political Participation in Five West European States. DPhil Dissertation for Purdue University.

[41] Horelli L. 1998. Creating child-friendly environments: case studies on children's participation in three European countries. Childhood 5 (2). pp. 225-239.

[42] Horelli L. 2001. A comparison of children's autonomous mobility and environmental participation in Northern and Southern Europe: the cases of Finland and Italy. Journal of Community Applied Social Psychology 11. pp. 451-455.

[43] Hu B. 2005. The system of water environmental standard in European Union. (in Chinese) Research of Environmental Sciences 18 (1). pp. 45-48.

[44] Hurtak DE. 2007. The European Union and the Environmental Movement in Europe: An Analysis of Insider Status. DPhil Dissertation for The New School for Social Research.

[45] Joss S. 2010. Accountable governance, accountable sustainability? A case study of

accountability in the governance for sustainability. Environmental Policy and Governance 20. pp. 408-421.

[46] Khagram S, Ali SH. 2008. Transnational transformations: from government centric inter-state regimes to multi-actor, multi-level global governance? In Park J, Conca K, Finger M (eds.). The Crisis of Global Environmental Governance. London: Routledge. pp. 209-249.

[47] Majoor S, Jorgensen J. 2007. Copenhagen Orestad: public partnership in search of the market. In Salet W, Gualini E (eds.). Framing Strategic Urban Projects. London: Routledge. pp. 172-198.

[48] Mares C, Gilia C. 2011. E-Environment as Part of the National Strategy E-Romania. In Gasco M (ed.). Proceedings of the 12th European Conference on e-Government. http: //academic-conferences.org/pdfs/ECEG_2012-Booklet.pdf (Accessed 25/10/2014).

[49] Mehnen N, Mose I, Strijker D. 2013. Governance and sense of place: half a century of a German nature park. Environmental Policy and Governance 23. pp. 46-62.

[50] Mulgan R. 2003. Holding Power to Account: Accountability in Modern Democracies. Basingstoke: Palgrave.

[51] Naughton HT. 2010. Globalisation and Emission in Europe. The European Journal of Comparative Economics 7 (2). pp. 503-519.

[52] Omanga E, Ulmer L, Berhane Z, Gatari M. 2014. Industrial air pollution in rural Kenya: community awareness, risk perception and associations between risk variables. BMC Public Health 14 (377). pp. 1-28.

[53] Peters M, Turner RK. 2004. SME environmental attitudes and participation in local-scale voluntary initiatives: some practical applications. Journal of Environmental Planning and Management 47 (3). pp. 449-473.

[54] Pickett STA, Cadenasso ML, Grove JM, Nilon CH, Pouyat RV, Zipperer WC, Costanza R. 2001. Urban ecological systems: linking terrestrial ecological, physical, and socioeconomic components of metropolitan areas. Annual Review of Ecology and Systematics 32. pp. 127-157.

[55] Poloni-Staudinger LM. 2005. How to act and where to act: the dynamic determinants of environmental social movement activity in the United Kingdom, France and Germany. DPhil

Dissertation for Indiana University.

[56] Rauschmayer F, Paavola J, Wittmer H. 2009. European governance of natural resources: participation in a multi-level context. Environmental Policy and Governance 19 (special issue). pp. 141-147.

[57] Reed MS. 2008. Stakeholder participation for environmental management: a literature review. Biological Conservation 141. pp. 2417-31.

[58] Robinson D. 1993. Public participation in environmental decision making. Environment and Planning Law Journal 10 (5). pp. 320-340

[59] Roggero M. 2013. Shifting troubles: decision-making versus implementation in participatory watershed governance. Environmental Policy and Governance 23. pp. 63-74.

[60] Ronis DL, Yates JF, Kirscht JP. 1989. Attitudes, decisions and habits as determinants of repeated behaviour. In Pratkanis AR, Breckler SJ, Greenwald AJ (eds.). Attitude, Structure and Function. Hillsdale, NJ: Erlbaum. pp. 213-239.

[61] Ruggie JG. 2004. Reconstituting the global public domain: issues, actors and practices. European Journal of International Relations 10. pp. 499-531.

[62] Rydin Y. 1999. Environmental governance for sustainable urban development: a European model? Local Environment 4 (1). pp. 61-65.

[63] Slocock B. 1996. The paradoxes of environmental policy in Eastern Europe: the dynamics of policy-making in the Czech Republic. Environmental Politics 5 (3). pp. 501-521.

[64] Squires H, Renn O. 2011. Can participatory modelling support social learning in marine fisheries? Reflections from the invest in Fish South West Project. Environmental Policy and Governance 21. pp. 403-416.

[65] Stoll-Kleemann S, Welp M (eds.). 2008. Stakeholder Dialogues in Natural Resources Management: Theory and Practice. Berlin: Springer.

[66] Swatuk LA. 2005. From "project" to "context": community based natural resource management in Botswana. Global Environmental Politics 5 (3). pp. 95-124.

[67] Town and Country Planning Act 1990. http://www.legislation.gov.uk/ukpga/1990/8/contents. (Accessed 25/10/2014).

[68] Tsang S, Kolk A. 2010. The evolution of Chinese policies and governance structures on

environment, energy and climate. Environmental Policy and Governance 20. pp. 180-196.

[69] United Nations Economic Commission for Europe (UNECE). 1998. Aarhus Convention. http: //www.unece.org/fileadmin/DAM/env/pp/documents/cep43e.pdf. (Accessed 25/10/2014)

[70] United Nations Environment Programme (UNEP). 2007. Ensuring Stakeholder Engagement in the Development, Implementation and Updating of NBSAPs, Training Module B-5, Version 1.

[71] van Laerhoven H. 2014. When is participatory local environmental governance likely to emerge? A study of collective action in participatory municipal environmental councils in Brazil. Environmental Policy and Governance 24. pp. 77-93.

[72] Wolchik S. 1991. Czechoslovakia in Transition: Politics, Economics and Society. NY: Pinter Publishers.

[73] Zeiter AA. 2000. Environmental Democracy: An Analysis of Brownfields Policy-Making. DPhil Dissertation for Purdue University.

[74] Zhu Y. 2007. Research on the Public Participation System in Environmental Legislation. (in Chinese) L.L.M Degree Thesis for East China University of Politics and Law.

环境保护的自愿参与视角：来自欧盟的经验

陈文山*

导言

从传统上来说，政府通常被认为是设计环境政策并执行政策的主体。在 OECD 国家特别是美国和欧洲国家中，这种以政府为中心的对于治理的理解在不断地受到挑战。无论是在学术领域还是在政策制定者中，社会治理行为和政府机构的管理之间一直被看成为相互区分的事情。治理行动可以确定为通过不同的利益相关方为了公共责任（如保护自然）分享而采取的公共行动。这一定义是把治理进程看做是对那些在社会中有影响力但分散的公共行动者的协调过程。在此过程中，政府仍然保留着重要地位，但它们已经不再是指挥和控制这一进程来保持集权化和最终的权威性。相反，公民、公民社会组织和企业能够在这一进程中构建起这种协调的权威性。自愿的环境参与视角能够很好地满足我们在超越传统政府管理的视角上来理解治理含义的需要。该视角时常为非政府的行动者（如公司和公民社会组织）所倡导，并在规范和执行上强调公民社会组织参与的意义。

采用公民自愿参与环境治理的视角具有以下积极的意义：

a）加强合法性。在全球环境治理（如联合国机构）的背景中，一个社会中因缺乏民主所形成的缺点可以部分地通过公民社会和民间组织的积极参与来进行弥补。

b）增进协调和运作。通过强调不同的利益群体的利益，公共部门和私营部门之间的联系将会更为密切。在许多情况下，私营部门也可以对于履行政府治理目

---

*陈文山，德国发展政策研究所博士；本文原文为英文，中译者为李骅（浙江大学公共管理学院）。

标起到积极的作用,并缓解政府的官僚主义倾向。

c) 扩大知识领域。即使掌握信息资源最全面的政府也难以拥有全能全知的官员来制定理性的决策。相反,他们的政策领域和议题讨论常常受限于有限的知识运用。通过与民营部门进行合作,政府可以从中得到专门的知识。

d) 资源的积累。除了具备更丰富的专业知识资源的能力,非政府部门也能够为追求公共产品提供更多的资源,他们能够降低政府在管理和运作过程中的成本(如 Arora and Cason, 1996)。

尽管具有这些潜在优势,自愿参与的视角也常常受到各种批评。特别是在国家和地方层面上,运用自愿参与的方式会提升环境治理的私有化导向(Cashore, 2002)。其次,某些政府管理方式会削弱其原有的民主制衡功能,也会影响到自主的私立部门。另外,污染企业涉及那些私营的和公私营污染企业所具有的绿色计划,它们的核心商业部门仍然具有污染性。第三,诸如"洗绿"的行动削弱了自愿环境治理机制的有效性。第四,对于企业来说,履行环境责任会导致较高的成本和不断增长的管理负担,例如,社会组织的行动者要求企业提供更多的环境信息,并向公众进行报告。

本研究的目的在于通过揭示公共机构治理环境的行为来激励私营部门采取更多对环境负责的行为。本文考察了欧盟的环境自愿视角并回顾了欧盟及其成员国在支持环境自愿行为方面的政策,还对这一视角在欧洲之外的可运用性进行简略的讨论,特别是对于中国来说的适用性。

典型的个案和政策研究都强调了他们从自愿的环境参与视角来讨论并确立对政策和个案的选择,这种政策和个案的选择可以根据以下的标准:

讨论欧盟流行的经验和教训;

概括与自愿参与环境行为有关的政策来获取经验,特别是绿色公共采购、自愿协议和环境报告等;

讨论那些超越于欧盟背景的个案及其相关性,特别是在中国背景中(例如,中国比较多的国有企业和传统的政府管制部门);

专业选择要与成功经验以及发展战略和措施相关。

在本文中,我们特别关注以下视角:政府报告、生态标签、谈判过程、环保的公共采购标准。

本研究强调支持自愿环境参与的视角所采取的公共政策，尽管这些公共政策是被来自于民间的动力所驱动。本研究控索的公共议题是公共部门在指引企业的积极性并采取对环境负责的企业行为。

报告

环境报告是指社会组织特别是私营企业发布的有关环境状况的信息报告。这类报告的目的在于增强环境行为的透明性、可度量性及可记录性。增强透明性意味着允许公共机构检查和公共关系介入，对于投资者来说，也意味着区分和评估环境风险和机会。与财务信息的发布不同的是，营利组织通常不具有非财务信息发布的义务。然而，越来越多的企业意识到发布非财务性报告的好处。环境报告是企业维护和建立与利益相关者、（潜在的）投资者和社会利益的相关者（如工会和非政府组织）之间信任的机制。此外，环境报告还有利于改善企业管理并带来较好的风险管理，也是价值增值的有效工具。因此，有一些企业将以超越法律要求发布环境报告列为其预期的竞争优势。

欧洲的企业在企业环境报告方面的工作长期居于领先地位。根据 2012 年 6 月可持续性信息披露数据库的数据显示：在全球报告倡议组织（Global Reporting Initiative，GRI）登记的 10 034 份报告中，有 4 626 份报告来自于欧洲地区的各种组织（不包括欧盟），这一数据占总数的 46%。其他与欧盟相关的组织构成了另外的 41% 的报告。在与欧盟相关的企业提供环境报告的行为十分普遍的情况下，新型工业化国家中的各种组织所提交的环境报告数量在近年来也快速增长。

各种支持性的举措也增进了环境报告的质量和可信度。下面我们将简单地讨论全球报告倡议组织（GRI）、联合国全球契约及担保标准。

驱动力 I：全球报告倡议组织（GRI）

全球盛行的环境报告提出了有关报告形式的问题。如果不规范报告的形式和标准，那么将很难对不同报告进行比较。因而来自于各方行动者的披露需求便无法满足。虽然目前不存在国际通用的报告标准，但全球报告倡议组织的指标体系是最为接近通用标准的。1997 年 4 月，这一指标体系由 CERES 和联合国环境规划署（United Nations Environment Program UNEP）共同发起。该指标设立了三条底线，即经济、环境和社会。2000年，第一版指导书发表。2002年，第二版指导书递交到了在约翰内斯堡召开的"可持续性发展世界首脑大会"。同年，该组织成为了一个独立的组织并在荷兰阿姆斯特丹设立秘书处。这一组织继续与联合国机构合作，特别是联合国发展署和联合国全球契约（Global Compact）。其目标是增进社会与环境责任方面的可持续性环境报告，或者说是非财务的责任的报告。2006年，新版的可持续性报告指南出版。目前，该指南已更新到第四版（GRI2012）。该指南的更新与技术发展相联系，为了更好地适应国际标准并采纳更多的网络报告。可持续性报告指南包括报告内容、确保报告信息质量和披露标准的"绩效指标"和其他项目，并对具体技术问题进行指导。组织可采用此 GRI 框架标准申请其分类：A、B、C，其中 A 代表着最高级的报告。

驱动力 II：全球契约和对话进程

全球契约（UNGC）是 2004 年由联合国发起的旨在履行千年发展目标的活动。它包括联合国成员国和国际组织约定在 2015 年实现的 8 个目标。千年发展目标将环境可持续性作为其发展目标，联合国全球契约也鼓励私营企业介入运行这些国际约定的目标中。截至 2012 年，该项目已有一万多名参与者，并成为世界上最大的社会和环境自愿协作的企业合作组织。

全球契约没有一个明确的固定格式，只要求其参约者每年与那些企业的利益相关者直接对话。对话形式并不固定，但它要确保企业决策者愿意续约，并达成可见的且结果是可以被测量的实践行动。企业在参与的前两年可以免去履行对话进程的要求。在此基础上这些参约者被界定为全球协定的积极行动者。这就意味着参约者符合对话行动的要求，或者成为全球协议的高级行动者，即参约者所做的已超过全球协议的基本要求。但如果达不到这些要求，参约者就要以见习者的身份提交一份新报告以满足契约的要求。

驱动力 III：担保标准

自愿的自我报告同样有报告的可靠性问题。由外部组织做担保是一个选择，它提供了一个能够被利益相关者所信任的机会。在可以由别的机构做担保的情况下，GRI 提供了自我担保的选择。环境报告的担保提供了几种方案，例如 ISAE 3000 和 AA1000AS。国际标准 ISAE3000 是由国际审计和担保标准委员会发展形成，它强调报告的担保要超出财务信息。这些议题包括环境绩效、企业治理、内部协调、利益相关者的介入和其他企业责任。ISAE 3000 的特点是处在有限的担保和合理的担保之间。有限担保是一个否定性描述（如果没有证据能证明报告是无效的），合理担保包括了有效性的正面的表述。

AA1000 担保标准于 2003 年发布，它是一种评估是否能够以及在多大程度上确保可靠性的标准。这些原则包括：包容性（在回应可持续性问题上，利益相关者在发展企业战略和获得回应中的积极参与），成熟性（确立一个议题对于一个组织和利益相关者的相关性和重要性），责任性（一个组织要回应那些影响组织可持续性的与利益相关者的议题）。在 2008 年版的 AA1000AS 标准中（Accountability, 2008），要求开放公共信息的这一标准涉及 ISAE3000 作为可资借鉴的方法。这一标准超越了对于资料可靠性的评估，要求担保提供者评估他们所遵循的一系列原则。根据 AA1000AS 标准的可靠性，把那些对于高风险的担保者和对于中度风险的担保者区分出来。要求担保提供者要获得充分的证据来支持（尽管没有低到零）尽可能地降低风险（Accountability, 2008, 11）。AA1000 的标准体系包含利益相关者的介入标准，这提供了确保利益相关者介入的分析框架。在使用可持续性报告担保的问题上，各个区域区别很大。2008 年，在美国只有 3%的企业风险报告具有担保，而在欧洲这一比例是 45%（Brown, de Jong and Levy, 2009）。

驱动力Ⅳ：碳披露计划

碳排放计算和信息的披露问题与气候政策相关，这也是在不断增长的国际报告中出现的。最显著的驱动力或许是"碳披露计划"（Carbon Disclosure project, CDP）。该计划于 2000 年启动，致力于在投资者和管理者之间建立更透明的关系。试图引起投资者对于气候变化的关注，并告知投资者关于公司在气候变化方面所造成的风险。"碳披露计划"向那些全球大公司发送由机构投资者提出的问卷，以搜集他们的答复，或让他们能够在线回复。进一步说，CDP 按照各个公司的指标和得分来进行排名，继而决定哪些公司在碳信息披露指标排名最高（CDP, 2013）。碳披露计划的成功在于使投资者和公司能够同时介入并使得许多投资者参与了每年的问卷调查。比如在 2003 年的一次问卷发放中，有 35 个机构的投资者回答了问卷，这些公司具有 45 亿美元的资产。到 2011 年例行调查问卷调查时，CDP 代表总资产达 710 亿美元的 551 个投资者向全球 500 强公司发送问卷，同一时期答复率也从超过 235 家企业增长到超过 3 000 家企业。

欧盟及其成员国推进报告的政策

2001 年，欧盟委员会就企业的环境报告发表了免责指南。在第四次年度披露的统计中（European Commission，2003a），企业被要求发布环境和就业相关的信息以确保企业的运行。然而，成员国可以免除中小企业的信息披露。有些成员国已经要求其披露机制要高于欧盟的规定（例如荷兰、英国、瑞典、西班牙、丹麦和法国）。此外，还有一些针对环境状况的特殊指南和环境宣言。例如，欧洲化学工业理事会已经在 1998 年对欧洲化学企业发表了指南。目前，欧盟并不提供单一的环境报告框架，而是为这些报告提供法律基础，以便在整个欧洲的法律体系和法令中，特别是在欧盟企业可持续责任战略和社区环境管理和审计计划的规则中体现出来。同时欧洲单一市场法案也承诺各个部门的企业都需在社会和环境信息透明化方面提供法律提议。

尽管存在着许多相关的规则和国际标准，但在发布非财政信息方面仍然存在着不透明的问题（许多企业不愿意公布这些信息）以及所公布的信息质量较低的问题。作为回应，欧盟在 2013 年 4 月 16 日制定了主要针对大公司的社会和环境信息透明化的法令（European Comission，2013），这些法令修改了以往的法律（据第四和第七有关年度统计），要求有 500 人以上的大公司在他们的年度报告中需包含信息材料的披露。那些在自愿的基础上提供非财务报告的公司可以免除这一提供报告的责任。这一法令能够确保自愿报告的框架能够延续。

欧盟影响环境报告的政策 1：社区环境管理和审议计划

社区环境管理和审计计划（Community Environmental Management and Audit Scheme，EMAS）是评估管理和不断提高单个组织履行环境义务的工具。它要求提供明确的环境政策的运作程序和环境报告，其环境标准具有一定的真实性。

EMAS 和其他报告（如 ISO 14001）的区别在于，其他的环境报告要求企业在登记注册前，由执行环境保护的部门对其环境绩效和数据进行评估，再由第三方验证（认可的验证者）在组织（或组织的特定部分）注册。成员国可指定注册登记主管机关。

主管机构也负责监督和认证的情况（成员国可指定许可机构）。成员国主体机构每年都会召开年会，以确保 EMAS 通过建议以影响程序（如经济部门和地理区域）。只有发表了环境验证声明，并向有关登记机构登记公示（通常是州或地方机构）才能得到成立组织（公司、子公司、生产基地）的认证。EMAS 已经成功地吸引了越来越多的网站和组织。注册的网站已经从 1998 年的 2 140 个增加到 2010 年的 7 794 个，而 2003—2010 年的组织（企业）数量也从 3 055 个增加到 4 542 个（Słonimiec and Świtała, 2013）。目前，已有超过 4 600 多个组织参与注册。注册数量多少都与公开程度和环境验证情况相关，它可以被视为一个大型的报告格式，虽然报告并不是 EMAS 的主要目的。

欧盟影响环境报告的政策 2：企业社会责任

有关企业社会责任（Corporate Social Responsibility，CSR）的第一个信息是欧盟于 2001 年发表的绿皮书。绿皮书强调需要一个（公共）政策框架，同时也必须以自愿行动为原则。自愿行动准则有助于促进国际劳工标准的提升，然而，这也取决于正确的实施（欧洲经济共同体委员会，2001：13-14）。委员会强调"整体欧洲框架的规定"（同上：6）。然而，2002 年有关企业社会责任的描述则是：企业对可持续发展作出的贡献（欧洲经济共同体委员会，2002b）。它强调企业的"自愿贡献"和经济效益……把对社会和环境问题的考虑纳入他们的日常工作并从中受益（同上：14）。企业社会责任标准的应用可能"源于市场的失真"（同上：8）。最新消息显示：2011 年 11 月发布的"2011—2014 年企业社会责任欧盟战略"（欧盟委员会，2011b）强调，"企业社会责任需要平衡多方利益和商业行为的市场回报之间的关系，在更大程度上保持透明度和更加关注人权"（Martinuzzi, Krumay, and Pisano, 2011：21）。该战略将企业社会责任重新定义为"企业对于社会的影响"，而以前的定义是自愿。从对企业社会责任的重新定义为自愿行为来看，每一个企业对企业社会责任和环境报告都将采取更多的强制性措施。下面是一些成员国的发展趋势，它们说明了强制性方法已变得越发普遍。

欧盟影响环境报告的政策3：气候政策

也许欧洲最大的强制性政策就是在气候政策框架下实行的保证程序。但从环境的角度来看，气候（以及排放）的焦点是有限的，其框架中的指标、协议及报告标准都比较明确。欧盟碳排放交易体系（EU Emission Trading System，EUETS）成立于2005年，是碳排放限额的交易规模最大的国际体系。它覆盖27个欧盟成员国以及列支敦士登、冰岛和挪威。越来越多的能源密集型企业要求加入到该体系中（例如能源、金属、矿物和纸浆及造纸行业）。2012年，该计划扩展了航空领域。2013年又新增了石油化工、氨和铝产业。列入的行业需要根据该计划获得一个减少温室气体的排放许可证，从而逐步降低排放量。为实现和欧盟碳排放交易这一体系，对温室气体排放的监测和报告是十分必要的。从航空活动（2009/339/EC）（欧盟委员会，2009）的报告和温室气体排放监测指南（2007/589/EC；2004/156/EC）（欧盟委员会，2004，2007）的结果来看，该体系决定在已公布的监测和报告指南中再加入二氧化碳地质储存量（2007/589/EC）（欧盟委员会，2007）。2004年，监测和报告指南由委员会确立。2007年又发布了一个修订版指南（欧盟委员会，2007）。此外，该委员会还从其监测中公布了吨/公里排放标准和每年航空活动的碳排放量。

欧盟成员国影响报告的政策1：瑞典

早在2002年，瑞典政府就推出一个全球责任单位。2007年11月，瑞典政府成为世界上第一个要求本国国有企业参照全球报告倡议组织（GRI）的标准发布可持续性发展报告的国家。瑞典强制国有企业按照GRI的要求发布报告的措施也激发了挪威和荷兰在其国家采取相应的政策。采用强制性框架的主要原因是，瑞典政府有很大一部分国家股份，它可以通过此途径在50多个国有企业施展经济影响（例如采矿、房地产、金融和赌博业）。此外，由于国有企业的资产最终是每个瑞典纳税人的财产，政府因而交给国有企业一项特殊的权利和义务。NGOs提出在本国的企业可能不会对人权和环境负责任的情况发生。而采用GRI的标准能够确保瑞典企业在国际市场上的竞争力。显然，瑞典要采用最广泛认可的标准。自

2008年以来，该国国有企业都必须遵守"服从或解释"的原则，即发布其可持续发展报告或者有足够的理由解释为什么他们不报告。GRI标准的使用，也满足了外部的发展。

根据乌普萨拉大学商学部的研究（Frostenson，Windell，and Borglund，2010），该政策不仅增强了企业发布可持续性报告的意识，而且各企业的董事也重视起GRI的要求，纷纷加强企业内部的可持续性建设，也以更加结构化的方式管理着公司。最受影响的是环境政策、采购、人力资源和沟通部门。根据调查中的一些受访者所述，利益相关者对话机制得到改进，新的采购政策已被采纳并增加了对供应商的要求。然而，一些公司认为"GRI指南可能导致重大的改变，这有助于改善可持续发展的过程"（Borglund，Frostenson and Windell，2010），然而实际上GRI指南并没有发生重大的组织变化。最大的问题是只有个别有经验的企业能够提供可持续发展报告，他们已经制定出了新的政策和程序。有些企业抱怨GRI指标不适合他们的行业，或测量太难。尽管如此，大多数受访者都表示该政策是积极的。应用GRI不仅能够促进环境报告的发展，也可以在可持续发展的过程中不断改进。

报告的要求不过是一个工具，它需要因地制宜地被运用到本土。据瑞典企业部的数据显示，2010年，有96%的国有企业已发布报告。与此同时，国有企业为其他公司树立了榜样（Didong，2010）。私人公司的报告也增多了，有约30%的私人公司也已按照GRI报告标准发布了环境报告。同时，瑞典的政策也激发了荷兰等国家。荷兰拥有约30家有关基础设施、博彩业、电气业等公司。自2009年以来，荷兰的国有企业也需要根据GRI指南和荷兰透明度基准报告对企业社会责任进行报告。这一政策的施行离不开2008年荷兰政府信用危机的影响（Schaay，2010）。这场危机揭示了对利润的短期焦点是灾难性的。企业社会责任报告有助于使企业采取更平衡的方式发展经济。这一政策的最终目标是将企业社会责任作为一种主流政策内化于企业中。

欧盟成员国影响报告的政策2：法国

在欧洲企业非财务信息披露的专家组会议上，法国曾表示希望欧盟能发布欧洲报告标准。标杆管理和协调在欧盟以外是很常见的，虽然更多的是施行在英美国家的政策和企业社会责任报告中，但法国是企业社会责任报告的先驱（Antal and Sobczak，2007）。它在1977年就出现了强制性的企业社会报告（Code du travail，partie legislative，livre iv，titre iii，article L438）（Gouvernement Française，1977）。法律要求报告企业有关的134项内容。然而，不同于其他的是这种报告一般不提供给公众，它只向政府机构汇报。这个事实不仅说明了法国政府对经济的干预，也证明了该国中央集权的历史渊源。国家在经济方面的作用在20世纪70年代逐渐缩小，甚至在以后的几十年中慢慢消失。

然而，"立法仍被认为是鼓励企业社会责任的主要工具"（Antal and Sobczak，2007）。例如在法国，国际影响了养老基金透明化改革。同时，除了立法之外，政府一直鼓励企业参与自愿性环境行为。在全球契约组织中，法国公司的注册量从2003年的8个增加至2012年的801个，占注册公司总数量的7.8%。Antal和Sobczak描述了总统鼓励其参与全球契约文件签署的场景："2004年1月，希拉克总统邀请参与签约的公司领导人前往爱丽舍宫与科菲·安南共同签署了全球契约在法国举办的友好论坛文件……法国公司签署的全球契约被认为是正式加入全球契约治理的标志性事件"（Antal and Sobczak，2007）。通过构建网络，法国公司的企业社会责任参与率已显著提高。2001年，新立法通过了环境和社会标准可以体现在公司年度报告中这一要求。

法国是第一个以三重底线为上市公司公开股票报告的国家（Gouvernement Française，2001）。公司依据报告要求设定了涉及员工问题的定性和定量指标，他们工作的地理位置以及提供环境保护方面的信息，例如提高能源效率、减少损害，符合法律的证据，环境管理体系及补偿第三方造成的损害。环境和社会指标与GRI标准部分重叠。相比于GRI，该指标在劳动关系方面更详细，而人权问题是GRI体系中所没有的。因为立法并没有提供制裁措施，也不需要立法之外的审计。法国的企业社会责任报告案例比其他国家的有趣的地方在于，它说明了历史、文化和社会经济背景的重要性，虽然它与国际标准趋同（如GRI）。此外，法国的经验说明了国家作用的重要性，即建立网络和确保合法性和促进企业发布社会责任报告。同时，法国的情况还表明，尽管有着强烈的民族传统背景，但其仍免不了要受到国际影响力的冲击。报告是由环境指标纳入证明，它包括GRI要求的一些要素。

生态标签

欧洲环保生态标签的增长可以作为自愿方式成功的案例。然而，大量的环保标签也混淆了消费者对于不同标签的区分。

欧盟政策的影响标签：欧洲生态标签计划

欧洲生态标签计划致力于提高透明度和公信力，这是欧盟最突出的支持性政策。其成立于2000年，目的在于"减少产品在其整个生产周期对环境的影响，并为消费者提供准确的、非欺骗性的和基于科学的产品环境影响信息（欧洲议会和欧盟理事会，2009）。标记过的产品应有10%~20%与环境有关的产品绩效。绩效是衡量在产品整个生命周期中对环境领域的影响，包括废弃物水资源的利用、废气排放、能源利用效率、材料的利用率和生物多样性。它的标准由具有相似特性的产品/服务产品组起草。如果产品含有某些有毒有害化学物质，则在无权添加生态标签，不过当产品有着显著的环境绩效时，该委员会可以允许使用该标签作为例外情况。

欧洲生态标签计划创建标签时，会考虑到它能够应用在广泛的产品中。当生态标签的使用涉嫌虚假或有误导性，消费者可以（匿名）投诉指定的主管机构。主管机构可以调查产品是否仍符合产品标准组，并最终禁止使用的标识。

从理论上讲采用欧盟生态标签可能使过程简约化，因为消费者将不需要了解不同的国家和不同的标签组的产品。

谈判协定

谈判协定（NAs）是环境政策的替代工具。其特点是私人行为者的参与程度高，为了公共利益（特别是一个更好的环境绩效）而和公共权利组织谈判。在欧洲，谈判协定已经是一些国家政策的重要部分，尤其是荷兰和德国。欧盟委员会还认识到谈判协定是环保政策重要和首选的实施工具。谈判协定更有效的原因是它包括行业里最了解环境方面的人；通过紧密合作，参与企业和公共机构交流知识，并在这一过程中相互学习。与分化公共机构和私人行为者不同的是，谈判协定能加大政策支持并为各方实现学习的机会。然而，谈判协定的使用也伴随着风险：因为行业是自利的，它并不以环境绩效为发展目标。同时，它也很难确定谁应该参加，谁不应该参加（提出问题的合法性）；当一些当事人不信守协议时，自由参与便成为一个问题。因此，下面将介绍一些强制性内容，如批准或采用监管的方式带来更好的环境绩效。

欧盟政策对 NAs 的影响：环境协定的沟通

谈判协议沟通于 1996 年发布（Commission of the European Communities, 1996）。该协议主要针对一些成员国，虽然它也被称为欧盟层面的环境协定架构。根据协议，环境协定应该是非强制性的，但它们可以被委员会通过信函或推荐交换。同时，它们应该解释定性标准、量化目标和中期目标。它们应该声明独立监管，并确保部长理事会和欧洲议会都知晓。最新的环境协议采用的是 2002 年的版本。它规定：意为简化和完善监管环境的"行动计划"（欧洲经济共同体委员会，2002a）并不特别针对环境协定。相反，它也适用于法律文书，确保立法的质量并简化相关规则。

委员会区分了两种类型的谈判协定，在欧盟的法律框架下，第一类是自我承诺的行业实施自愿协议。这样的谈判协定虽然可能被认可，但欧盟仍可以决定是否启动正式规则。原则上这种自愿协议是完全自愿的，因为委员会认为该协议是一种自律形式，没有法律层面的约束力。然而，欧盟可以通过一些简单的变更承认环境协议。此外，委员会承认额外的监控决策。在这种情况下，委员会可以推荐协议，并结合议会和理事会共同监管。第二类被称为有限的规定，同意其在法律的框架内。这需要以指令和法律程序的形式设置条件；制定环境目标、时间安排及监测情况。强制性的规定可能会导致协议失败。OECD 国家中对于协议自我承诺和协商上之间的区别相应于行业工业和政府之间的谈判。当然，完全自愿和协商的跨度范围很大。

谈判协议在欧盟成员国之间的使用很不平衡。根据 1982—1986 年的研究，欧盟形成了 23 项新的自愿环境协议，而 1992—1996 年，则形成了 123 项协议（karamanos, 2001）。70 年代早期，大部分的（根据 1997 年的数据，大约占总人数的 3/4）法国人和英国人定居在德国和荷兰（Dalkmann et al., 2005）。在欧洲层面，只创建起了少数的谈判协议，例如关于能源效率（电视和 DVD 播放器[2003]，冰箱[2002]，洗衣机[2002]）。几乎三分之一的谈判协定是关于化学工业的，另一个大方面是食品产业。其成员国内部的能源及排污管理部门已大多采用这一标准（Dalkmann et al., 2005）。

欧盟政策对 NAs 的影响：荷兰

荷兰利用环境协定（NAs）建立了领袖谈判协议环境政策（De Vries, Nentjes, and Odam, 2012），这不仅因为该国的谈判协定数量多，而且也由于这一政策的成功应用。例如 OECD 报告中提到的，在荷兰："所有行业都在提高环保性能，特别是通过环境协议（如契约）和环境管理及审计"（OECD, 2004; Bressers, De Bruijn and Lulofs, 2011）。协议（或 convenanten，荷兰语）在政府部门之间协商（通常为环境部和一个部门或行业组织的代表）（见 Bressers, De Bruijn, and Lulofs, 2011）。该协议是为了落实国家环境政策计划。他们通过提供定量指标细化政策目标。政府和行业协商如何实现目标并设定时间表。

这一程序适用于荷兰的政治文化，它被描述为"浮地模式"。它是政府和产业、劳动和利益组织代表之间的谈判模式。这种文化的特点是通过协商一致的方式进行政治挑战。谈判中，合作伙伴被认为是负责的行动者，政府试图通过政策的制定和实施强化这些组织的责任。这个过程不是完全自愿的。根据协议，义务被分配到（单个企业）并与他们的许可证相连。由此，投机取巧将会被制裁。如果企业拒绝参与其中，他们将发现自己的处境很艰难。

Bressers et al.（2011）认为，荷兰个案说明："契约就是谈判，以灵活的立场谈判；包括行业组织及其会员的正式立场。另一方面，如果达成协议，那么就表示这些设计在原则上是可执行的。这样的设计清楚地表明用"志愿的"和"强迫的"两分性具有误导性"（2011：190）。他们通过分析 59 个协议发现，社会政治环境（例如政治文化）的不同实际上与绩效相关，一些协议甚至比其他做得更好（203）。这意味着如果对协议设计、政府压力以及实际的成本花费有足够的关注，协议可以同样在另一些国家实现。

范例：建立标杆契约

1999 年，基于本国的工业，荷兰签署了世界领先的能源排放标准。签约起草能源效率计划（Energy Effihciency Plan，EEP），宣布他们与世界先进水平的距离并解释他们将如何追赶。2003 年，共有 233 个组织签署契约。另有 276 个符合参与标准的组织，这远高于预期。工业部门参与企业 1999 年的总能源消耗为 704PJ，相当于能源密集型产业总消耗量（768PJ）的 92%。参与计划的公司需提交他们的能源效率计划（EEP），由鉴定机构出具建议并评估他们的主管机关。其目的是达到世界能源密集型企业的先进水平。由主管部门考核，参与者和主管者之间达成的协议便是 EEP。2003 年，共有 111 项能源效率计划由主管机关同意。

计算结果表明，到 2012 年二氧化碳排放量减少 5.1×10^9 kg 是可能的。这大大超过了之前 $3.2\times10^9\sim4.0\times10^9$ kg 的估计值。标杆管理委员会提供了一个协定的广泛概述。管理办公室及主管机关也会以年度监测报告来监测整体排放量（Commissie Benchmarking，2004）。

绿色公共采购

绿色公共采购（Green Public Procurement，GPP）是促进企业自愿展开环境行为的重要措施。通过公共采购合同标准，政府可以增加对绿色产品和服务的需求，同时也促使企业报告产品和服务的环境标准。2002 年，欧盟公共采购的份额约为国内生产总值的 16%（Steurer et al.，2007），这一数字增加到 2009 年的 19.9%（Renda et al.，2012）。这个庞大且不断增长的公共采购体系冲击市场，它成为刺激绿色产品、服务和技术发展的好机会。

欧盟影响 GPP 的政策

2001 年，在解释沟通会上（欧洲共同体委员会，2001b），欧盟委员会首次提出了绿色公共采购（COM，2001），并讨论了如何将奖励机制考虑进绿色公共采购的过程。2003 年，"通信集成产品政策"（欧盟委员会，2003b）将公共采购视为持续环境改善的核心工具，并鼓励各成员国制定国家行动计划。到 2010 年，已有 21 个成员国实施行动计划或类似的政策，同时另有 6 个成员国正在推进计划制定。例如，GPP 计划对于葡萄牙和德国的木材行业和捷克的 IT 行业有法律约束力。2004 年发布的两项指令，2004/17（欧盟，欧洲议会和理事会，2004）和 2004/18（欧盟，欧洲议会和理事会，2004）。在这里可以反映在采购过程的早期阶段对社会和/或环境问题的考虑，并"阐明政府如何为环境保护和促进可持续发展作贡献，同时又能确保经济效益最大化"（2004/18/EC l134/114）。

这两项指南特别包含了生态标签和环境管理措施等环境要求。2008 年，欧盟委员会提出了"利用公共采购创造更好环境"的要求（欧盟委员会，2008）。它包括对政府绿色采购的目标的确立：到 2010 年绿色公开招标程序达到 50%；符合欧盟的绿色公共采购在建筑，运输和服务的标准。委员会为 GPP 设立了产品和服务及服务过程的标准。该标准是基于现有的欧洲和国家生态标签标准，并邀请了来自民间社会和行业的利益相关者等来共同制定。一旦由成员国签署，这些标准都会包含在国家的指导方针和行动计划中。最初的标准是针对 10 组产品制定，它们目前存在于 21 类产品中。为确保更高的绿色标准，标准会定期修订。欧盟区分核心标准（适用于任何成员国的缔约当局）用以解决影响环境问题的关键问题（尽量减少需要增加的成本）和实际需求（希望购买市场上最好的产品则需要额外的成本投入，或与其他产品相比成本略有增加）。

2011 年，一项由欧洲政策研究中心和欧洲学院受委员会委托的研究发现，所规定的 50% 的目标并没有达到，其中，2009—2010 年，只有 26% 的采购合同达到 GPP 标准，而 55% 的合同只参照 GPP 的一些标准（Renda et al.，2012）。然而，尽管 GPP 的核心标准在不断提高，但是 2009—2010 年仍只有 38% 的采购包括某种形式的 GPP 标准，大约价值为 1 175 亿欧元。核心标准主要和两方面有关，即节能表现和二氧化碳排放量。虽然 GPP 的标准水平不断提高，但在欧洲的表现则不同。比利时、荷兰、瑞典、芬兰和丹麦是主要领先的成员国。2010 年 AEA 为委员会进行的研究指出在 9 个成员国和挪威产品和服务标准数目发展情况不同（AEA，2010）。一项发现是联邦政府国家（比利时、德国）的标准更少，并且比 GPP 范围小。然而这些附加政策有时也是很有远见的（例如，比利时的佛兰德地区的目标是 100% 地将 GPP 标准列入采购）。

欧盟成员国影响 GPP 的政策：英国

可持续公共采购和框架的案例证明，英国在考虑到中央和地方政府财政预算的情况下，每年 1 500 亿英镑的年度公共采购预算目标是必要的（Fletcher, Duisterwinkel et al., 2009）。英国的情况也表明监测和管理系统的需要，也要考虑到能力建设和最低标准之外的要素，如产品周期的考虑也是必要的。

2002 年，可持续发展世界首脑级会议提出了促进公共采购政策、鼓励环保商品和服务的发展等要求。英国政府表示，其目标是 2009 年成为欧盟可持续采购的领先者。2003 年的政府采购标准，对各种采购物品的最低标准是"速效方案"（Quick Wins）。它们在中央政府部门和机构内部强制进行，且不包括其他公共部门。2008 年底，这一标准覆盖了 9 个重点产品群中的 46 种产品。同年，这一标准又进行了更新，它包括了自愿实施标准和强制性标准。

然而，速效方案的标准是基于市场内部的平均标准，而非对整个生活环境的影响分析。2007 年 8 月，21 个中央部门中的 15 个表明他们符合这些标准，但仍有 6 个部门没有系统测量。近年来，政府采购标准（Quick Wins 的新名称）已与欧盟委员会的政府绿色采购计划结盟。2005 年，可持续发展战略重申要在 2009 年将成为欧盟可持续采购的领先者。它制定了一个灵活的框架来评估可持续采购进展。该框架确定了 5 个主题："政策，策略和沟通""测量和结果""采购过程""供应商"与"人"的等级。它分为 1~5 五个等级，其中 5 代表最先进。自 2006 年以来，政府部门在所谓的"政府资产可持续经营"的框架下报告进展，该框架规定了所有中央政府部门都需要降低运作影响。该框架包括 14 项经营绩效目标，它包括办公室和道路车辆的碳排放、能源利用效率、废物排放及回收利用、生物多样性、水的消耗和可再生能源发电。另外，它也增加了部门的 8 项达标要求，以提高监测水平和报告质量，并得到环境管理体系的认证。

2007 年，可持续采购行动计划作为欧盟可持续发展战略目标实现的英国路线图，重申了框架，并增加了 5 个可持续发展承诺：领导和问责制（常任秘书长要对其部门的进展及适当激励负责任）；预算与会计（审查预算安排和绩效框架以解决可持续解决方案的障碍）；能力建设（制定行动并考虑柔性框架的使用）；市场参与和鼓励创新（确保供应商有降低碳足迹的计划）；提高标准（停止购买速效方案标准以下的产品)。

在 2009 年国家审计署的调查中发现,该框架所要求的成为欧盟领先水平的目标并没有实现,其原因之一是缺乏目标的量化。此外,灵活的框架会破坏评估机制,让评估变得没有保证。而进步并不是基于自我评估,因为框架需要外部的衡量机制。另一个问题是,政府工作人员对于成本和收益并不清楚,且不确定它们会不会低于最低标准。国家审计署发现,并非所有部门都遵守速效方案(Quick Win),他们往往缺乏监测和绩效的管理系统。

指数化

列出企业名录是改善其环境行为的一个外部方法,同样,点名批评也能够达到该效果。这种方法假定通过提供多个企业的环境绩效信息,来促使企业改善环境行为。这种方法可能会以指标形式(环境绩效指数),正向排名(比如在投资者的投资组合)和负面列表(NGO 经常做的)出现。欧盟不提供详细清单和指数化方案,而由欧洲的几项大计划(或具体部门)负责。

市场指数公司已经根据不同的主题划分了公司,包括环境领导力。比如 FTSE4Good Leaders Europe 40 Index(富时指数)是由一些公司组成,它们"运用环境策略来控制企业的环境风险和影响,并减少其环境足迹。公司必须发布关键的环境绩效数据来证明他们的环境战略和管理体系,并评估风险和影响"(FTSE,2013)。道琼斯也已建立了类似的指标——道琼斯欧洲可持续性发展指数(Dow Jones Sustainability Europe Indexes,DJSI 欧洲),它由一个综合指数和几个子指数所构成(例如,不包括酒精、烟草、赌博、武器、枪支和成人娱乐相关的公司)。企业可持续性发展评估主要体现在经济、环境和社会维度;环境标准包括报告标准(保证、覆盖范围、定性和定量数据)和行业标准(例如 EMS、气候策略、生态效率)(DJSI,2014)。市场指数公司会回报那些表现良好的企业,而专业市场指数则通常排除那些表现不佳的企业。

通过从指数化中排除这些企业,投资者丧失了去激励这些企业改进其绩效的机会。环境跟踪是可以替代专门的指数化的另一种方法,它可以包括环境保护做得好的和差的两方面的公司。环境投资组织(Environmental Investment

Organization，EIO）是一个专注于生态投资的非营利组织，它开发了一种跟踪方法。该方法结合了信息披露、验证和市场指数的评估，并被描述为"旨在鼓励世界上最大的公司减少其排放量和提高信息透明度的水平"（EIO，2012）。环境跟踪指数中的市场指标根据其在碳排名上的位置加权。环境投资组织自2010年公布了几个区域的碳排名，包括欧洲300、亚太3000、金砖四国300、全球1000和全球800等。排名是基于绝对排放量，包括直接与间接排放。环境跟踪碳排名的目的是提供企业温室气体排放强度信息以及信息披露和验证的水平。

根据2011年的排名，欧洲的碳披露水平在全球领先。据EIO，欧洲有53%的上市公司提供完整的数据报告。这一数据在荷兰达到100%，意大利为88%，德国为84%（The Climate Group，2011）。如果发现排名不正确，企业可以上诉。

讨论

欧盟企业和欧盟成员国已经制定了政策和战略以鼓励企业自愿展开环保行动。欧盟的企业和组织在环境报告分享、披露机制和组织程度的水平都很高。然而，在一些新兴国家，例如中国，其企业越来越认识到参与自愿计划的好处。

对企业而言，自愿的企业环境行为可以是有益的：
（1）减少与监管机构的冲突；
（2）允许更大的灵活性，尤其是相对于传统的管理方法；
（3）公司可以通过绿色创新和强制性规定在竞争中取得优势。

对政府来说，加强私人企业的角色可以导致如下：
（1）降低实施成本；
（2）从目标行业获得更大的支持；
（3）更好地利用专业知识。

然而，对于自愿参与的方法人们也会做出如下批评：
（1）允许"放纵行为"，把那些污染企业列为绿色企业（"洗绿"）；
（2）在特定的工业利益驱使下，设置较低的标准；
（3）缺乏民主合法性；
（4）遭受排斥；

（5）在制定国际及全行业标准方面缺乏支持。

本文表明，为了增加企业效益并避免负面影响，自愿参与的方式需要精心设计。那些过于宏大的战略（如英国的 GPP 战略）在提供监控和保证体系方面常常会无效。研究还表明，自愿参与应该伴随着一些强制手段的采取，如制裁或提高违规成本。如在荷兰，NAs 是允许的实施系统。在许多情况下，它都有伴随着"阶层的影子"：当政府采取更严格的规章制度时，私人公司更有可能采取环保行动。自愿性与强制性之间的区别并不是一成不变的。例如一些自愿性质的计划，如 EMAS 和欧洲生态标签在欧盟法规和指令的支持下发生作用的，对于其程序过程和参与概念的界定已为欧盟委员会、欧洲议会和欧盟成员国所用。

在某些情况下，自愿计划也可能转化为强制性计划。比如作为可持续发展报告的 GRI 本是一个自愿的标准，但现已在瑞典和荷兰强制成为国有企业必须遵照的标准。对于国家案例的研究说明了两类公司在对环境负责时可以被授权的合理理由。国有企业由于是公有的，因此他们应该在影响环境、社会和经济方面承担更多。以法国为例，公开股票的上市公司必须披露其社会和环境信息。投资者需要股票上市公司公开更多的财务信息来评估风险和机会，以便作出明智的投资决定。

欧洲的经验和教训可能不适用于中国的情况；然而有几点可以探讨。考虑到高份额的国有企业在中国经济中的所占比例，像瑞典那样强制报告以增加报告公司的数量会较为有效。应该指出，这并不一定能导致更好的环境绩效。因为欧盟的报告往往是针对股东而不是社会团体（工会组织）。然而，法国的案例表明，由政府发布报告同样可以帮助政府了解企业环境污染的信息。

某种程度上来说，欧盟的环保标签是其自身的受害者。大量的标签计划迷惑消费者。欧盟正试图通过为多项产品和服务引入欧洲生态标签来解决这一问题。虽然中国没有欧盟这么丰富的标签体系，但它也可以采取相似的多产品和服务的方法。然而，标签体系的关键在于其可靠性，因此，这就需要精准的认证和监测系统，以及对滥用标签系统的制裁。

鉴于中国政府支出比重较高，在所有自愿参与的外部方法中，绿色公共采购似乎是很有前途的。但是应该保障监控和认证体系的理念会被贯彻到政府机构的运行中去。需要说明的是，环境协定的范围是有限的。荷兰是一个比较容易在谈判桌上聚集整个行业的小国家，如果是在一个在单个行业拥有众多企业的大国家则可能很麻烦。但是在中国，在省和地方特别是大城市，可能会有效。此外，协商一致的决策重点似乎与中国的和谐发展理念是相吻合的。

Voluntary Environmental Approaches
—— Experiences in the EU

Sander Chan*

• **Introduction**

Traditionally, governments are regarded as the primary actors in devising and implementing environmental policies. However, this state-centric understanding of governance has increasingly been challenged throughout the OECD world, in particular the US and the EU. Both in academia and among policy-makers the act of governance and the institution of government are increasingly seen as separate. The act of governance could be defined as the provision of public goods (e.g. the protection of nature) through the sharing of responsibilities among several stakeholders. This definition reimagines governance as a coordination process in a society that is fragmented in terms of authoritative public actors. Government remains important in governance, however, their centrality and their overarching authority – as in command-and-control coordination – is no longer presumed. Rather, citizens, civil society organizations and firms can constitute authoritative governance coordination. Voluntary environmental approaches fit in this understanding of authority beyond traditional government, as they refer to approaches which are sometimes devised by non-state actors such as firms and civil society organizations, and at least assume their participation in norm-setting and implementation.

Several advantages of voluntary environmental approaches for governments are:

---

* Sander Chan, Doctor and Researcher in German Development Institute (DEUTSCHES INSTITUT FÜR ENTWICKLUNSPOLITIK).

a) Greater legitimacy. In global environmental governance (and UN institutions), a lack of democratic representation could partly made up by the greater participation of civil society and private actors.

b) Smoother coordination and implementation. By taking into account interests among a larger group of stakeholders, public-private relations will feature less antagonism, and subsequent coordination and implementation of policies can proceed more smoothly. In some cases private actors can even fulfill operational implementation tasks, relieving government bureaucracy.

c) Greater knowledge. Even the most informed governments cannot have all-knowing bureaucracies to make rational decisions; rather they act on incomplete knowledge and with limited understanding of specific policy areas. Through collaboration with the private sector governments can benefit from expert knowledge.

d) Greater pool of resources. In addition to greater expertise, non-state actors can contribute with additional finance and other resources in the pursuit of public goods. This could potentially reduce administrative and transactions costs for governments (e.g. Arora and Cason, 1996).

In spite of the many potential advantages, voluntary environmental approaches have been criticized quite extensively. Especially at the local and national levels of governance, the application of voluntary instruments has raised concerns over the privatization of environmental governance (Cashore, 2002). Previously government administered governance functions are removed from democratic control, and relegated to unelected private actors. Moreover, polluting firms has been can refer to private and public-private "green schemes", while their core businesses remain polluting. Such green-washing practices undermine the credibility of voluntary environmental mechanisms. For businesses, environmental responsibilities can also cause higher costs and increased administrative burdens, for instance when civil society actors and demand more environmental information disclosure and reporting.

The purpose of this study is to inform about what public authorities could do to incentivize more responsible environmental behavior from private actors. This paper reviews voluntary environmental approaches in the EU. It gives an overview of different types of approaches, and considering policies by the EU and individual Member States to support the uptake of voluntary approaches. Moreover, the paper concludes with a brief discussion of the applicability of voluntary environmental approaches beyond the EU, in particular in the People's Republic of China.

The sheer amount of possible case studies and policies to highlight in an area as broadly defined as voluntary environmental approaches complicates a selection of policies and cases. The selection of policies and case studies was informed by the following criteria:

To include lessons and experiences from across the EU.

To include lessons from different policies related to voluntary corporate environmental behavior, in particular green public procurement, voluntary agreements and environmental reporting.

Cases that are relevant to appliance beyond the EU context, in particular the Chinese context (for instance, taking into account China's relatively large state owned enterprise sector and long tradition of a statist orientation).

The selection should encompass best practices, as well as ambitious strategies and policies.

In this paper, we particularly look at the following approaches:

Reporting

Eco-labeling

Negotiated agreements

Green public procurement

Indexing

The emphasis lies on public policies supporting voluntary environmental approaches, even when there are numerous private initiatives. The reason for the public policy focus is related to the purpose of this study: addressing what public authorities could do to incentivize more responsible environmental behavior.

- **Reporting**

Environmental reporting is the production of reports by organizations (in particular private businesses) that contain information about environmental performance. The aim of such reporting is to improve transparency, measuring, and tracking of environmental performance. Increased transparency allows for public scrutiny and public relations, but also for investors to identify and assess environmental risks and opportunities. Contrary to financial information disclosure, non-financial information disclosure is generally not mandatory. Nonetheless, more and more businesses recognize the benefits of non-financial reporting. Environmental reporting can be a communication mechanism to maintain, build (and restore) confidence with other shareholders, (potential) investors and social stakeholders (such as labor unions and NGOs). Reporting can also be instrumental to the improvement of business management, leading to better risk management and value creation. Some businesses will go beyond legal reporting requirements as they anticipate competitive advantages.

European businesses have long been the leader in corporate environmental reporting. According to the Sustainability Disclosure Database in June 2012, 4 626 reports out of 10 034 reports registered in the Global Reporting Initiative 3 have been drafted by organizations based in the European region (not exclusively EU) (46%), the share of EU based organizations is approximately 41%. While the share of Europe based companies in reporting remains high, reporting by organizations in emerging countries has increased dramatically in recent years.

Several supportive initiatives and developments have improved of the quality of reporting and the credibility of reporting; in the following we briefly discuss the Global Reporting Initiative, the UN Global Compact, and assurance standards.

Supportive initiative 1: Global Reporting Initiative (GRI)

The profusion of reporting globally raises questions about reporting formats. Without agreed reporting formats and standards, comparison between reports is difficult, and information disclosed may not match demands from various stakeholder groups. Although there is no agreed international standard for sustainability reporting, the Global Reporting Initiative (GRI) comes closest. In April of 1997 the Global Reporting Initiative was launched by CERES and the United Nations Environment Program, introducing a "triple bottom line" (economic, environmental and social) for reporting. The first version of the guidelines was released in 2000 ("GRI 1"), a second was presented at the 2002 World Summit for Sustainable Development in Johannesburg (WSSD) ("GRI 2"). That year the GRI also became an independent organization with its secretariat in Amsterdam (The Netherlands). The GRI continues to collaborate with UN agencies, in particular UNEP and the United Nations Global Compact. Its aim is to increase sustainability reporting on social, environmental responsibilities, otherwise known as non-financial responsibilities. A newer generation of GRI Sustainability Reporting Guidelines (G3) was launched in 2006. The GRI had now developed the fourth generation of standards (G4) (GRI, 2012), to keep up with technology, to better link to other international standards, and to adapt to more web-based reporting. The Sustainability Reporting Guidelines (the Guidelines) consist of "Principles" for defining report content and ensuring quality of reported information and "Standard Disclosures" containing "Performance Indicators" and other disclosure items, and guidance on specific technical topics. Organizations can apply the GRI framework to different degrees, and subsequently self-declare their "GRI Application Levels": A, B, or, C, where A stands for the advanced reporters.

Supportive initiative 2: Global Compact and "Communications of Progress"

The Global Compact (UNGC) is an UN initiative launched in 2004, aimed at the implementation of the Millennium Development Goals, a set of eight goals which member states and international organizations have agreed to achieve by 2015. MDG 7 sets environmental sustainability as a goal. The UNGC engages the private sector in the implementation of these internationally agreed goals. With over 10 000 participants (in 2012), it is the world's largest social and environmental voluntary corporate initiative.

While strictly not a reporting scheme, participation in the UNGC requires annual reporting in the form of a Global Compact Communication of Progress (COP) defined as a "direct communication from business participants to their stakeholders" (UN, 2011). The format is flexible but it should contain a "statement by the chief executive expressing continued support", a "description of practical actions", and "measurement of outcomes". Businesses can be exempt of submitting COP in their first two years of participation. On the basis of the COP, participants are designated as either "GC Active", indicating that participants have met all COP requirements, or "GC Advanced", indicating that participants have gone beyond requirements. When requirements are not met, participants are granted a one-time "Learner Grace Period" to submit a new COP that meets all the requirements.

Supportive initiative 3: assurance standards

Voluntary self-reporting has raised questions regarding reliability. Assurance by an external party is an option and could lead to higher reliability and trust with stakeholders. The GRI offers the option of assurance by the GRI itself, or by other external stakeholders. Several schemes have been established for the assurance of reporting, in particular: ISAE 3000 and AA1000AS. The International Standard on Assurance Engagements 3000 (ISAE 3000) (previously ISAE 100) was developed by the International Auditing and Assurance Standards Board, and addresses the assurance of reporting beyond financial information. Topics include environmental performance, corporate governance, internal compliance, stakeholder engagement and other of corporate responsibility. One of the key distinctions that ISAE 3000 makes is between limited assurance and reasonable assurance. A limited assurance statement is a negative statement (when no evidence is found that the content of a certain report is not valid), while reasonable assurance includes to positive validations.

The AA1000 Assurance Standard, introduced in 2003, is a methodology to evaluate whether and to which extent organizations keep to Accountability Principles. The three principles subject to assurance are: inclusivity ("participation of stakeholders in developing and achieving an accountable and strategic response to sustainability"), materiality ("determining the relevance and significance of an issue to an organization and its stakeholders") and responsiveness ("an organization's response to stakeholder issues that affect its sustainability performance"). The 2008 version of AA1000AS requires public disclosure. AA1000AS (Accountability, 2008) refers to ISAE3000 as a compatible methodology. AA1000AS goes beyond the assessment of reliability of data by requiring the assurance provider to evaluate the adherence to a set of principles. The standard distinguishes "high assurance" from "moderate assurance"; according to Accountability, which issues the AA1000AS standard, an "assurance provider achieves high assurance where sufficient evidence has been obtained to support their statement such that the risk of their conclusion being in error is very low but not zero", whereas "The assurance provider achieves moderate assurance where sufficient evidence has been obtained to support their statement such that the risk of their conclusion being in error is reduced but not reduced to very low but not zero" (Accountability, 2008, 11). The AA1000 family of standards also includes a Stakeholder Engagement Standard (AA1000SES) which provides with a framework to ensure stakeholder engagement processes. Regional differences in the use of assurance of sustainability reports appear to be high. In 2008, only 3% of reporting companies in the US assured their reports, while about 45% in Europe did so (Brown, de Jong and Levy, 2009).

Supportive initiative 4: Carbon Disclosure Project

Carbon accounting and disclosure related to climate policies are also addressed in a growing number of international frameworks. The most prominent initiative is perhaps the Carbon Disclosure Project (CDP). Launched in 2000 the CDP aims to further transparency between investors and managers, by voicing investors, concerns about climate change and by informing investors about companies, risks associated with climate change. The CDP sends questionnaires signed by institutional investors to the largest global firms, collects responses and makes them available online. Moreover the CDP also ranks companies by "carbon performance scores" and "disclosure scores", determining which top scoring companies qualify for a listing in the "Carbon Disclosure Leadership Index"(CDP, 2013). CDP's success at engaging investors and companies is illustrated by the number of investors that signed the annual questionnaires; the first questionnaire in 2003 was signed by 35 institutional investors representing 4.5 trillion USD in assets. By 2011 the CDP sent its annual questionnaire to Global 500 companies on behalf of 551 investors representing 71 trillion USD in assets. Responses in the same period increased from over 235 companies to over 3 000.

- **The EU and its Member states policies to stimulate reporting**

In 2001 the European Commission proposed non-binding guidelines on corporate environmental reporting. The Fourth Directive on Annual Accounts 2003/51/EC (European Commission, 2003a)decrees that enterprises have to disclose environmental and employee-related information "to the extent necessary for an understanding of the company's development, performance or position". However, Member States may exempt SMEs from disclosure. Some Member States already have requirements that go beyond EU legislation (i.e. the Netherlands, the UK, Sweden, Spain, Denmark, and France). Additionally some sector specific guidelines for environmental reporting exist, and guidelines for environmental statements. For instance, the European Chemical Industry Council (CEFIC) has published guidelines in 1998 for European chemical companies. Currently, the European Union does not put forward a single framework for environmental reporting, rather the legal basis for reporting is found in references throughout EU legislation, regulations, and directives, in particular in the EU's Corporate Sustainable Responsibility strategy, the Community Environmental Management and Audit Scheme (EMAS) regulation, but also the Single Market Act (SEC (2011) 467) (European Commission, 2011a) which promises "a legislative proposal on the transparency of the social and environmental information provided by companies in all sectors".

With many references in regulations and international standards, the problem remains that there is inadequate transparency with regards to non-financial information (many companies choosing not to disclose) and insufficient quality of disclosed information. In response, an EU directive was adopted on 16 April, 2013, to enhance "the transparency of certain large companies on social and environmental matters" (European Commission, 2013). The directive amends the Accounting Directives (Fourth and Seventh Accounting Directives on Annual and Consolidated Accounts, 78/660/EEC and 83/349/EEC, respectively) and requires large companies (with more than 500 employees) to disclose material information in the form of a statement in their annual report. Companies that prepare detailed non-financial reports on a voluntary basis are exempted of this reporting obligation. This stipulation ensures the continued relevance of existing voluntary reporting frameworks.

EU policy affecting environmental reporting 1: Community Environmental Management and Audit Scheme (EMAS)

The Community Environmental Management and Audit Scheme (EMAS) is an instrument to assess manage and continually improve the environmental performance of individual organizations. It requires explicit environmental policies, operational procedures and reporting in the form of Environmental Statements which are subject to validation by accredited verifiers. An important difference between EMAS and most other schemes (such as ISO 14001) is the fact that environmental statements, which contain data on environmental performance and results against set targets, should be verified by a third party (accredited verifier) before organizations (or parts of organizations [specific sites]) can be registered. Member States assign Competent Bodies in charge of registration and management of the registry.

Competent Bodies are also charged with the supervision and accreditation of verifiers (Member States can designate Licensing Bodies to license verifiers). Competent Bodies of Member States meet annually in a forum, to ensure the consistency of EMAS throughout the Union and to suggest improvements and more detailed procedures for specific clusters (e.g. economic sectors and geographic areas). Only upon verification of environmental statement, publication of the statement and registration with competent registration bodies (usually a Member State or local body) can an organization (company, subsidiaries, production sites) be certified. EMAS has been successful at attracting a growing number of sites and organizations. Registered sites have increased from 2 140 in 1998, to 7 794 in 2010, while the number of organizations (businesses) have increased steadily from 3 055 to 4 542 between 2003 and 2010 (Słonimiec and Świtała, 2013). Currently, more than 4 600 organizations have been registered. As the number of registrations more or less corresponds to the number of publicly available and verified environmental statements, it could also be considered as one of the larger reporting schemes, although reporting is not the primary purpose of EMAS.

EU policy affecting environmental reporting 2: Corporate Social Responsibility (CSR)

The first communication on CSR was the 2001 EU Green Paper. The Green Paper emphasized the need for a (public) policy framework, while also referring to the principle of voluntary action "While voluntary codes of conduct can contribute to promote international labor standards, their effectiveness, however, depends on proper implementation and verification" (Commission of the European Communities, 2001a 13-14). The Commission highlighted the provision of an "overall European Framework" (idem: 6). However, the 2002 Communication concerning Corporate Social Responsibility: A Business contribution to Sustainable Development (Commission of the European Communities, 2002b), emphasized corporate "voluntary contribution" and economic benefit for responsible corporations businesses ... would benefit from the inclusion of social and environmental issues into their daily operations (idem: 14). The application of CSR instruments standards and schemes, however, could be "a source of market distortion" (idem: 8). The latest communication, "Renewed EU Strategy 2011-14 for Corporate Social Responsibility" (European Commission, 2011b) issued in November 2011, stresses the need for a "balanced multi-stakeholder approach, market rewards for responsible business conduct, for more transparency, and for greater attention to human rights"(Martinuzzi,

Krumay and Pisano, 2011: 21). The strategy also redefines CSR as "the responsibility of enterprises for their impacts on society", whereas previous definitions referred to voluntariness. The redefinition of CSR from voluntary behavior to responsibility for every enterprise maybe a precursor for a more mandatory policy approach towards CSR and reporting, following the trend in some member states where mandatory approaches have become more common.

EU policy affecting environmental reporting 3: climate policy

Perhaps the largest European wide mandatory scheme which includes an assurance procedure is found in the framework of climate policy. While – from an environmental standpoint – the focus on climate (and emissions) is limited, the framework is relatively clear in terms of indicators, protocols and reporting standards. The EU Emission Trading System (EU ETS) established in 2005 is the largest international scheme for the trading of emission allowances, covering 27 EU member states and Liechtenstein, Iceland and Norway. A growing number of energy-intensive businesses are required to take part in the scheme (e.g. in the energy, metal, mineral and pulp and paper sectors). The scheme was extended to the aviation sector in 2012; the petrochemical, ammonia and aluminum industries followed in 2013. Industries need to acquire a decreasing number of greenhouse gas emissions permits under the scheme, thus lowering emissions over time. To enable the EU ETS, monitoring and reporting of greenhouse gas emissions is necessary. Under the scheme a series of Commission decisions have been published for monitoring and reporting guidelines with regards to greenhouse gas emissions from i.e. capture, transport, geological storage of carbon dioxide(2007/589/EC)(European Commission, 2007), from aviation activities (2009/339/EC) (European Commission, 2009), guidelines for reporting and monitoring of greenhouse gas emissions (2007/589/EC; 2004/156/EC) (European Commission, 2004, 2007). The Monitoring and Reporting Guidelines were established by the Commission in 2004. A revised version of the Guidelines (European Commission, 2007) was adopted in 2007. In addition, the Commission has published electronic templates for monitoring plans and reports for ton-kilometer data and annual emissions from aviation activities.

EU Member State policies affecting reporting 1: Sweden

The Swedish government launched a Global Responsibility Unit as early as in 2002. In November 2007, the Swedish government became the first in the world to decide that state-owned companies should present a sustainability report in accordance with the GRI guidelines. The Swedish mandatory GRI reporting by state owned companies has inspired similar policies in Norway and the Netherlands. The main reasons for adopting a mandatory framework are the fact that the Swedish government has a large stake and influence in the economy through their involvement in more than 50 state-owned companies (e.g. in the mining, housing, finance, gambling sectors), moreover, the government sees a special obligation and responsibility for state-owned companies, since they ultimately are the property of every Swedish tax payer. Moreover, NGOs have complained about the possibility that Swedish companies would act irresponsibly with regards to human rights, and, the environment. The choice of the GRI standard was informed by the fact that Swedish companies have to compete in an international market. Therefore it seemed obvious to adopt the most widely recognized standard. Since 2008 state-owned companies are required to report under the "comply or explain" principle, i.e. companies have to publish sustainability reports, or they should have good reasons explaining why they do not report. GRI standards are used, with the additional requirement of external verification.

According to research by the Department of Business Studies at Uppsala University (Frostenson, Windell, and Borglund, 2010), the policy has increased awareness of sustainability reporting and the GRI among chairmen of boards, strengthened the focus on sustainability issues in state owned companies, and, sustainability issues are addressed in a more structured way within companies. The most affected processes have been those related to environmental policy, purchasing, human resources and communications. According to some respondents in the survey, stakeholder dialogues have been improved, new purchasing policies have been adopted, and demands have been increased on suppliers. However, few companies think "that the GRI guidelines have resulted in major changes and improvements to their sustainability process" (Borglund, Frostenson and Windell, 2010). Major organizational changes have not occurred. The most significant impact has been on companies with little or no experience with sustainability reporting, they have developed new policies and procedures. Some companies complain that GRI indicators are not suitable for their industries, or that measurement is too difficult. Nonetheless, most respondents experienced the policy as positive. Applying GRI should not only result in better reporting but also changes in the sustainability process.

The reporting requirements are but one tool that needs to be flexible and adapted to local conditions. According to the Swedish Ministry of Enterprise in 2010, 96% of all state-owned companies were reporting. Also, state-owned companies seem to set an example for other companies (Didong, 2010). There is also an increase of reporting among private companies, about 30 percent are reporting according to GRI standards. The Swedish policy has also inspired other countries, such as the Netherlands, where the state owns about 30 companies (e.g. infrastructure, gambling, gas, electricity). Since 2009, Dutch state-owned enterprises are also required to report according to the GRI guidelines and the Dutch transparency benchmark on CSR. For the Dutch government the credit crisis (2008) has been a major impact (Schaay, 2010). The crisis revealed how a short-term focus on profits alone can prove disastrous. CSR and reporting is helpful to make corporations take a more balanced approach. The goal is to integrate CSR as a mainstream policy within companies.

EU Member State policies affecting reporting 2: France

In the meetings of the Expert Group on Disclosure of Non-Financial Information by European Companies, France has expressed the wish for the EU to define a European reporting standard. Benchmarking and harmonization is necessary vis-à-vis suppliers outside the EU. Although much more is published on Anglo-American policies and instruments of CSR and reporting, France has been a pioneer in corporate social reporting (Antal and Sobczak, 2007). Mandatory corporate social reporting was introduced in 1977 (Code du travail, partie legislative, livre iv, titre iii, article L438) (Gouvernement Française, 1977). The law required reporting on employment related responsibilities on 134 items. However, different from elsewhere, such reporting was not generally available to public; rather reports were addressed to government agencies. This is illustrative for the fact that in France government interference and centrality of government have historically been widely accepted. The role of the state and the narrow scope of the seventies have given somewhat away in subsequent decades.

However, "legislation is still considered a major tool in stimulating CSR in companies" (Antal and Sobczak, 2007). In France, international influences have inspired for instance reforms for more transparent pension funds. Also beyond legislation, business is engaging in voluntary environmental behavior. The number of French companies that have registered under the Global Compact increased from 8 in 2003 to 801 in 2012, 7.8% of total registrants. Antal and Sobczak describe the intervention by the President to encourage participation in the Global Compact: "Chirac invited the leaders of the signatory companies to the Elysée Palace in January 2004 to launch the Forum of the Friends of the Global Compact in France in the presence of Kofi Annan … the French network of companies having signed the Global Compact was integrated in one of the two working groups created to think about the governance of the Global Compact" (Antal and Sobczak, 2007). By building networks, the participation in CSR by French companies has been relatively high. In 2001, new legislation was enacted to mainstream integration of environmental and social criteria in annual reporting by listed firms.

France was the first country to mandate triple bottom-line reporting for publicly-listed companies (Gouvernement Française, 2001). Under the law companies have to report according to a set of qualitative and quantitative indicators relating to employee issues, the location/location they are operating in, as well as provide information in the area of environmental protection, for instance: improving energy efficiency and limit damage, evidence of legal compliance, environmental management system, and compensation to third parties for caused damage. Only part of the environmental and social indicators overlaps with the GRI standard. Compared to the GRI the regulation is more detailed in terms of employment relationships as indicators, whereas human rights issues, that feature large in the GRI is absent. The legislation does not provide sanctions in case of non-compliance, and neither does the legislation require external auditing. The French case of CSR and reporting is especially interesting for other countries, because it illustrates the importance of historical, cultural and social economic contexts, even though there is also convergence towards international standards (such as GRI). Moreover, the French experience illustrates the role of the state; the state plays a central role, i.e. in building networks and giving legitimacy and promoting CSR and reporting. At the same time the French case shows how in spite of strong national traditions there is a strong international influence. In reporting this is evidenced by the later inclusion of environmental indicators, including some elements of GRI reporting.

• Eco-labeling

The growth of environmental ecolabels in Europe attests for the success of labeling as a voluntary approach. However, the profusion of ecolabels has also led to confusion among consumers over what different labels indicate and which are the requirements under different schemes.

EU policy affecting labeling: European Eco-label scheme

The EU's most prominent supportive policy, to enhance transparency and credibility of labeling is the European Eco-label scheme. Established in 2000, the European Eco-label scheme was intended to "promote products with a reduced environmental impact during their entire life cycle and to provide consumers with accurate, non-deceptive, science-based information the environmental impact of products" (European Parliament and the Council of the European Union, 2009).

Labeled products should belong to the 10~20 percent of top environmental performers within a product group. Performance is measured throughout the life cycle of a product across key environmental areas: waste, water use, emissions, energy efficiency, material efficiency, biodiversity. Criteria are drafted by product group, a group of products/services with similar functional properties. Products containing certain hazardous and toxic chemical substances cannot carry the eco-label, although the Commission allows exceptions when products have a significantly higher overall environmental performance.

The European Eco-label scheme created one label that can be applied across almost a very wide variety of product groups. The Eco-label logo can be used on products and in promotional material for products awarded with the Eco-label, with a registration number; and optionally, with up to three environmental performance properties. When the use Eco-label is suspected of being false or misleading, complaints can be made (anonymously) to designated competent bodies. A competent body can investigate whether the product still complies with the product group criteria, and ultimately prohibit the use of the logo.

Theoretically the European Eco-label could lead to a substantial simplification; consumers would not have to learn different labels for different product groups across different countries. In practice, one could also argue that yet another label was added to a profusion of environmental labels.

• Negotiated Agreements

Negotiated Agreements (NAs) are alternative instruments for environmental policies. They are characterized by high involvement of private actors, the pursuit of public goods (in particular a better environmental performance), and negotiations with public authorities. In Europe, NAs have been an important part of the policy mix in several countries, most notably the Netherlands and Germany. The European Commission also recognizes NAs as important and preferred instruments for the implementation of environmental policy. NAs can be more effective because they include those industry actors who know most about the environmental aspects of their business; they allow more flexible (and cost-efficient) implementation by setting general targets, while leaving the actual methodology for implementation to individual firms. By working closely together, participating firms and public authorities exchange knowledge, enabling a mutual learning process. Instead of polarizing public authorities and private actors, NAs enable collaboration, increasing the chances that policy will be endorsed and implemented by all parties. However, the use of NAs is also accompanied by risks: industry may be self-interested, aiming at low environmental performance targets. It is also difficult to determine who should participate and who not (raising questions of legitimacy); free-riding can become a problem when some parties do not keep their end of the agreement. Introducing some mandatory elements, such as sanctioning, or the threat of adopting binding regulation are strategies employed by public authorities to achieve better performance.

EU policy affecting NAs: Communication on Environmental Agreements

A Communication on negotiated agreements was published in 1996 (COM (1996) 561 final) (Commission of the European Communities, 1996). The Communication was mainly directed at member states, although it also referred to a framework for EU-level NAs. NAs, according to the communication, should always be non-obligatory; however they may be recognized by the Commission through an exchange of letters or a recommendation. NAs should explicate qualitative criteria, quantifiable goals, intermediate goals, they should be published, independently monitored, and keep the Council of Ministers and the European Parliament informed. The latest Communication on Environmental Agreements was adopted in 2002. The 2002 Communication: "Action plan 'Simplifying and improving the regulatory environment'" (Commission of the European Communities, 2002a) is not especially targeted at NAs, however, it reviews the use of legislative instruments, the quality of legislation, and tries to simplify regulation. One of the actions stipulated in the Communication was to "make it easier to choose the most appropriate instrument or combination of instruments (of both a legislative and non-legislative nature) from the wide range of options available (regulation, directive, recommendation, co-regulation, self-regulation, voluntary sectoral agreements, open coordination method, financial assistance, information campaign)" (Commission of the European Communities, 2002a).

The Commission distinguishes two types of NAs, self-commitments agreed by industry to implement voluntary agreements are outside of any legal framework at the EU. Such NAs may be recognized, but the EU can still decide to initiate formal regulation. In principle such voluntary agreements are completely voluntary; the Commission considers the agreement as a form of self-regulation that does not have legally binding effects at the Community level. However, the EU can acknowledge environmental agreements, by a simple exchange of letters. Alternatively, the Commission can acknowledge with an additional monitoring decision. In this case the Commission can recommend an agreement while combining it with a Parliament and Council decision on monitoring. The second type is referred to as co-regulation, which is agreed within a legal framework. This takes the form of a directive, a legal procedure, to set conditions; stipulating environmental objectives, time schedules, and monitoring conditions. Mandatory regulation can follow when an agreement fails. This distinction between self-commitments and co-regulation echoes the OECD categorization between "unilateral commitments" and "negotiated agreements" (OECD, 2003), where the former is a voluntary commitment by firms and industries and the latter refers to a negotiated outcome between industry and public authorities. Of course, the range between completely voluntary and negotiated is large. Generally, VAs tend to be negotiated when there is some kind of "shadow of hierarchy", the possibility/threat that public authorities enact binding regulation (Börzel, 2010).

The use of NAs among EU member states is very uneven. According to Karamanos (2001) between 1982 and 1986, 23 new voluntary environmental agreements were formed in the EU, while between 1992 and 1996, 123 agreements were formed. While early adopters in the seventies were France and the UK, most (according to a 1997 survey, ¾ of all) NAs are in Germany and the Netherlands (Dalkmann et al., 2005). At the European level, only a few NAs have been created, for instance with regards to energy efficiency (televisions and DVD Players [2003], Refrigerators and Freezers [2002], and Washing Machines [2002]). Almost a third of all NAs concluded are in the chemical industry, another big sector where many NAs are concluded is in the food industry. Within member states level most agreements have been adopted in the energy and waste management sectors (Dalkmann et al., 2005).

Member State policies affecting NAs: the Netherlands

The Netherlands is a leader in the use of negotiate agreements (NAs) in environmental policy(De Vries, Nentjes, and Odam, 2012), not only because the high number of NAs, but also because of the relative successful applications of NAs, for instance the OECD reports that in the Netherlands: "overall industry has been responsive and often proactive in improving its environmental performance, particularly through environmental agreements (e.g. covenants) and environmental management and auditing" (OECD, 2004; Bressers, De Bruijn, and Lulofs, 2011). Agreements (or "convenanten" in Dutch) are negotiated between public authorities (usually the Ministry of Environment) and a sector or industry organization that represents firms in that sector or industry(see: Bressers, De Bruijn, and Lulofs, 2011). The agreements are meant to implement aspects of the National Environmental Policy Plan. They refine policy targets, for instance through providing quantitative targets. Government and industry then negotiate how to achieve targets, and set timelines.

The procedure fits well with Dutch political culture, which has been described as a "polder-model", a negotiation model between government and representatives of industry, labor and interest organizations. This culture is characterized by a consensual approach to political challenges. Partners in negotiations are regarded as, and expected to be, responsible actors, the government tries to enhance responsibilities by engaging these organizations in policy making and implementation. The process is not entirely voluntary. Upon agreement, commitments are allocated(to individual firms)and linked up with their licenses. Free riding will be sanctioned. While firms can refuse participation, they will find themselves in a confrontational position vis-à-vis other firms in the sector, and they will likely be subjected to new restrictions through conventional regulatory (permitting) instruments.

Bressers et al. (2011) therefore argue that, in the Dutch case: "Covenants are thus negotiated, with flexibility regarding positions taken during negotiations; the formal position of the sectoral organization and its membership, in other words, is substantially greater than what is conventionally seen in a standard regulatory, rule-setting process, wherein interested parties can seek to influence authoritative actors but cannot stop action by withdrawing from or resisting the process. On the other hand, if and when an agreement is reached, it is in principle enforceable. This design clearly demonstrates that crude distinctions between "'voluntary' and 'coercive' approaches are misleading"(2011: 190). In their analysis of 59 agreements, they found that socio-political context (for instance political culture) is not necessarily related to performance, some of the agreements perform better than others (203). This means that agreements can be implemented in other countries as well, if there is due attention to the design of agreements, governmental pressure, and practical attention to costs.

Example: benchmarking energy efficiency covenant

In 1999 Netherlands based industrial companies signed up to match world leaders in the area of energy performance. Companies that signed up drafted Energy Efficiency Plans (EEP) declaring their distance from the world leaders, and, explicating how they would catch up. By 2003, 233 organizations signed the covenant. 276 organizations qualified for participation, which was well above expectation. The joint energy consumption of participants from the industrial sectors was calculated, equaling 704 PJ in 1999, corresponding to about 92% of total consumption (768 PJ) by the energy-intensive industry. The participating companies submitted Energy Efficiency Plans (EEP) while a Verification Agency issued recommendations about the plans, and the competent authority then assessed them. The aim was to match the best performers in the world among energy intensive companies. After the assessment by the competent authority, the EEP formed the basis of an agreement between the participant and the competent authority. In 2003 111 EEPs were agreed upon by competent authorities.

Calculations indicate that a reduction 5.1 billion kg of the CO_2 emissions is possible by 2012. This is considerably more than the 3.2 to 4.0 billion kg reduction which was estimated before the covenant. A Benchmarking Committee provides an extensive overview of the interim results of the covenant. A managing office and the competent authority receive annual monitoring reports to monitor overall performance (Commissie Benchmarking, 2004).

• Green Public Procurement

Green Public Procurement (GPP) is an important measure to stimulate voluntary corporate environmental behavior. By including criteria in public procurement contracts, public authorities can increase demand for green products and service, but also incentivize reporting on the environmental aspects of products and services. The share public procurement in the EU amounted to about 16 per cent of the gross domestic product(GDP)in 2002(Steurer et al., 2007), and increased to 19.9% in 2009 (Renda et al., 2012). This large and growing share of public procurement provides with a great opportunity to impact the market and stimulate green products, services and technologies.

EU policy affecting GPP

The European Commission first addressed Green Public Procurement in 2001 in an Interpretative Communication (COM, 2001) (Commission of the European Communities, 2001b), discussing how contract awarding processes could take into account GPP. In 2003 a "Communication on Integrated Product Policy" (European Commission, 2003b) addressed public procurement as a central tool for continuous environmental improvement and encouraged every member state to draw up national action plans. By 2010 21 member states have implemented an action plan or equivalent policy, while action plans were underway in another 6 member states. Aspects of GPP have become legally binding in e.g. Portugal, Germany (for wood) and the Czech Republic (for IT). In 2004, two directives, 2004/17 (European Parliament and Council of the European Union, 2004) and 2004/18 (European Parliament and Council of the European Union, 2004), where published which respectively allow the consideration of social and/or environmental issues at an early stage of the procurement process, and "clarifies how the contracting authorities may contribute to the protection of the environment and the promotion of Sustainable Development, whilst ensuring the possibility of obtaining the best value for money for their contracts" (2004/18/EC L134/114).

The two directives specifically address the possibility to include environmental requirements such as the use of eco-labels and environmental management conditions. In 2008 the Commission presented a communication "Public Procurement for a Better Environment" (European Commission, 2008). The Communication included a voluntary EU-wide target for Green Public Procurement: by 2010 50% of all public tendering procedures should be green, that is: compliant to EU GPP criteria drafted for product and service groups such as construction, transport, and services. The Commission sets GPP criteria for product and service groups which authorities can include as criteria in tendering procedures. The criteria are based on existing European and national eco-label criteria and also collected from stakeholders from civil society and industry. Once endorsed by member states, these criteria are included in national guidelines and action plans. Initially criteria were formulated for ten product groups; currently they exist for 21 product groups. Criteria are regularly revised, to keep up with higher green standards. The EU distinguishes core criteria (those suitable for use by any contracting authority across the Member States) and address the key environmental impacts for the product in question (designed to be used with minimum additional verification effort or cost increases), and comprehensive criteria (those procurers wishing to purchase the best products available on the market that may require additional verification efforts or a slight increase in cost compared to other products).

A review in 2011 by the Center for European Policy Studies and the College of Europe commissioned by the Commission found that 50% the target has not been met, instead only 26% of all procurement contracts signed in 2009-10 included all GPP criteria, while 55% of all contracts did refer to some GPP criteria (Renda et al., 2012). Nonetheless uptake of core GPP criteria is increasing. In 2009-10 38% of total procured value included some form of GPP criteria, representing a value of about 117.5 billion Euros. Uptake was especially high with regard to core criteria concerning double printing, IT energy performance, and transport CO_2 emissions. While the overall level of uptake of GPP criteria is increasing, performance in Europe varies significantly; Belgium, Netherlands, Sweden, Finland and Denmark are leading member states. A 2010 study for the Commission by AEA also indicated differences in development of criteria for the number of product and service groups among 9 member states and Norway (AEA, 2010). One finding was that countries with federal governments(Belgium, Germany)generally define fewer criteria, and the scope of GPP is narrower, however, at the level of constituent states there are additional policies which are sometimes quite ambitious (for instance, the Flemish Region in Belgium aims at 100% GPP criteria inclusion in procurement).

EU Member State policy affecting GPP: UK

The case of the UK and sustainable public procurement and frameworks are illustrative for the ambition level as its aim to become a leader in public procurement given especially considering the large of the annual procurement budget of central and local government which is estimated at £150bn a year (Fletcher, Duisterwinkel et al., 2009). On the other hand the case of the UK also demonstrates the need for monitoring and management systems, knowledge capacity-building and standards beyond the minimum that e.g. take into account product cycles.

Following the call at the World Summit on Sustainable Development in 2002 for promoting public procurement policies to encourage development and diffusion of environmentally sound goods and services, the UK government stated its goal to be amongst the leaders in Europe on sustainable procurement by 2009. In 2003 Government Buying Standards, minimum environmental standards for the procurement of a variety of goods were introduced as "Quick Wins". They were mandatory within the central government departments and agencies, but not with other parts of the public sector. By the end of 2008 the standards covered 46 types of product in nine priority product groups, in that year standards were updated to include voluntary best practice standards as well as mandatory ones.

However, the Quick Wins standards have been based on average standards within the market and not on whole-life environmental impact analyses. By 2007/08, 15 out of 21 central departments reported that they complied with these standards, but six of the departments did not have systems to measure compliance. Over the years Government Buying Standards (the new name for Quick Wins) have been aligned to the European Commission's Green Public Procurement initiative. The 2005 Sustainable Development Strategy restated the ambition to be amongst the sustainable procurement leaders in the EU by 2009. It set out a "Flexible Framework" to assess progress in sustainable procurement. The Framework identified five themes for measuring progress on: "policy, strategy and communications"; "measurements and results"; "procurement process"; "engaging suppliers"; and "people", by rating levels from 1 to 5, where 5 stands for leading. Since 2006 government department report against progress under the so-called "Sustainable Operation on the Government Estate (SOGE)" framework, which sets out mandatory requirements for all central government departments to reduce operational impacts. The framework includes fourteen operational performance targets covering carbon emissions from offices and road vehicles, energy efficiency, waste, recycling, biodiversity, water consumption and renewable energy generation. Eight requirements were added, covering the steps that departments should take to reach performance targets, and to improve monitoring and reporting, and to enable accreditation of environmental management systems.

In 2007 the Sustainable Procurement Action Plan, UK's roadmap to the implementation of objectives set within the EU Sustainable Development Strategy, reiterated the Framework, it added five commitments to sustainability: leadership and accountability (permanent secretaries are accountable for their department's progress and appropriate incentives); budgeting and accounting (resolving barriers to sustainable solutions by reviewing budgeting arrangements and performance frameworks); building capacity (setting out actions and considering the use of the Flexible Framework); market engagement and capturing innovation(ensuring suppliers have plans to lower carbon foot prints); raising standards(stop buying products that are below the Quick Wins standards).

In a review the National Audit Office (2009) found that the goal to become a leader in the EU has not been realized because of a lack of quantifiable targets (p.7). Moreover, while the Flexible Framework would provide a basis for assessment against progress, there is no assurance in place, rather progress is based on self-assessments which are not necessarily consistent, because the Framework did require outcome measurements.Other problems related to uncertainty with government staff to which it was not clear what the costs and benefits were of going beyond the minimum standards. The National Audit Office found that not all departments complied with the mandatory Quick Win standards, and they often lacked management systems for monitoring and performance.

- **Indexing**

An external approach to improve environmental behavior is the listing of corporations, or the naming and shaming of companies. This approach assumes that by providing information on companies, environmental performance of multiple companies, companies are incentivized to improve their environment performance. This approach may take the form of indexes (environmental performance indexes), positive listings (for instance in investor portfolio's) and negative listings (often done by NGOs). While the EU does not provide with listing and indexing schemes, several

European-wide (and sometimes sector specific) schemes are active.

Market index companies have identified companies on specific topics, including environmental leadership. For instance the FTSE4Good Leaders Europe 40 Index is made up of companies that "are doing more to manage their environmental risks and impacts whilst reducing their environmental footprint. Companies have to demonstrate they have environmental strategies and management systems, assess risks and impacts, and need to publish key environmental performance data" (FTSE, 2013). Similar indexes have been established by Down Jones, the Dow Jones Sustainability Europe Indexes(DJSI Europe), consisting of a composite index and several subset indexes(for instance, one excluding companies that related to alcohol, tobacco, gambling, armaments, firearms and adult entertainment). Corporate sustainability is assessed across economic, environmental, and social dimensions; environmental criteria include reporting criteria (assurance, coverage, qualitative and quantitative data) and industry specific criteria(relating to e.g. EMS, climate strategies, eco-efficiency)(DJSI, 2014). While market index companies reward better corporate performers, specialized market indexes typically exclude companies that are underperforming.

By excluding these companies from the indexes, investors miss an opportunity to incentivize these companies to improve their performance. Environmental tracking is an alternative to specialized indexing, which includes both high and low environmental performing companies. The Environmental Investment Organization (EIO), a non-profit organization aiming at ecological investment, has developed an Environmental Tracking methodology, a combination of assessment of disclosure and verification and market indexing, which it describes as a "market mechanism designed to incentivize the world's largest companies to reduce their emissions and improve their levels of transparency" (EIO, 2012). In ET indexes market indexes are reweighted according to their position in a carbon ranking. Since 2010 EIO publishes ET Carbon Rankings for several regions, among them the ET Europe 300, Asia-Pacific 3000, BRIC 300, and the ET Global 1000 and ET Global 800. Rankings are based on absolute emissions, direct and indirect. The purpose of the ET Carbon Ranking is to provide information on a company's greenhouse gas emissions intensity as well as its level of disclosure and verification.

According to the 2011 rankings, Europe is the global leader in carbon disclosure. According to EIO, 53% of all listed companies in Europe provided complete reporting data. Disclosure is especially high in the Netherlands (100%), Italy (88%) and Germany (84%) (The Climate Group, 2011). Companies can appeal when they find they have been inaccurately represented in the ranking. Appeals have been successfully filed by the Veolia and Carnival companies in 2011, moving them up in the indexes.

• Final discussion

EU based companies and EU Member States have been pioneering in policies and strategies incentivizing voluntary corporate environmental behavior. The share of EU based firms and organizations in environmental reporting, disclosure schemes and environmental assurance is high. However, emerging countries, such as China, are catching up as corporations increasingly recognize the benefits of participating in voluntary schemes.

For corporations voluntary corporate environmental behavior can be beneficial as:

(1) conflicts with regulators is reduced;

(2) greater flexibility is allowed, especially compared to traditional regulatory approaches;

(3) companies can gain a competitive edge over competitors through green innovation and anticipating future mandatory regulations.

For governments, an enhanced role of the private sector can lead to:

(1) lower implementation costs;

(2) greater support from targeted industries;

(3) better use of expert knowledge.

However, voluntary approaches are also criticized for:

(1) allowing free-riding, for instance presenting a firm as green when it continues

to pollute ("green-washing");

(2) setting lower standards, under the pressure of special industry interests;

(3) lacking democratic legitimacy;

(4) being exclusive;

(5) lacking the broad support to set international, industry-wide, standards.

The overview in this paper indicates that, to increase benefits under voluntary approaches and to avoid negative effects, voluntary approaches have to be well-designed. Strategies that are ambitious (such as the UK's GPP strategy) fail to deliver when monitoring and assurance systems are not in place. The study also shows that voluntariness should be accompanied by more coercive approaches that sanction or increase the cost of non-compliance. For instance, in the Netherlands, NAs are accompanied by permitting systems. In many cases there is "a shadow of hierarchy": private companies are more likely to act when government threat to enact (stricter) binding regulations. The distinction between voluntary and mandatory, moreover, is not set in stone. Some voluntary schemes, for instance EMAS and the European Eco-label are backed by European regulations and directives. Definition of the instruments, its procedures, and the terms of participation are outlined in detail by the European Commission, European Parliament and Member States.

In some cases voluntary schemes become mandatory. For instance the GRI as a voluntary sustainability reporting standard has become mandatory for state-owned companies in Sweden and the Netherlands. The country case studies highlighted two types of companies where responsible environmental behavior can be mandated on grounds of public legitimacy. State-owned companies are ultimately owned by the people and should therefore be more accountable in terms of how their operations impact the environment, society and the economy. In the case of France, publicly listed companies have to disclose social and environmental information. Investors in publicly listed companies often need more than financial information alone to assess risks and opportunities, and to make informed investment decisions.

European experiences and lessons learned may not readily translate in to policy recommendations for the Chinese context; however a few points can be made. Considering the high number and share of state-owned enterprises in China's economy, mandatory reporting – not unlike the case of Sweden – could be particularly effective at increasing the number of reporting companies. It should be noted that this does not necessarily result in better environmental performance. Reporting in the EU has often been directed to shareholders and rather than to social groups (labor unions and NGOs). However, as the case of France shows, reporting can be addressed to governments as well, helping the government to keep informed on the role of companies in environmental pollution.

Environmental labeling in the EU is, to some degree, a victim of its own success. The large number of labeling schemes has confused consumers. The EU is trying to counter this by introducing the European Eco-label for multiple product and service groups. A similar multi-product and service approach may be adopted more effectively in China, since labeling is not yet as profuse as in the EU. However, key to any labeling scheme is reliability, which requires well designed accreditation and monitoring systems, and credible sanctions against abuse of labeling.

Among voluntary external approaches, GGP seems to be promising, given the high share of government expenditure in China. However monitoring and assurance systems should be in place, and there should be broad awareness and support throughout government bureaucracies. The scope for NAs seems to be limited. In a small country like the Netherlands it is relatively easy to gather an entire sector around the negotiation table, in a larger country with many more companies in a single sector such a process may prove cumbersome. However, provincial and local level applications within China, especially in the larger cities may be effective. Moreover, the emphasis on consensual decision making seems to fit well with China's concept of harmonious development.

References

[1] Accountability. 2008. AA1000 Assurance Standard 2008.

[2] AEA. 2010. Assessment and Comparison of national Green and Sustainable Public Procurement Criteria and Underlying Schemes. Brussels.

[3] Antal, Ariane Berthoin, and André Sobczak. 2007."Corporate Social Responsibility in France: A Mix of National Traditions and International Influences." Business Society no. 46(9): 9-32.

[4] Arora, Seema, and Timothy N. Cason. 1996. "Why Do Firms Volunteer to Exceed Environmental Regulations? Understanding Participation in EPA's 33/50 Program." Land Economics no. 72 (4): 413-432.

[5] Borglund, Tommy, Magnus Frostenson, and Karolina Windell. 2010. Increasing responsibility through transparancy? A study of the consequences of new guidelines for sustainability reporting by Swedish state-owned companies. Stockholm: Regeringskansliet.

[6] Bressers, Hans, Theo De Bruijn, and Kris Lulofs. 2011."Negotiation-based Policy Instruments and Performance: Dutch Covenants and Environmental Policy Outcomes." Journal of Public Policy no. 31 (2): 187-208.

[7] Brown, Halina Szejnwald, Martin de Jong, and David L. Levy. 2009. "Building institutions based on information disclosure: lessons from GRI's sustainability reporting." Journal of Cleaner Production no. 17 (6): 571-580. doi: 10.1016/j.jclepro.2008.12.009.

[8] CDP. 2013. CDP Global 500 Climate Change Leaders 2013. https://www.cdproject.net/en-us/results/pages/leadership-index.aspx, accessed: 10-01-2015.

[9] Commissie Benchmarking. 2004. Convenant Benchmarking enegie-efficiency. Avalable at: http://www.benchmarking-energie.nl/pdf_files/covtned.pdf (Dutch), accessed: 07-06-2012.

[10] Commission of the European Communities. 1996. Communication from the Commission to the Council and the European Parliament on Environmental Agreements. Brussels: Commission of the European Communities.

[11] Commission of the European Communities. 2001a. Green Paper. Promoting a European Framework for Corporate Social Responsibility. Brussels.

[12] Commission of the European Communities. 2001b. Commission Interpretative Communication on the Community law applicable to public procurement and the possibilities for integrating

environmental considerations into procurement. Brussels: Commission of the European Communities.

[13] Commission of the European Communities. 2002a. Communication from the Commission. Action plan "Simplifying and improving the regulatory environment". edited by Commission of the European Communities. Brussels.

[14] Commission of the European Communities. 2002b. Communication from the Commission concerning Corporate Social Responsibility: A business contribution to Sustainable Development.Brussels.

[15] Dalkmann, Holger, Daniel Bongardt, Katja Rottmann, and Sabine Hutfilter. 2005. Review of Voluntary Approaches in the European Union. Feasibility Study on Demonstration of Voluntary Approaches for Industrial Environmental Management in China. In Wuppertal Report. Wuppertal: Science Centre North Rhine-Westphalia, Institute of Work and Technology.

[16] De Vries, Frans P., Andries Nentjes, and Neil Odam. 2012. "Voluntary Environmental Agreements: Lessons on Effectiveness, Efficiency and Spillover Potential." International Review of Environmental and Resource Economics no. 6: 119-152.

[17] Didong, Jenny. 2010. Learn how State-Owned Companies Take the Lead in Reporting. In The Amsterdam Global Conference on Sustainability and Transpancy. Rethink. Rebuild. Report, Amsterdam.

[18] DSJI. 2014. Dow Jones Sustainability Indices. Methodology. http://www.djindexes.com/mdsidx/downloads/meth_info/Dow_Jones_Sustainability_Indices_Methodology.pdfEuropean Commission. 2003a. Fourth Directive on Annual Accounts Brussels: European Commission, accessed 15-01-2015.

[19] EIO. 2015. EIO Background and ET Concept. http://www.eio.org.uk/pdf/EIO_Background_and_Further_Reading.pdf, accessed 15-01-2015.

[20] European Commission. 2003b. Integrated Product Policy: Commission outlines its strategy to stimulate greener products. Brussels: European Commission.

[21] European Commission. 2004. Commission Decision of 29 January 2004 establishing guidelines for the monitoring and reporting of greenhouse gas emissions pursuant to Directive 2003/87/EC of the European Parliament and of the Council. Brussels: European Commission.

[22] European Commission. 2007. Establishing guidelines for the monitoring and reporting of greenhouse gas emissions pursuant to Directive 2003/87/EC of the European Parliamsnt and of the Council. edited by European Commission. Brussels.

[23] European Commission. 2008. Communication from the Commission to the European Parliament, the Council, the European Economic and Social Committee and the Committee of the Regions. Public procurement for a better environment. Brussels.

[24] European Commission. 2009. Commission Decision of 16 April 2009 amending Decision 2007/589/EC as regards the inclusion of monitoring and reporting guidelines for emissions and tonne-kilometre data from aviation activities. edited by European Commission. Brussels: European Commission.

[25] European Commission. 2011a. Communication from the Commission to the European Parliament, the Council, the Economic and Social Committee and the Committee of the Regions. Single Market Act. Twelve levers to boost growth and strengthen confidence. "Working together to creat new growth". edited by European Commission. Brussels: European Commission.

[26] European Commission. 2011b. Communication from the Commission to the European Parliament, The Council, The European Economic and Social Committee and the Committee of the Regions. A renewed EU strategy 2011-14 for Corporate Social Responsibility. edited by European Commission. Brussels: European Commission.

[27] European Commission. 2013. Commission moves to enhance business transparency on social and environmental matters. Press release. 16 April 2013. http: //europa.eu/rapid/press-release_IP-13-330_en.htm, accessed 10-01-2015.

[28] European Parliament and the Coucil of the European Union. 2004. Directive 2004/17/EC of the European Parliament and of the Council of 31 March 2004 coordinating the procurment procedures of entities operating in the water, energy, transport and postal services sectors. Brussels: European Union.

[29] European Parliament and the Council of the European Union. 2004. Directive 2004/18/EC of the European Parliament and of the Council of 31 March 2004 on the coordination of procedures for the award of public works contracts, public supply contracts and public service

contracts. Brussels: European Union.

[30] European Parliament and the Council of the European Union. 2009. Regulation (EC) No 66/2010 of the European Parliament and the Council of 25 November 2009 on the EU Ecolabel (Text with EEA relevance). http://eur-lex.europa.eu/legal-content/EN/TXT/? uri=celex: 32010R0066, accessed 15-01-2015.

[31] Frostenson, Magnus, Karolina Windell, and Tommy Borglund. 2010. Mandatory sustainability reporting in Swedish state-owned companies: Perspectives and Consequences. Paper read at Earto 2010, 19 May 2010, at Uppsala University.

[32] FTSE. 2013. Groundrules for the management of the FTSE4Good Environmental Leaders Europe 40 Index. Available at: http://www.docstoc.com/docs/155191447/FTSE4Good-Environmental-Leaders-Europe-40-Index-Ground-Rules, accessed 10-01-2015.

[33] Global Reporting Inititiative (GRI). 2011. Sustainability Disclosure Database. edited by Global Reporting Initiative (GRI). Amsterdam.

[34] Gouvernement Française. 1977. Loi n° 77-769 du 12 juillet 1977 RELATIVE AU BILAN SOCIAL DE L'ENTREPRISE. edited by Gouvernement Française. Paris: Gouvernement Française.

[35] Gouvernement Française. 2001. Nouvelles Régulations Economiques (NRE). Paris.

[36] GRI (2012) Sustainability Disclosure Database. Beta 0.9. Available at: http://www.database.globalreporting.org, accessed 01-06-2012.

[37] Karamanos, Panagiotis. 2001."Voluntary Environmental Agreements: Evolution and Definition of a New Environmental Policy Approach." Journal of Environmental Planning and Management no. 44 (1): 67-84. doi: 10.1080/09640560124364.

[38] Martinuzzi, André, Barbara Krumay, and Umberto Pisano. 2011. Focus CSR: The New Communication of the EU Commission on CSR and National CSR Strategies and Action Plans. In ESDN Quarterly Report Vienna, Austria: European Sustainable Development Network (ESDN).

[39] OECD. 2003. Voluntary Approaches for Environmental Policy: Effectiveness, Efficiency and Usage in Policy Mixes. Paris.

[40] OECD. 2004. Environmental Performance Review of the Netherlands, Executive Summary. Paris: OECD.

[41] Renda, Andrea, Jacques Pelkmans, Christian Egenhofer, Lorna Schrefler, Giacomo Luchetta, Can Selçuki, Jesus Ballesteros, and Anne-Claire Zirnhelt. 2012. The Uptate of Green Public Procurement in the EU 27. Brussels: Centre for European Studies (CEPS), College of Europe.

[42] Schaay, Jean-Paul. 2010. Learn how State-Owned Companies Take the Lead in Reporting. The Amsterdam Global Conference on Sustainability and Transpancy. Rethink. Rebuild. Report.

[43] Słonimiec, Justyna, and Jakub Świtała. 2013. European development of eco-management and audit scheme (EMAS) in the European Union. Management Systems in Production Engineering 4 (12): 28-32.

[44] Steurer, Reinhard, Gerald Berger, Astrid Konrad, and Andre Martinuzzi. 2007. Sustainable Public Procurement in EU Member States: Overview of government initiatives and selected cases. Final Report to the EU High-Level Group on CSR. In Analysis of national policies on CSR, In support of a structured exchange of information on national CSR policies and initiatives. Vienna: RIMAS - Research Institute for Managing Sustainability, Vienna University of Economics and Business Administration.

[45] The Climate Group. 2011. ET Global Rankings: Europe Leads on Carbon Disclosure, Boasts Lowest Carbon Intensity. http://www.theclimategroup.org/what-we-do/news-and-blogs/et-global-rankings-europe-leads-on-carbon-disclosure-boasts-lowest-carbon-intensity, accessed: 15-01-2015.

[46] UNGC. 2011. Online Application Guideline Business. http://www.unglobalcompact.bg/wp-content/uploads/2014/05/Online_Application_Guideline_Business.pdf, accessed: 15-01-2015.

[47] 1 Parts of this paper are based on a study commissioned by the EU-China Environmental Governance Programme and written by the present author in collaboration with Patrick Schroeder at the China Alliance for NGO Collaboration (Chan, Sander and Patrick Schroeder (2013) Comparative Policy and Practice Study 4 (EU). Voluntary Corporate Environmental Behavior. Prepared for EU-China Environmental Governance Programme, June 2012, Beijing).

[48] 2 Sander Chan. PhD researcher at the German Developpment Institute/Deutsches Institut für Entwicklungspolitik (DIE), Bonn, Germany. E: sander.chan@die-gdi.de/T: +49 (0) 22894927-293.

英国的环境公众参与概况

辛里奇·沃斯*

导言

  环境污染和环境恶化是大多数国家面临的严峻问题（Ycelp and Ciesin，2014）。一些国家环境信用系统都发现，无论发达国家还是发展中国家都存在着水、土壤和空气的污染，而日益严重的污染问题也威胁着生物多样性和动物的生存空间。此外，温室气体的排放也应该受到人类的重视，因为其本身就是一个重要的污染源（IPCC，2014）。尽管所有国家都面临着如何解决环境问题的困境，但发展中国家在强调经济增长和实现经济赶超的情况下会面临着更大的挑战。不过，对发展中国家而言，经济发展与保护环境相比具有更加重要的意义，因为实现经济和财富的高速增长是国家发展的根本。

  对企业发展来说，企业领导人的自由任期制促使企业基于利润最大化原则去指导其生产经营活动，把关注的重点放在投资开发、生产销售和提供商品服务上。只有当环境问题对其追求商业成功造成直接影响时，企业才会积极考虑在环境保护上的支出费用。在这种情景下，企业就容易把环境污染的外部效应转嫁给社会来承担。正因为环境污染对公众社会的负面效应，全球经济长期持续增长的神话也开始遭到更多的质疑。因此，如何在环境保护和经济增长之间找到平衡点是一个长期以来困扰着我们的问题。缓解这一困境的一种方法是将各个利益相关者聚集在一起，并适当赋予他们参与环境决策的权利。

---

*辛里奇·沃斯，利兹大学商学院讲师，BSc IB 项目的副主任和利兹大学孔子商学院执行主任。本文原文为英文，中译者为易龙飞（浙江大学公共管理学院）。

公众参与环境治理和建立公众协商机制能为社会大众构建一个表达利益诉求的渠道。这一渠道的设立不仅能够让公众向利益相关者申诉其在生活中已经或即将受到的影响，也使决策者和投资者意识到这些问题的潜在后果。在这一过程中，不同社会利益群体之间的误解会得到有效缓解，从而形成一个理性运转的社会。

如果基于排污企业的视角看待经济增长与环境保护的二难问题，实质上就变成了什么时候让他们参与决策过程的这一问题。公众可能对环境议题的民主参与感兴趣，但并不会对每一个议题都如此。也就是说，排污企业必须意识到一系列利益相关者在不同情景下的重要性。下面笔者将以英国为例来分析和说明环境问题的公众参与如何运作。

通过排污企业的视角看待公众参与环境问题，与研究公众参与的机制以及研究更为广泛的环境污染问题是同等重要的。如果公众参与仅仅涉及那些"受到影响的群体（affected party）"，那么将会忽视其排污企业的自身约束力。本文主要从组织或企业的视角出发分析组织或企业如何进行公众参与，以及这一公众参与的机制是如何建构和相互联系，组织或企业又如何看待这种机制以及公众要求他们改变行为的要求。

环境保护公众参与的概念框架背景

环境保护的公众参与可以追溯到 1970 年的美国《国家环境政策法》(National Environmental Policy Act，1970)（Morgan，2012）的制定和 Arnstein（1969）对公众参与环境保护概念的界定。这一概念旨在增进受到组织决策或组织发展潜在影响群体的参与权利，以及促进这些群体在组织决策制定过程中的参与（Bickerstaff, Tolley and Walker, 2002）。公众参与同样遵循着民主原则，要求政府回应公众的诉求以及公众对政府施政目标的意见，因此民主的思想要求公众在决策过程中积极参与（Shepherd and Bowler, 1997）。通过这种参与，民主思想本身得到重申并焕发生机。一个民主的决策应该是充分考虑到受影响群体相关诉求之后的决策，正因为公众有权参与决策的制定，所以其给予了公众了解排污企业更多细节的机会。因此，公众也会更加理解和支持排污企业的进一步行动。

《奥胡斯公约》（Aarhus Convention）规定了公众参与的基本原则，即获取信息是一项重要的公民权利，不仅限于公共机构（Mason，2010）。通过信息的获取，公众可以了解环境问题对他们所在社区的发展和活动造成的影响，因此也能更好地了解他们生活的环境在多大程度上会受到水、土壤、空气和噪声污染的危害。也就是说当企业决定建立可能会产生噪声和空气污染的新工厂时，或者当企业打算引进可能会增加化学物使用的新生产线时，或者当企业决定搬迁现有工厂以更新生产设备时，公众有权了解相关信息并提出意见。

公众参与并不会随着决策过程的终止而停止，而要延伸到企业的日常运作中。受到企业决策或运营影响的群体可以通过公众参与环境治理的机制为自己发声，并采取必要的措施迫使排污企业改变其运作方式（Lee et al.，2013）。通过环境问题的公众参与，排污组织和企业也能获得更多的信息以便使项目运作更加丰富完善（Shepherd and Bowler，1997；Lee et al.，2013）。环境问题公众参与过程中的关键问题在于如何界定谁是"公众"；"公众"的范围被确定后，关于公众何时参与，参与到什么程度的问题随之出现。

Bickerstaff 等强调，在公众参与问题的讨论中我们有必要明确界定"参与（participation）"、"协商（consultation）"、"信息（information）"这三种方式（Bickerstaff et al.，2002）。"参与"是一个企业和公众之间的互动过程，这一互动能够促进企业决策的形成。公众有权利和影响力对企业施加干预，企业的决策必须整合这些来自公众的意见。"协商"是一个询问公众意见和建议的过程，它并不硬性规定企业的最终决策必须采纳这些意见和建议。最后，在公众参与作为一种实践"信息"公开的情况下，公众仅会被告知即将发生的事情以及他们可能受到的影响。

以上是三种常见的公众参与的方式，它们往往同时存在。在英国，一旦法律规定了公众参与的过程，那么就要确定谁是可以参与的公众。Bickerstaff 等进一步指出，英国的交通基础设施项目中存在着不同界定边界的"公众"：公众首先是指当地居民、企业、交通运营商和用户，至于弱势群体（如有特殊需求的老年人）的代表人、医疗服务和教育的提供者、环保组织等则没有清晰的界定（Bickerstaff et al.，2002）。尽管清晰的界定深受不同政党的欢迎，但依然会产生一些问题。因为对公众概念的界定过于固定有可能会把受环境问题影响最深的那部分群体排除在外，也不能反映社会经济环境的变化，因此需要不断微调。

《奥胡斯公约》指出：公众参与应当是尽早且有效的。这就意味着参与的过程越早越好，并且这个过程应该是包容的、开放的、互动的、连续的和互惠的（Bickerstaff et al.，2002；Hartley and Wood，2005）。但是 Hartley 和 Wood（2005）及 Morgan（2012）则强调公众对规划、法律和授权等知识的匮乏，以及信息的沟通不顺畅甚至误解往往会导致公众参与的实际效果大打折扣，而民众对 HS2 建设项目给环境、经济和社会造成的影响所知甚少也制约着公共协商机制的建立。公众对协商过程的回应表明，在进行任何咨询协商之前应当充分向公众提供涉及环境、经济和社会等方面的信息。但事实却是，信息公开尚未使用简单通俗的语言，一些敏感的信息也会被不同程度地保留（Crompton，2013）。

使用"组织"这一概念是为了区别公众参与环境问题强调的是公众和企业以及公众和其他组织（如公共组织、政府和非政府组织等）之间的关系，下文将主要讨论公众与企业（或相关公共组织）之间的关系。

1. 利益相关者的识别与管理

由于文章将从企业的视角考察公众参与环境治理，因此笔者会采用与企业活动密切相连的分析框架作为分析背景。在企业经验与管理的学术研究中，公众参与这一概念最早来源于利益相关者（stakeholder）管理的概念（Freeman, Madsen and Ulhøi, 2001；Verbeke and Tung, 2013；Helin, Jensen, and Sandstroem, 2013）。与前文介绍的公众参与的定义类似，利益相关者被定义为"对公司财富增长做出自愿或非自愿贡献的个人和团体，他们是潜在的受益者或风险的承担者"（Post, Preston, and Sachs, 2002：19）。财富增长或财富贬值的潜在风险承担者是受到环境污染影响的广义上的公众。这一定义也包括其他可能受到影响的企业或组织、组织的雇员、投资者及政府。这种企业和公众之间交互关系的框架在商业领域的学术研究中是十分常见的，同时也是企业展示自己的典型方式。公众参与和利益相关方参与都是描述和评估社会参与的路径，但它们基于不同的分析角度和现实观念。这种方法灵活地定义了"公众"这一概念。

公众参与和利益相关者管理的目的之一是基于组织合法性或组织特定的规则来建立理解和信任的关系（Carroll and Shabana，2010）。利益相关者和组织之间的信任指的是他们知道该组织承认并尊重法律法规。组织既没有逃避监管，也没有剥削劳动力，更没有违反环保规定，相反它对社会作出了贡献。利用利益相关者对组织进行管理的实践表明组织愿意接受公众的意见,愿意倾听他们的担忧。组织问责和信息公开机制的缺乏会严重阻碍公众对组织的信任。因此,为了建立这种信任关系并取得组织合法性,组织在日常活动中的问责与透明机制便显得很重要,尤其要确保对所有利益相关者一视同仁。透明度涉及组织是如何履行环保义务,并且多大程度上公开其决策结果,它包括对问题的详细说明,以及如何回应特定的公共问题。

在利益相关方参与的过程中,组织必须认识到并不是所有的利益相关者都拥有同样的地位,他们话语权的大小和受重视的程度是不一样的。为了区分不同利益相关者的时空差异以及重要性,Mitchell、Agle和Wood提出了"重要利益相关者(salient stakeholder)"这一概念,当组织内部与外部利益相关者人数过多时,他们应当被区别对待(Mitchell,Agle and Wood,1997)。Mitchell等的推动使这一概念愈发流行,它不仅确定了重要利益相关者,而且根据重要程度把这些重要利益相关者再排序。这一概念把权力(power)、合法性(legitimacy)和紧迫性(urgency)作为重要性排序的指标。权力指利益相关者能够把自身意志强加给组织的能力大小;合法性指利益相关者的行为是否能符合社会预期并被社会认可;紧迫性则描述了利益相关者要求的时限性。

最重要的利益相关者应该同时具备这三个特征(见图1)。具体到环境污染问题上,最重要的利益相关者可能是一个水源地正在遭受着上游有毒污染物威胁的社区,组织无法回避这些重要利益相关者的意见。当然,也会出现不具备以上特征的"利益完全不相关者":比如一个人想要把该组织社会责任报告中一张树的照片更改为水的照片。这两个极点之间的"显著利益相关者"和"利益完全不相关者"的区别在于他们是否可以合法地提出治理要求、合法性和/或紧迫感。介乎于重要利益相关者和利益完全不相关者这两个极端之间的是普通利益相关者,他们也日益成为组织关注的焦点。因为他们可以正当地提出自己的诉求并且具有一定的权力、合法性和紧迫性。

随着时间的推移,利益相关者的重要程度会发生显著变化。环境污染问题注定具有很强的紧迫性,因为该问题直接影响到利益相关者的日常生活并有可能导致潜在的严重后果。环境保护被视为是维护生态平衡和实现可持续发展的重要举措。然而,这种环境保护的理念在以牺牲环境为代价换取经济增长和社会福利的地区中很难被主流社会所接受。这种动态的发展需要组织定期评估究竟谁才是真正的重要利益相关者。

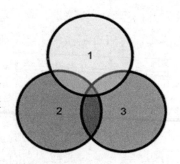

1. 权力：强加自己意愿的能力
2. 合法性：社会接受的或期望的行为
3. 紧迫性：时间约束和建议的尖锐性

来源：Mitchell 等，1997。

图 1　重要利益相关者模型

2. 跨越生命周期的利益相关者管理

当我们考虑一个组织的生命周期时，主要利益相关者的时空性质就会变得清晰。利益相关者的管理与公众参与是一个贯穿组织生命周期始终的过程（Verbeke and Tung, 2013）。生命周期视角将从特定活动（如新建基础设施）的规划过程开始，走向法定的审批流程，继而是开展施工和组织的日常运营，最后将以组织设施的退役而告终。然而，组织的商品和服务可以比组织本身享有更长的寿命。通过这些商品和服务，利益相关者可以继续参与公众活动（Shepherd and Bowler, 1997）（见图 2）。

图 2　跨生命周期的利益相关者管理

组织生命周期的各个阶段表明，普通利益相关者可能会试图介入组织生命周期发展阶段的一个特定决策过程中。随着生命周期的发展，组织会遇到各种利益相关者通过公众参与或其他重要利益相关者的介入施加压力的情况。某些特定的阶段，普通利益相关者会变得愈发突出，组织需对未来可能涉及的利益相关者做好准备。如前所述，积极主动是增强信任的好方法，组织需要及时寻求这些信息。

3. 公众参与的渠道

环境污染包括水、空气和土壤污染等不同种类的污染。同样，公众参与的渠道和方式也会根据环境污染的种类及其发生在组织生命周期的阶段而各不相同。公众参与的法律义务（即告知、咨询和交流）贯穿于组织生命周期的始终。《奥胡斯公约》（1998）第一条提出："为保证当代人和未来几代人能够健康富足地生活在一个环境优美的地球，各方必须确保公众的信息知情权、决策参与权以及符合本公约规定的司法权利。"作为《奥胡斯公约》的签署国，英国和欧盟已经把《公约》中的相关规定纳入自身的规定条款中（European Directive，2003/4/EC）。在英国，《环境信息规定》赋予了公众获取政府环境信息的权利（ICO，2014）。

英国政府网站（www.gov.uk）显示，一些政府部门参与了环境与气候变化相关议题的咨询。自2008年以来，网站列出了182个与"环境"相关的问题和160个与"气候变化"相关的问题展开磋商与咨询。涉及的政府部门包括：环境、食品和农村事务部，环境总局，能源与气候变化部，交通运输部，社区与地方政府部，英格兰自然环境署，海洋管理组织，国家森林公司，国防部和小排量汽车管理办公室。多个政府部门开展环境问题的公共咨询证明了大范围开展公众参与环境治理活动具有一定的可能性。除此之外，一些公司也被要求加入到公共参与的过程中来进一步实现环境民主。

同时，当公共协商的渠道已经不起作用时，可以采用法律的手段加以解决。比如英国的一些大型交通基础设施建设项目（高速列车和机场扩建）实施的过程中，公共协商失效之后随之使用法律来解决。较为典型的法庭案件是公众对施工工具和过程的抱怨，当居民日常生活受到当地工厂或建筑工地噪声的严重干扰时，这些投诉是可以被受理的。

纵观组织的整个生命周期，传统媒体和社交媒体是十分常见的投诉渠道。通过媒体的参与，环境污染问题可以得到迅速的传播和广泛的曝光。这一工具有助于信息的传递，也有助于把相关各方召集到谈判桌上来有针对性地讨论各自关注的问题。换句话说，媒体的参与使利益相关者拥有了更多的权利和砝码，因此他们有可能从非重要的利益相关者成为重要的利益相关者。

公众参与的实践

下面,笔者将以英国作为案例来研究组织在整个生命周期中如何处理与利益相关者的关系(Tsang,2013)。基于英国不同行业和不同地区的真实案例,我们可以研究公众参与的程度,也可以在英国特定的社会经济、政治和文化背景下研究如何管理重要利益相关者(Morgan,2012)。尽管英国是遵守欧盟法规及国际公约的欧盟成员国(见图3),但欧盟和国际法规与英国本地法律法规的融合方式仍具有其独特性。

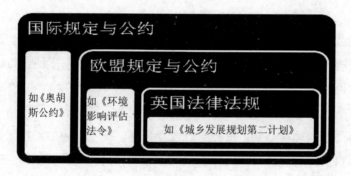

图3　英国公众参与的制度框架

1. 规划过程中的公共参与

项目的规划过程是公众可参与的第一个阶段。这一阶段不会形成具有约束力的决定,因此组织可以广泛咨询公众的不同意见并把相关信息告知公众。公众参与在重大基础设施建设项目实施的前期十分常见,比如最近伦敦机场的扩建项目。当伦敦市长 Boris Johnson 对在伦敦东部泰晤士河口建设新机场表示公开支持时,现有机场和其他群体也建议政府扩建希思罗机场和盖威克机场的航班容纳能力(Mulholland,2009；Topham,2013)。无论是新建还是扩建都会给当地带来严重的环境问题,政府在多大程度上允许公众参与机场建设项目以及最终的结果如何都还有待观察。

围绕机场扩容建设和 2 号高速铁路①（High-Speed 2）建设讨论的核心是项目建设规划初期利益相关者的参与问题，以及随着项目进展如何保持这种参与机制的问题（Bickerstaff，Tolley and Walker，2002；Dooms，Verbeke and Haezendonck，2013；Lee et al.，2013；de Luca，2014）。20 世纪 90 年代末期，公众对地方交通基础设施规划的参与构成了整个英国公众参与环境问题的基础（Bickerstaff et al.，2002）。

高速铁路

从伦敦出发经由伯明翰到曼彻斯特和利兹的高速铁路项目（2 号高铁线，HS2）是重要利益相关者管理的第一个案例。2009 年政府首次提出了这条高铁线路的建设计划，但最终通过项目立项是在 2014 年（Dft，2009）。如果项目在 2017 年正式开工，那么最终建成预计到 2030 年。不过，2017 年能否开工建设将取决于公众参与过程何时结束，以及政府能否顺利解决所有阻止项目建设的问题。Bolden 和 Harman（2013）认为 HS2 高铁线路的建设需要有充分的理由和目的，因为该线路将采取政府外包的方式由一家名为 HS2 高铁公司（High Speed Two Ltd.）的商业公司具体负责建设。HS2 高铁公司虽然为商业公司，但其实际是由政府建立并代表政府利益的集团。目前，整个高铁线路的建设估计总成本为 400 亿～800 亿英镑（Wellings，2013）。因此有反对的声音认为这一庞大的建设计划成本过高，为了避免资金不足而不应该过早开工。

尽管公众强烈反对该项目的实施，但中央政府一直没有放弃 HS2 的建设计划（Synnott，2013）。2011 年，政府的公众咨询显示有 90%的受访者反对该项目的建设，相似的情况同样出现在 2012 年和 2013 年（Crompton，2013）。公众主要担心两方面的问题：第一个是人们担心随着伦敦和伯明翰之间铁路建设的项目进展速度，沿线特别是奇尔特恩地区（Chilterns）的生态环境将遭到不可复原的破坏；第二个是高速列车运行所带来的环境成本（Miyoshi and Givoni，2013）。这些忧虑都需要通过利益相关者管理和公众参与来解决。奇尔特恩地区是一个当地居民引以为豪的自然风景区，并被誉为是一个"无与伦比的自然景观"（Tomaney and Marques，2013）。该地区的特殊价值应当受到保护，特别是要防止出现土地污染和噪声污染（Synnott，2013）。此外，奇尔特恩地区也是伦敦和伯明翰市民重要

---

① 1 号高铁线路是指连接伦敦和巴黎之间的铁路，由欧洲之星（Eurostar）公司运营。

的旅游休闲区，伯明翰政府也利用该区域的旅游资源获得了大量财政收入。

许多行动团体（action group）对已经形成的 HS2 计划表示反对，这些行动团体是由多种反对 HS2 建设为目的的利益相关者组成。行动团体多由铁路沿线的居民、地方政府、环保 NGO 和其他组织（如致力于保护英国名胜古迹和文化遗产的国民托管组织 National Trust）所建立，反对组织包括为地方政府充当保护伞的直接利益相关者（如 51m 组织），也有为个人和私人组织充当保护伞的团体（如 StopHS2 组织和 HS2 行动联盟）。这三个组织由于力量太小而被当局所忽视，因此约 70 个当地反对团体加入其中，形成了反对高铁 2 号线建设的团体联盟（Federation of Action Groups against High Speed Two），各个团体相互协作彼此支持迫使政府终止 HS2 的建设计划（Synnott，2013）。反对团体中也包含那些并没有受到直接利益影响但在政治上反对执政党的政党组织，如英国独立党（UK Independence Party）和绿党（Green Party）。其中，后者和其他一些反对组织不仅出于环境保护的考虑，而且还以经济成本计算的不确定为由反对项目的建设。

不同群体均已表达出了对 HS2 建设计划的正式与非正式担忧（见图 4）。这些团体被中央政府和 HS2 高铁公司纳入正式的公众咨询，这是一种法定的自上而下的协商方式，也是一个通过现有框架向公众提供信息和咨询并回应他们问题的参与方式（cf. Fraser et al.，2006；Crompton，2013）。由于铁路沿线社区的环境脆弱性评估由 HS2 公司完成（Crompton，2013），因此，公众力量对 HS2 项目的改变在很大程度上被限制在这一环节中。显而易见，这一过程侧重于公众咨询而非公众参与。

非正式过程是指一种自下而上的公众参与方式。通过这一方式，当地社区和居民不仅可以接触 HS2 项目的决策者，也可以接触到英国其他公众的广泛意见。非正式过程允许个人和地方政府对决策者施加压力并可以保证个人和地方政府的诉求不被决策者忽视，这一过程可以被视为是弥补正式参与不足的一个途径。

当然，在我们关注公众以环境保护为由反对 HS2 项目建设，并通过各种方式表达他们不满的同时，我们还要注意到支持项目建设的那部分"公众"。伯明翰市政厅、伯明翰商会、伯明翰机场和一些组织组成了"Hands up for HS2"、"Go-HS2"等支持组织；HS2 铁路沿线较大城市伯明翰、布里斯托、加的夫、爱丁堡、格拉斯哥、利物浦、曼彻斯特、纽卡斯尔、诺丁汉、利兹和谢菲尔德等市政厅，以及三个主要政党（保守党、工党和自由民主党）则组成了"HSR 英国"组织支持高铁项目的建设。

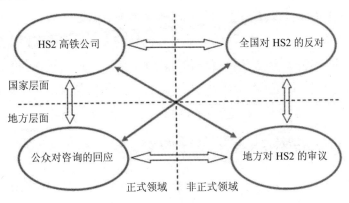

资料来源：Crompton，2013。

图 4　HS2 项目的公众参与

无论是支持团体还是反对团体，他们都经过了独立分析，把整个参与过程向公众公开并试图向公众解释和说明他们的工作流程等的一系列过程。Bolden 和 Harman（2013）公开对高速铁路项目表示支持，但却质疑 HS2 项目未来对社会产生的效益。基于政府的预测，Miyoshi 和 Givoni 表示 HS2 项目的建设会减少温室气体的排放。事实上，到 2033 年温室气体的减少量微乎其微，其仅仅占 2007 年英国国内交通工具尾气排放量的 0.1%。而他们强调在铁路运输过程中温室气体仅占交通运输排放总量的 0.4%，因而所产生的影响也十分有限。

英国政府正在以 HS2 项目会提升公共福利为由努力博得民众的支持。英国政府向议会提交议案，希望铁路建设项目能获得通过，并希望议会赋予政府更大的权力（Bolden and Harman，2013）。作为议案的一部分，政府可以对环境问题置之不理。该议案仿照了之前中央政府通过的在连接伦敦与巴黎的英吉利海峡上建高铁线路的议案，其对环境问题的关注不是很多。2013 年，该议案的政治进程得到了上诉法院的支持，未来 HS2 项目的实施将不会再有来自法律上的阻碍。

应对气候变化的基建项目

公众参与的另一个领域是旨在减少温室气体排放的基础设施建设。它以取代化石燃料的能源基地为目的，但问题的关键在于该项目的选址（Lee et al.，2013）。而英国公众考虑的是该项目的建设是否必要，因为其会产生直接或间接的社会成本。由于英国是一个岛国，而且其有着发展风能、太阳能、潮汐能源和碳转化技术的多种选择。实现公众参与的路径有着国家层面、欧盟层面和国际层面的法律依据，但却没有较好地实施。目前公众参与的普遍途径是在国家投资重大基础设施项目的执行过程中，通过法律规划和环境评估的方式加以干预。但在实际操作中，由于公众咨询力度有限，所以这种参与方式的影响范围也十分有限，因此企业和公众双方都很难实现自己的预期（Lee et al.，2013）。

2. 审批过程中的公众参与

为了保证项目执行的合法性，项目规划后的审批需要得到当地政府的许可。而为了获得政府批准，公众参与环境影响评估磋商则是必需环节。它包括组织必须遵守有毒废弃物和化学废弃物的处理规定、必须获得温室气体排放许可的规定、必须符合放射性物质管理的规定等。这些规定涵盖了综合污染物的防控、垃圾和废弃物的管理、工矿企业的废水排放等方面。环境总局在颁发许可之前，必须设置 12 周的环境公众咨询时间。此外，排污组织必须回应居民因担心建设项目对身体健康和环境质量等带来负面影响而滋生的"不要建在我家后院（Not In My Back Yard）"心理，避免当地居民产生情绪化的集体反对甚至抗争等行为。

以美国为例，一家名为"Airport Watch"的公民社团（www.airportwatch.org.uk）集聚了众多反对力量向英国机场建设项目施压，他们反对的理由主要是机场建设可能带来的噪声污染、空气污染以及温室气体排放量增加。不同派系之间对待项目建设的不同行为见表 1，第一象限中组织关注的是项目获得所有政党的支持；第二象限是项目获得批准的最佳选择，其既得到了重要利益相关者的支持，也有公共利益的支持；第三象限是目前 HS2 项目遭遇的情景，虽然有公共利益的支持但却遭到重要利益相关者的反对。

表 1 重要利益相关者管理和规划考虑的二维矩阵

	支持项目的重要利益相关者	反对项目的重要利益相关者
支持项目建设的规划考虑	批准项目	基于私利反对公共项目建设
反对项目建设的规划考虑	牺牲私利促进公共项目建设	拒绝项目

除了可以支持和反对项目生命周期中的环节，公众还可以通过环境影响评估

的方式参与环境治理。它包括社会影响评估、人体健康评估、可持续发展战略评估、气候变化评估、文化影响评估和累积效应评估等（Morgan，2012；Lee et al.，2013）。该评估涉及环境影响的程度及如何解决环境的负面影响等（Weston，2000）。例如，美国的环境评估实践允许公众审阅相关的环境评估报告并提出意见，公众也可以通过法律途径对环境影响评估施加影响（Shepherd and Bowler，1997）。在英国，《城镇和国家规划法则》（Schedule II of the Town and Country Planning Regulation，1988）这一法令允许地方评估机构（LPAs）和项目开发商一同评估项目的实施是否会产生重大环境问题，其实质就是通过环境影响评估来确定潜在环境风险（Weston，2000）。

公众参与和环境影响评估面临的最大挑战是由于相关部门并没有诚意邀请公众加入其中，而导致其参与过程流于形式（Shepherd and Bowler，1997）。由于各种限制，在审批阶段的公众第一次参与通常不会产生实质影响。在这一过程中会披露较多的项目细节，公众参与也由之前的相互过程变成了一个信息发布的单向过程（Shepherd and Bowler，1997）。公众对于 HS2 项目的反应表明，政府确定最终解决方案之前并没有询问公众对铁路系统更新是否还有其他的替代方案。

Facebook 和微博等社交网络的使用可以在正式的环境评估之外获得与公众更为广泛的联系。因为有了社交媒体和网络，更多的负面新闻和有损组织形象的信息可以迅速传播，这也使得组织面临着来自于法律之外的更大压力。这一途径不同于请愿书，因为法律规定后者提交之后必须得到回应并采取相应行动。在英国，下议院（House of Commons）建立了一个快速回应公众诉求的电子请愿网站，通过该网站公众可以影响政府的政策进程。为了引起下议院的重视，该网站在一年内收集了至少 100 000 个签名，而最多的一个请愿书中得到了 304 214 个签名。

英国政府的电子请愿网站除了对公共政策产生影响之外，还会影响到私人交易的审批进程。一家名为 Midland Pig Producers 的公司申请在德比郡（Derbyshire）建造一个大规模的养猪场，如果获批其将成为英国最大的养猪场，计划周屠宰量将达到 1 000 头的水平，如此规模的养猪场在英国十分少见。德比郡以农村景观而闻名，因此当地居民认为这样的养猪场并不利于当地生态环境的保护，况且大规模的屠宰行为有悖于人道。他们还担心养猪场的存在会导致一些传染性疾病的爆发，而且有可能增加空气和水的污染程度并对当地交通状况造成影响。截止到

2014 年 5 月 8 日，电子请愿网站已经得到了来自全球 32 269 名支持者的签名，其中也包括一些国会议员和环保团体。

请愿活动由当地的利益相关者介入，他们要弄清楚当中有多少请愿者并不居住在英国（从斯堪的纳维亚到美国都有大量请愿者提交请愿书）。如瑞典的"Stop Britain's Largest Pig Farm"得到 567 人的支持、"Stop Plans for Britain's Largest Pig Factory Farm"得到 657 人的支持，美国的"Factory Farm Trying to Silence its Critics...Don't Let It!"得到 571 人的支持，除此之外还有来自于环保组织网站和社交网络平台的支持者。这些反对意见导致该养猪场的建设至少推迟了 3 年，建设方必须要向环境总局详细阐述环境污染问题的应对措施。Midland Pig Producers 在 2010 年暂停了原有的申请计划，并在 2013 年根据环境公众咨询结果调整了申请计划，但德比郡南部区议会最终否决了该项目的建设计划。从这个例子中可以看到通过非正式渠道的公众参与同样能够产生巨大影响。

3. 建设过程中的公众参与

除了空气和噪声污染之外，建筑工地施工也会不可避免地影响到周围水和土壤的质量。在英国，施工方必须设法监控周围环境以减少污染的产生，公众也可以通过信息公示了解工程的建设规划和施工细节，包括噪声水平和持续时间、评估指标和安全标准等。公众对噪声和空气污染的理解可能与法律的界定不同，这里所指的噪声污染是指超出人们预期并可能对周围居民造成影响的污染。此外，为了确保施工方担负起长期的环保责任，还需要评估污染水平是否会随着项目建设而增加。为了减少光污染，施工通常在白天进行以便保证周围居民可以在晚上有合理的睡眠质量的保证。施工方提前向公众告知施工活动的潜在污染，公众可以据此评估是否需要临时搬迁来避免这些影响。这种信息公示的方法其实是最低程度的公众参与，即所有附近居民都被公平对待，所有积极主动的参与都必须代表公众利益。

4. 项目日常运作过程中的公众参与

项目运作过程中出现的任何问题都需要及时与公众和重要利益相关者沟通。预警机制固然重要，但企业和有关部门也必须具备处理突发事件的能力，必须学会灵活面对新出现的重要利益相关者。噪声污染、水质污染、有毒物质泄漏等问题可以潜在于任何行业。如餐馆在烹饪过程可能产生空气和噪声污染；大学、音乐厅、金融服务机构等的通勤班车可能产生空气污染。一些生产活动也可能间接地引发社会问题和环境问题，如生产生物燃料可能影响到食品价格，生产转基因产品可能影响到生物多样性并威胁生态平衡。居民关心的是这些活动是否违背了环保规则和相关法律，因此这些问题都必须要依法处理。例如，伦敦希思罗机场的扩建工程遭到公众重大抗议。多年来围绕是否增建第三条跑道的问题一直处于谈判中。大量当地居民和环保 NGO 强烈反对修建第三条跑道，理由不仅是因为噪声污染问题，而且他们认为扩建工程可能进一步影响到气候变化。公众对未来环境问题的担忧深深影响到希思罗机场建设项目的进展。

中小企业的公众参与

利益相关者管理和公众参与并不局限于规模庞大的企业和组织中，中小型企业（SMEs）也是公共参与重要的实践主体。和大多数国家一样，中小型企业对英国的经济增长也发挥着至关重要的作用。Peters 和 Turner 研究了中小企业自愿的公共参与活动，其活动通常受制于有限的人力和财政资源，因此加强公众参与对这些公司来说是不小的负担。他们还发现东英格兰（East Anglia）的中小企业更有兴趣加入公众参与环境治理的志愿活动，他们可以从这种与当地政府和利益相关者的互动中获得收益（Peters and Turner，2004）。这些志愿活动还显示，公众参与并不应该在环境危机出现时或法律有明确规定时才被动进行，积极主动地参与环境问题可以给各方带来更多的好处（Aragon-Correaand Sharma，2003）。

参与型模式的污染

公众参与涉及公共信息的采集和验证。布里斯托、谢菲尔德和约克三个英国城市建立了空气质量管理系统，其要求公众参与系统的测试工作，即确定官方收集的空气质量数据的正确性和可靠性（Yearley, et al., 2003）。空气质量相关的信息涉及空气污染、噪声、气味和灰尘等，识别和认识这些更广泛的信息非常重要，它可以帮助组织了解公众对环境感知的预期。同样，经过这些磋商可以确定哪些是在公众的眼中受影响最严重的领域并立即采取行动（Yearley et al., 2003：252），尤其生活在该环境中的人对这一特定区域的环境测量和数据更为敏感。

虽然通过信息共享和更广泛的民主参与可以增加参与型模式和监测污染的包

容性，但它也有局限性。信息收集和评估过程缺乏公众参与使这一模式并不能实现全面的公众参与，它仅仅限于信息收集和共享（Conrad and Hilchey，2011）。McDonald 等（2002）发现有部分伦敦市民难以理解收集和分析空气质量数据的科学方法，这就限制了他们参与政策决策的能力。因此，有效的环境监测和决策过程中的公共参与必须保证公民拥有良好的知识储备和专业技能。

5. 项目终止过程中的公共参与

最后一个生命周期是项目的终止或退役，无论是写字楼、工厂、交通基础设施，还是业务服务流程，都有一个时间期限，而技术进步和革新也需要对现有设备更新换代。一般来说，所有的终止和退役过程都会经历以下几个阶段（Fairley and Tilling，2012）：

在第一阶段，组织需要收集相关信息，并需要评估终止或退役之后可能对环境产生的潜在影响，还要履行法律和合同的义务来处理各种善后事宜。而利益相关者的识别需要遵循一个协商的过程，在此过程中对待重要利益相关者的方式与在前期规划以及建设阶段类似。至于公众参与，则必须找到一个各方都能够接受的方案。设备的退役并不意味着将其摧毁，它可以被回收翻新之后重新利用。工业革命时期保留下来的工厂、教堂等建筑经过重新装修可以用作办公室或商场。这种设施和空间的循环再利用要求相关方必须彻底清理有毒材料，并保证翻新过程符合现代建筑标准。暂时无法翻新的建筑和设备可以暂时封存，等到日后技术和条件成熟时再做处理（Bond，Palerm and Haigh，2004）。这个阶段还必须考虑谁应该为这些清理工作负责，这需要经过详细的计算才能得出最终结论。

在第二阶段，即将退出的企业必须与环境保护部门和承接商签署协议，还需要对环境影响做出评估，保证当地环境能够得到彻底的清理和恢复。化学工厂退出前要处理好危险的有毒废料并防止造成污染土壤，工矿企业退出前要拆除所有设备并保证自然景观的复原，而涉及放射性核废物的处置必须严格遵守英国和欧盟的相关法规。Bond 等特别指出，核电站退役过程中的公众参与必须保证环境评价是客观公正的，而环境信息发布是绝对透明的（Bond，Palerm and Haigh，2004）。

在第三阶段，需要进一步调查以确定环境修复的最终标准，并最终开展环境的清理工作。公众参与和咨询在最终形成结果前一直会持续下去。接着，最终结果确定后开始进入实施阶段（阶段四），随后是保证实施结果符合利益相关者预期并达到相关法规标准的第三方认证阶段（阶段五）。

Tetley 啤酒厂

位于利兹市中心的 Tetley 啤酒厂是嘉士伯啤酒公司的一个工厂，当嘉士伯公司公开宣布关闭该工厂后，新的投资商准备在原址建一个本地啤酒工厂。建厂前，周围社区的居民就被告知老厂拆除的施工方案和噪声污染水平。居民担心这些活动可能会影响他们的正常生活，因此打算进一步联系该公司并获得更加详细的信息。居民的集体行动延缓了该项目的施工，最终的妥协方案是保留部分原建筑并将其翻新成为一个展览空间，而其他区域则在老厂搬迁后重新施工。

核电站

Bond 等研究了之前位于北威尔士 Trawsfynydd 核电站退役过程中公众参与问题的案例。在向地方政府提交正式退役和拆除计划之前，核电站咨询了雇员、工会组织代表和核电站附近 30 km 范围内的居民这三个利益相关群体的意见。除了当地居民的就业问题，利益相关群体关心的是任何可能导致污染的因素。此外，环保 NGO 组织也表达了他们对核泄漏等问题的担忧。最后，所有这些意见和担忧都得到了核电站方面的高度重视并得以解决。

结论与政策建议

环境问题的公众参与贯穿了整个组织生命周期的始终，在任何一个阶段公众都可以通过特定的渠道和方式表达自己的利益诉求，公众主动要求加入咨询活动的情况也十分常见。无论被动的参与还是主动的加入，这种参与机制都能有效避免多个利益相关者间潜在的矛盾冲突。组织和企业需要赢得公众的信任从而获得项目执行的合法性，对社会和环境任何不透明和不负责任的行为都将使合法性丧失。文章汇总了英国的众多案例，并具体分析了公众参与不同组织生命阶段的具体形式和特点，并得出一些有意义结论。

首先，对一般企业而言，重要利益相关者会随着时间推移而发生变化。其既有好处也有不足，不足来源于新凸显的利益相关者对现有组织关系和运作模式的了解，他们关注的焦点可能因此变得更加分散；而好处则在于这一模式必须不能依靠常规思路来回应利益相关者的诉求，必须通过不断创新来应对新凸显的利益相关者。

其次，利益相关者可能聚集起来并形成地方性的、区域性的、全国性的甚至国际性的行动。根据情况的危机程度，一些利益相关者会相互联系并形成联盟，因而可以拥有在全国范围内甚至国际上更大的话语权，同时能够更大程度上影响和监督相关环境协议的签署。一些利益相关者的活动也获得了法律的授权，而法律也明确要求企业和组织回应利益相关者的合法诉求。

最后，公众参与往往同时存在这两种方式，自上而下的方式最终落脚到参与社区的层面，同样自下而上的方式最终把问题反馈到企业和组织。我们无法简单确定哪种方式更为有效，但只要企业和组织以真诚的态度及早地回应公众诉求，就可以保证公众参与的成功实现（Petts，1995；Shepherd and Bowler，1997；Ng and Sheate，1997；Bond et al.，2004）。

政策启示

中央和地方政府支持公众参与需要更加开放和透明参与的机制，并且要鼓励排污组织和企业代表的积极参与。回顾以往公众参与的实践，任何降低污染排放的措施其代价都是高昂的，珍贵的原生态环境受到无可挽回的破坏，保护环境的总花费甚至可能超过当初的建设成本（例如塞拉菲尔德核电站环境清理的总费用超过了 700 亿英镑）（Macalister，2013）。但是，如果对环境污染不作为势必会更影响公众健康，久而久之也将导致公众对政府、私人企业或组织的不信任，这种不信任也有可能影响到其他政策的实施。

对新兴国家的启示

英国以及欧盟的环境公众参与实践启示新型国家需建立多种参与机制和渠道以保证各方之间利益的平衡。个人、团体、企业、组织和政党可以在相互妥协和谈判的过程中进行公众参与形式的创新，从而保证公民个人、企业组织和自然环境的和谐发展。

Environmental Public Participation in the UK

Hinrich Voss[*]

Introduction

Environmental degradation and pollution are serious matters in most countries (Ycelp and Ciesin, 2014). Several studies and rankings on the environmental credentials of countries have found severe problems in developed and developing countries alike, ranging from water, soil, and air pollution to decline in biodiversity and over usage of animal stock, and extensive emissions of greenhouse gases (IPCC, 2014). While all countries face the difficulties of how to address environmental concerns, developing countries in particular are confronted with the additional challenge of when to address these concerns during their economic catching-up. Prioritizing economic growth over environmental protection can achieve significant economic growth rates and generate wealth quickly.

When leaving businesses free reign to conduct their activities in profit-maximizing fashion to their best ability, their efforts will be invested in the development, production and marketing of goods and services. Expenditures on environmental protection will be considered where and when the environment has an immediate impact on their business success. The externalities of any environmental degradation are passed on to society in this scenario. The success and long-term sustainability of economic growth is put into doubt because of the detrimental effects environmental loss can have. Addressing environmental and economic growth issues

---

[*] Hinrich Voss, Lecturer in Leeds University Business School, the deputy director of BSc IB programmes, the executive director in Business Confucius Institute at the University of Leeds (BCIUL).

and finding a balance that benefits both is delicate and has been a long-standing concern. One approach that can help mitigate some of the issues of this dilemma is to bring the concerned parties together and give the affected parties a voice in the decision-making process.

Public participation and public consultation on environmental matters gives society a channel to articulate their concerns and detail how their lives have been or will be affected. Effective and inclusive participation can increase the social quality of a society. Insights from participatory actions can inform policy makers and investors as they become aware of the potential of unintended consequences. Through this process, society will also be better informed about the rational of particular activities which can alleviate misconceptions and misinterpretations.

Considering the economy-environment development dichotomy from the perspective of a polluting organization, the question becomes why involve anyone outside the organization and, if this is necessary, when to involve whom in the decision making process. The public that could be interested in the process and become actively involved will not be same for every project, that is, the organization has to be aware of a range of stakeholders that gain importance at different instances.

Looking at public participation on environmental matters through the lens of a polluting organization is important as the mechanisms and tools of public participation, and of the broader issue of environmental pollution, cannot only be considered from the perspective of the group that wants to be heard and seeks involvement. Analyzing the processes and their usefulness from an "affected party" perspective only would neglect the constraints and options the polluting organization has and the selections it has made in order to engage with the public.

Taking an organizational perspective we use the example of the United Kingdom to illustrate and analyses how organizations engage in public participation, how the mechanism are structured and interlinked, how organizations perceive this mechanism and the request to change their behavior. The next section introduces the conceptual background which is followed by a section on environmental public participation cases from the UK. Finally, we conclude and provide policy recommendation.

Conceptual background for environmental public participation

Environmental public participation goes back to the National Environmental Policy Act 1970 in the USA (Morgan, 2012) and the conceptualization of public participation by Arnstein (1969). It seeks to facilitate the involvement of those (potentially) affected by an organization's decision or an organization's development and to facilitate the involvement of those affected in the decision-making process where possible (Bickerstaff, Tolley and Walker, 2002). Public participation also follows from the principles of a democracy according to which public should have a voice in how it is governed and what objectives the government pursues (Shepherd and Bowler, 1997). Shepherd and Bowler (1997) thus state that the very idea of a democracy asks for the public representation in decision-making processes. Through this involvement, the democracy itself is revitalized and reaffirmed. It also follows from this that within a democracy decisions should be suitable to the people who are affected by the decisions. Public participation brings with it the advantages of the public having perceived ownership of the project because of its involvement in the decision-making process. It can thus become more supportive of the project.

The underlying principles of public participation have been established in the international "Convention on Access to Information, Public Participation in Decision-Making and Access to Justice in Environmental Matters" (the "Aarhus Convention") which states that access to information is a key right for citizens; although this is restricted to public authorities (Mason, 2010). Through access to information, the public can be informed about the environmental impact of developments and activities in their neighborhood. The public thus has the possibility to understand better if and to what extent they and the environment could be affected through water, soil, air or noise pollution. That means to say that the public is informed and consulted when, for example, an organization is deciding to set up a new factory which could increase noise and air pollution, when an organization is deciding to set up a new production process which involves a new set of chemical inputs, or when an organization is deciding to relocate certain activities and is considering the decommissioning and rescaling of current facilities.

Public involvement doesn't stop with the decision making process, but it also extends to the daily operations of the organization. All of those who may be affected by an organization's decision and operations should be able through the environmental public participation to have a voice that is heard, taken serious and has a chance of changing the way a polluting organization is operating (Lee et al., 2013). Through calling upon public participation it is also possible to gain access to insights and expertise that was previously not accessible or unknown thus making the project overall stronger (Shepherd and Bowler, 1997; Lee et al., 2013). Key questions in this process are who is the public that is asked to be involved, and once "the public" is identified, questions arise about when the public is be informed, to what extent it is, and if the public wants to engage.

Bickerstaff et al. (2002) highlight that within public participation one can differentiate between "participation", "consultation", and "information". "Participation" is an interactive process between the consulting organization and the public to inform and shape the decision-making process. The public has the power and influence to imprint its objectives and concerns on the decision-making and these have to be incorporated by the organization. "Consultation" refers to a process that asks the public about its opinion but doesn't need to incorporate any suggestions and concerns in the decision-making process. Finally, where public participation is practiced as an "information" exercise, the public is merely told what is about to happen and how they may be affected.

All three approaches of public participation are common, often exist in parallel next to each other and can be found across the UK as will be shown below. The public participation type chose has obvious implications for the perceived social quality. These three participation offer varying degrees of participatory action (empowerment) and inclusiveness which can be an expression of social cohesion and affect the economic security of the public (Abbott and Wallace, 2012). The "participatory" public participation process arguably offers the greatest social quality because its provision decision-making involvement goes furthest.

Once it is decided, or legally prescribed, how the public can participate in a project, questions arise about who the public is. Bickerstaff et al.(2002)further indicate that different boundaries for the "public" exist with regards to transportation infrastructure projects in the UK. The public refers in the first instance to local people, businesses, transportation operators, users and their respective representative bodies in case of vulnerable (for example, elderly and people with special needs), health and education providers, and environmental organizations. But the geographical boundaries are not clearly defined (local, regional, national or international) nor are the weightings of the different public groupings. Although a clearer definition should be welcome because it gives clarity to the different parties, defining the public clearly can, however, cause problems. Introducing hard boundaries and fixed definitions can exclude those groups that are most affected by a project. It would also not account for changes in the socio-economic environment and would thus require a constant fine-tuning. The management literature on stakeholder identification and management may provide a solution for this as explained below.

The Aarhus Convention suggests that public participation should be early and effective which should mean that the process is inclusive, open, interactive, continuous, early, and reciprocal (Bickerstaff et al., 2002; Hartley and Wood, 2005). But Hartley and Wood(2005) and Morgan(2012) highlight that the public participation is often curtailed by poor knowledge of the public on planning, legal and licensing issues and lack of legal advice, by poor provision of information and mistrust towards the industry concerned, poor execution of the participation process with limited possibility to influence the decision-making process. The case for the public consultation is constrained by the limited knowledge the public has about the complex issues surrounding the environmental, economic, and social impact and interdependencies of, for example, the large-scale British infrastructure project "HS2".

Responses by the public to the consultation process indicated that greater efforts to inform the public comprehensively on environmental, economic, and social issues before embarking on any consultation would be required. Where this is not the case, or information is not provided in a format that is accessible in plain language to be understood by the lay person, a perception of information being withheld arises (Crompton, 2013).

The abstract notion of "an organization" is chosen here in order to clarify that the environmental public participation addresses the relationship between the public and companies, and the relationship between the public and any other organization such as public organizations, the government, and non-governmental organizations. One of the aims of public participation is to build understanding and trust among stakeholders from which organizational legitimacy, or legitimacy for a particular behavior, can follow (Carroll and Shabana, 2010). Trust between the stakeholders and the organization refers to knowing that the organization is acknowledging and respecting laws and regulations, that the organization is contributing to the society and not only exploiting local labor and regulatory conditions such as weakly written and enforced environmental regulations.

The organization that engages in stakeholder management, that is, public participation from an organizational perspective, indicates that it is concerned with what the public has to say and that it is listening to their concerns. In order to build the trust relationship and achieve legitimacy, it is important for the organization to be accountable and transparent in its activities and communication as well as being consistent in its behavior across stakeholders. Transparency involves the communication on the outcomes of decision, how the organization is aiming to implement processes on environmental concerns. This includes detailing how it is responding point by point to particular public concerns. Falling short of accountability and transparency or being pressured to reveal relevant information does hardly increase the trust towards the organizations.

In stakeholder engagement, it is important for the organization to recognize that not all stakeholders carry equal weight and require the same attention. Mitchell, Agle, and Wood (1997) developed the concept of the "salient stakeholder" in order to capture the spatial and temporal differences in importance across stakeholders. While the list of stakeholders that are internal or external to the organization is rather long, not all of them require the same level of attention at a given place and time (Verbeke and Tung, 2013). The approach popularized by Mitchell et al. (1997) is to identify the level of salience a stakeholder has and to order them accordingly. The characteristics they proposed for this are power, legitimacy and urgency: Power refers to the extent a stakeholder has the means to impose its will in its relationship with the organization. Legitimacy refers to the socially accepted and expected structures or behavior of the stakeholder. Urgency describes the time sensitivity or criticality of the stakeholder's claims.

The most salient stakeholder combines all three characteristics at the same time (Figure 1). With regards to environmental pollution, a salient stakeholder could be a community that is expose to toxic water pollution upriver which is threatening their livelihood. These are stakeholders the organization cannot avoid to deal with and to acknowledge and react to their concerns. At the other extreme are stakeholders that exhibit none of the characteristics. An example of that could be an individual who would like the photos in the organization's corporate social report to be changed as he prefers tree over water photos. Between these two poles of salient stakeholders and irrelevant stakeholders are stakeholders that matter to an increasing degree as they can rightfully claim stronger levels of power, legitimacy and/or urgency.

Power (1)
the extent a party has means to impose its will in a relationship

Legitimacy (2)
socially accepted and expected structures or behavior

Urgency (3)
time sensitivity or criticality of the stakeholder's claims

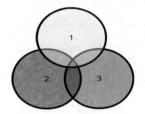

Adapted from Mitchell et al. (1997).

Figure 1 Salient stakeholder model

From this description it becomes clear that over time, the level of salience varies and with it the status a stakeholder enjoys. Environmental pollution is predestined to raise the urgency of a stakeholder's concerns as the pollution can affect him directly with potentially severe consequences. Environmental protection is often also seen as the socially correct behavior as nature is what society enjoys and benefits from. This relationship may though be questioned in cases where economic growth and welfare creation at the expense of the environment enjoys stronger acceptance. This dynamic and evolving relationship requires a constant awareness and regular check on behalf of the organization to assess who the salient stakeholders are (Figure 2).

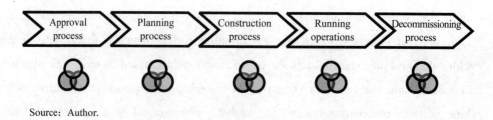

Source: Author.

Figure 2　Stakeholder management over the life-cycle

Channels for public participation

Environmental pollution can appear in different shapes and forms, including water, air, and soil and greenhouse gas pollution. Likewise, the channels for public participation vary by pollution and by the stage of life-cycle where they occur. Legal obligations for public participation (that is, to inform, consult, and interact) with the public exist across of stages of the life-cycle. The Aarhus Convention (1998) has brought this forward through Article 1 that says "In order to contribute to the protection of the right of every person of present and future generations to live in an environment adequate to his or her health and well-being, each party shall guarantee the rights of access to information, public participation in decision-making, and access to justice on environmental matters in accordance with the provisions of this Convention." The UK, and the European Union, as signatories of the Aarhus

Convention, have turned the Convention into EU directives (European Parliament and of the Council, 2003: European Directive, 2003/4/EC) and British regulation. In the UK, the Environmental Information Regulations gives the public thus the right to request environmental information from the government (ICO, 2014).

Several governmental departments are involved in environmental and climate change related consultation, according to the British government's website (www.gov.uk). It lists 182 "environment" and 160 "climate change" consultations since 2008 with the involvement of the following governmental departments: the Department for Environment, Food and Rural Affairs (Defra), the Environment Agency (EA), Department of Energy and Climate Change (DECC), Department for Transport (DfT), Department for Communities and Local Government (DCLG), Natural England, the Marine Management Organisation (MMO), the National Forest Company, Ministry of Defence (MoD), and the Office for Low Emission Vehicles. The range of government departments carrying out public consultations on environmental issues illustrates the wide range of environmental public participation possibilities with public authorities. In addition to these, companies are requiring to engage in public participation, offering further possibilities for environmental democracy.

Where public consultations are not required or have been exhausted, legal court cases can be pursued. Large transportation infrastructure projects in the UK, such as the high-speed train development and the airport expansions, have seen public consultations followed by court cases. Typical court cases are nuisance complaints that can be filed during ongoing operations of a facility. Nuisance complaints would be applicable, for example when the neighborhood feels upset by the noise levels of a local factory or construction site.

Across all stage of the life-cycle, the usage of traditional media and social media are common channels to be pursued. Through the involvement of the media, concerns about environmental pollution can be exposed quickly and widely. It is one tool to help convening a message and raise support for the concern with the prospect of bringing the targeted organization to the negotiation table. In other words, the involvement of the media can increase the level of power the stakeholder enjoys. He may therefore move from a peripheral to a salient stakeholder. Applying the conceptualization of stakeholder management into practice, we are looking now at examples in the United Kingdom.

Using the UK as a case study (Tsang, 2013), we investigate how organizations engage with stakeholders across the life-cycle of the organization. Using real-life examples across different industries and different regions in the UK, we can identify to what extent public participation is occurring within the UK. Staying with a single country case also acknowledges the particular socio-economic, political and cultural context of the UK and its implications for salient stakeholder management (Morgan, 2012). Even though the UK is a European Union member state and is following EU regulations as well as international conventions, the way these directives are adapted into UK law is likely to differ from other countries (Figure 3).

Source: Author.

Figure 3　Institutional framework for public participation in the UK

Stakeholder management in the UK

Stakeholder management in the development and approval stage

The first stage of a business where the public can get involved is the planning process. At this stage nothing is decided or fixed yet and various views and opinions could be heard and inform the decision process. Involvement at this stage is in particular common at larger infrastructure projects. A recent example of this is the consideration of expanding London's airport capacity. While the Mayor of London, Boris Johnson, has publicly favored the construction of a new airport east of London in the Thames estuary (Mulholland, 2009), existing airports and different other groups have proposed the expansion of Heathrow, Gatwick or one of the other airports (Topham, 2013). Either plan has significant environmental implications. The extent to which the invitation to get involved in the process of developing new airport capacity is shaping the decision-making process and the ultimate outcome remains to be seen. Key to the discussion around the case study of "High-Speed 2" below, is the stakeholder engagement earlier on in the planning cycle of large scale infrastructural projects and the continued engagement with them as the project is implemented and comes into operation (Bickerstaff, Tolley, and Walker, 2002; Dooms, Verbeke and Haezendonck, 2013; Lee et al., 2013; de Luca, 2014). Public participation in local transport planning became a cornerstone in the UK in the late 1990s (Bickerstaff et al., 2002).

The case of High-Speed Railway and its analysis

The first example of salient stakeholder management thus refers to a current infrastructure project in the UK, the construction of a high-speed railway lines from London via Birmingham to Manchester and Leeds. The development of the "High-Speed 2" (HS2)[①] train line has been first proposed by the government in 2009 (DfT, 2009) and a final decision on the construction of the line is expected for 2014. Construction of the HS2 train line could be finished around 2030, if it commenced by 2017. If construction can commence by 2017 or not, will depend on whether and when the public participatory processes have been satisfactorily concluded and any resistance

---

① High-speed 1 refers to the only other high-speed train line which links London with Paris and is operated by the Eurostar.

against the project has been seriously addressed. The reasons for building HS2, Bolden and Harman (2013) argue, need to rest on a clearly identified purpose and rationale.

The high-speed railway line is a national government contracted project that is executed by the business organization called *High Speed Two Ltd*. High Speed Two Ltd has been established by the government in order to create it on behalf of the government. The costs for the railway line are currently estimated to be around £40~80 billion (Wellings, 2013). But opposition had been formed to challenge this massive infrastructure project very early on and is now further supported because of the cost implications. A public consultation by the government about the proposed line found that 90 per cent of the respondents opposed the infrastructure project in 2011. This was followed by public consultations in 2012 and 2013 (Crompton, 2013). Despite this hefty opposition and concerns about the environmental of HS2 (Synnott, 2013), the national government has continued the planning of HS2.

There are two aspects of environmental concerns people of various sorts of life have expressed that need to be addressed through stakeholder management, that is public participation: The first is the perception and concerns about the destruction of the Chilterns between London and Birmingham, that is the environmental destruction during the construction and the irrevocable change it make to the landscape, and the second is the environmental cost for running high-speed trains (Miyoshi and Givoni, 2013). The Chilterns are a green belt that is praised by the local community and beyond for its landscape and is a designated "Area of Outstanding Natural Beauty" (Tomaney and Marques, 2013). The value of this particular stretch of land therefore should be protected against environmental destruction and increases in noise levels (Synnott, 2013). It also serves as a recreational area for the populations from London and Birmingham thus bringing revenues to the area through tourism.

A number of action groups against the HS2 have been formed. These action groups are formed by a wide range of different kinds of stakeholders that have been drawn together by their opposition against the HS2. Action groups have been established by individuals and local governments along the proposed train track, by environmental NGOs, and by other organizations such as the National Trust which is working towards protecting England's national and cultural heritage. Opposition groups include directly affected stakeholders like the umbrella organization for local authorities, the 51m group, and the umbrella organizations for individual and private opposition, StopHS2 and the HS2 Action Alliance. All three umbrella organizations indicate that they are too small on their own and would thus not be heard. Across them, some 70 local action groups have joined up to form the AGAHST, the Federation of Action Groups against High Speed Two (Synnott, 2013) .Working in coalitions, each group supports each other's causes as well as the overall objective of stopping HS2. Not directly affected but politically opposing groups are the national political parties the UK Independence Party (UKIP) and the Green Party. The latter and other organizations oppose the HS2 not only on environmental grounds but also on the basis of uncertain economic calculations.

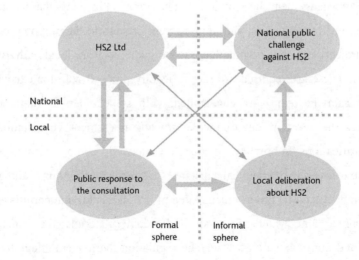

Source: Crompton (2013).

Figure 4　Public participation in the HS2 project

The concerns of these diverse groups have been expressed formally and informally. Formally, these groups are involved in the consultation process the national government and High Speed Two Ltd have instigated. This consultation process is very much a legally required top-down approach through which High Speed Two Ltd has sought to inform the public and consult with them about their opposition and how their concerns could be addressed within the given framework(cf. Fraser et al., 2006; Crompton, 2013). Communities along the railway line have also been are assessed on their environmental vulnerability from the HS2 (Crompton, 2013). As expressed earlier, the power of the public to make considerable changes to the HS2 are constrained in this process which focuses on consultation rather than involvement. The informal process indicates a bottom-up approach as the part of the public participatory process. Through this process local communities and individuals can reach out to the decision-makers of HS2, to the wider public in the UK and beyond to make their standpoint clear. The informal process allows individuals and local government to exert pressure and create a dynamic that cannot be ignored by the decision-makers. The informal process can thus be regarded as a tool to rectify engagement shortcomings of the formal public participation process.

Moving on from planning and development process to the approval process of a particular project, specific permits have to be gathered from local authorities by the organization. These are required in order to legally proceed with the execution of the project. To gain the approval, engagement with stakeholders is required and this usually takes the form of environmental impact assessment consultations. It also includes clarifications on how the organization intends to comply with (toxic) waste disposal, the recovering of chemical substances from waste(REACH), and greenhouse gas emission permits. The Environment Agency (2010) states that permits are requires for emissions to air, water and land, radioactive material, the management of radioactive and non-radioactive waste. The regulation that covers integrated pollution prevention and control installations, waste management licensed facilities, waste from extractive industries, and water discharge consents. In order to issue a permit, the Environment Agency consults with stakeholders for twelve weeks. Besides opposition against the project out of persuasion (for example, people should not fly), and opposition due to serious environmental concerns, the organization will in this process need to address NIMBY ("Not In My Back Yard") opposition. The NIMBY attitude towards projects expresses an opinion whereby stakeholders welcome the development of new projects, but would rather like to the development and the follow-on operations with all its benefits and challenges elsewhere.

In cases of such fundamental opposition it is difficult to carry out compromise-focused consultation processes as there is nothing the opposition could compromise on. The civil society group "Airport Watch" (www.airportwatch.org.uk) is an organization that is bringing together different oppositional forces against the development of UK's airports. Some of its members are profoundly against any airport extension on environmental grounds, mainly due to a perceived increase in noise levels, an increase in air pollution, and the contribution of air traffic to climate change, while others in this group just don't want particular airports to expand further. The opposing factions can lead to constellations whereby the public and the public interest agree or disagree on the development of a particular project. In the interest of the organization is it to position the project on the first quadrant to ensure the support from all possible parties. Second best options are where either the salient stakeholder agrees with the project or there is a strong public interest in its development. The strong support from either fraction can ensure that the project is approved. The above mentioned development of the HS2 falls into the quadrant where the British government has identified a strong public interest in the development of the project but salient stakeholders along the proposed track corridor oppose it (Table 1).

Table 1　Salient stakeholder management and planning consideration

	Salient stakeholders support development	Salient stakeholders oppose development
Planning considerations support development	Approval of project	NIMBY and selfish rejection of project which is in public interest
Planning considerations oppose development	Charge of self-interest in promoting the project	Rejection of project

Adapted from: Lee et al. (2013).

The challenge with public participation and with environmental impact assessment studies is that they are too often used to rubber stamped a project by having a notional public participation and don't allow, or call for, a sincere engagement with the public (Shepherd and Bowler, 1997). If public participation is considered at this, the approval stage, for the first time, there is limited scope for the public to have a fundamental influence. Typically, the kind and shape of the project is normally set at this stage and so are the location and the function of the project; shortcomings that will be highlighted in the case study below. Thus, the stakeholder engagement project turns from an interactive and iterative process into one that is aiming to sell the project to the public with limited or no objective to integrate their insights (Shepherd and Bowler, 1997). The public responses to the HS2 consultation reflect this sentiment. Rather than asking the public about alternative approaches to renew the rail stock in the UK and increase capacity, the government presented a final solution and consulted on the impact of this particular one.

A novel way to foster support for an action, and one that resides outside the formal environmental impact assessment, is the usage of social media. Social media site like Facebook, Twitter or Google+ can be used to inform and stay in touch with a large audience. Through social media, pressure can be developed for an organization to change course. This though occurs because of the negative reporting and image the pressure generates, not because the organization would be legally required to do so. This is different with petitions. Petitions can be collected and filed with organizations and individuals who are then legally required to respond and take action. The threshold level by which a petition is successful differs by country. The House of Commons in the UK, runs is very own electronic petition (e-petition) website through which the public can aim to influence government policies. In order to be considered by the House of Commons, an e-petition needs to collect at least 100 000 signatures within one year. One successful petition aimed to reconsider the culling of badgers which received 304 214 signatures.

Stakeholder management in the construction stage

Construction sites produce, among other things, air and noise pollution and can affect the water and soil quality of the surroundings. Elimination of these pollutants is near impossible at this stage. The aim of organizations is therefore the monitoring, reduction and minimization of any pollution and to introduce mitigating procedures where required. It is therefore common in the UK that the public is informed about forthcoming construction activities in their neighborhood by the operating organization. The information includes details about, for example, where and when increases in the noise levels are to be expected and for how long. They should ideally also contain information about the assessment criteria employed for measuring the impact and the thresholds used for measuring the impact. Legal thresholds do exist, but the public perception of what constitutes noise or air quality pollution can differ from these. The nature and the kind of noise is often detailed as well as when the created noise is of unexpected kind so not to surprise and frighten the neighborhood. This includes detailing how any pollution will develop over time. This requires continuous and clearly allocated on-site responsibilities. To reduce light pollution, construction activities are normally kept to daylight in order to reduce the impact during the night to a minimum and allow residents of a good night's sleep. Informing the public about construction activities sidesteps potential noise nuisance complaints and gives the public an opportunity to take pollution evading measures. This includes measures from closing windows to temporary relocation.

There may occur noise level, water pollution, toxic leaks problems that are problems which are attached to a particular business activity. Potential problems can be identified across any industry. Restaurants generate fumes and noise through the kitchen activities and the dining guests that frequent the location. Large-scale service providers like universities, concert halls, all the financial services generate pollution through the commute of its employees. Other operations generate indirect environmental concerns such as the creation of biofuels which is now considered to affect food prices in other regions of the world or the production of genetically modified products which could affect organisms locally and potentially threaten the ecological balance. The production and disposal of hazardous materials is a further key concern. They need to be managed and taken care of in a responsible manner. Common across these types of pollution is that they are concerned with the enforcement of rules and regulations. The organization has received approval to operate in the previous stage of its life-cycle, and it is currently concerned with complying with the approval and the local and national laws as well as the stakeholders' concerns.

Public participation can involve the public for data gathering and data verification. The three British cities of Bristol, Sheffield and York designated Air Quality Management Areas following legislative changes. Subsequently, these cities trialed to engage the public, citizens in this case, in establishing to what extent the officially collected data on air pollution is correct and to what extent the right data is collected (Yearley et al., 2003). Air quality to the involved public comprised air pollution, noise levels, air odor, and dust. The wider definition as brought forward by the public, is important to recognize. Understanding how the public perceives environmental categories allows to address them and to manage the public's expectations. Similarly, the "spatial representation of local knowledge" (Yearley et al., 2003: 252) these consultations deliver, identifies the areas that are most affected in the eyes of the public and therefore require most immediate action. These areas can overlap with official data but this is not necessarily the case as Yearley et al. (2003) have shown. People who are living and working in a particular area are more sensitive to environmental changes than measurements.

Although the participatory modelling and monitoring of pollution increases environmental democracy through inclusiveness, information sharing and a broader engagement of the community, it has its limitations. Lack of public engagement and insufficient expertise in collecting and assessing data makes this approach to public participation difficult. Moreover, it does not necessarily lead to a comprehensive engagement with the public, but could be used for information collection and sharing only (Conrad and Hilchey, 2011). McDonald et al. (2002) show for a group of citizens from London that they struggled to understand the scientific methods employed to collect and analyses air quality data. Their constructive involvement in any decision-making was thus limited. McDonald et al. (2002) thus propose that the aspired public participation in the environmental monitoring and decision-making should involve a well-informed public.

Conclusions and policy implications

Mechanisms for environmental public participation exist across the whole life-cycle of any organization. At any point in time, the public has access to channels through which it can express its concern or disagreement with particular decisions and activities. It is also common that the public is proactively consulted and informed about changes in their vicinity. Both, the reactive and the proactive, mechanisms aim to reduce any latent conflict and address the interests of multiple stakeholders. Organizations require the publics' trust to gain legitimacy for their operations and existence. Where they fall short of transparency and accountability of their engagement with the environment and the society, they threaten to lose both. In the process of stakeholder management, activities are taking place across UK, and beyond, and along the complete life cycle of the organization. For any organization the salient stakeholders vary over time. This weakens and strengthens the stakeholders at the same time. The weakness derives from a lack of an existing rapport with the organization and established modes of operandi. It is therefore more resource intensive for the new salient stakeholder to exploit the potential of saliency.

The strength of this system is that the organization finds it more difficult to employ routine responses in the engagement with the stakeholder. (Salient) stakeholders can form action groups with local, regional, national and even international peers supporting the cause and working for a closely related one. Depending on the industry and on the issues at stake, some of the stakeholders will be interconnected and thus work together to form alliances and allegiances. These have larger nationally (internationally) a greater voice, greater impact, and greater resources to monitor the compliance by the organization to the agreement. Some stakeholder engagements are legally supported and enforced in the sense that organizations are legally required to engage with stakeholders. The top-down approach reaches out to and engages communities. Similarly, the bottom-up approach sees the public reach out to the organization. Neither process is necessarily the more effective one, or provides greater social quality. What is driving successful stakeholder management, however, is the early and sincere involvement of the public (Petts, 1995; Shepherd and Bowler, 1997; Ng and Sheate, 1997; Bond et al., 2004).

The public participation mechanisms in the UK, and the European Union, suggest that the existence of multiple engagement channels supports public participation. Individuals, groups of individuals, organizations and political parties can start initiatives for being heard and consulted on environmental concerns. This requires acceptance from all parties involved that negotiations have to take place and compromises have to be found that take into consideration the citizens and their well-being as well as the well-being of nature. National and local governments that support public participation need to work towards open and transparent participation mechanisms that involve a significant degree of proactive engagement on behalf of the polluting organization. Retrospectively engaging the public and implementing corrective measures to reduce any pollution would be costly on two grounds. Precious and pristine environmental habitat can often be irretrievably destroyed. Even occurrences with less severe immediate damages to nature can be very costly as the example of the environmental cleaning costs in excess of £70 bn by Nuclear Management Partners at Sellafield, UK shows (Macalister, 2013).

Besides monetary and environmental costs, inaction towards environmental pollution can affect public health and support mistrust towards public and private organizations and question to extent to which they are servants of public interest and work towards social quality. Such mistrust has the potential to make it more difficult to pursue other policy goals. Developing and implementing effective environmental public participation procedures can help to moderate these tensions and contribute to greater social quality.

References

[1] Aarhus Convention. 1998. Convention on Access to Information, Public Participation in Decision-Making and Access to Justice in Environmental Matters. United Nations Economic Commission for Europe (UNECE). http://live.unece.org/fileadmin/DAM/env/pp/documents/cep43e.pdf.

[2] Abbott, P. and C. Wallace. 2012. Social Quality: A Way to Measure the Quality of Society. Social Indicators Research, 108 (1): 153-167.

[3] Arnstein, S.R. 1969. A Ladder of Citizen Participation. Journal of the American Institute of Planners 35 (4): 216-224.

[4] Bickerstaff, K., Tolley, R., and Walker, G. 2002. Transport planning and participation: the rhetoric and realities of public involvement. Journal of Transport Geography, 10 (1): 61-73.

[5] Bond, A., Palerm, J., and Haigh, P. 2004. Public participation in EIA of nuclear power plant decommissioning projects: a case study analysis. Environmental Impact Assessment Review, 24 (6): 617-641.

[6] Bolden, T., and Harman, R. 2013. New development: High speed rail in Great Britain—its rationale and purpose. Public Money and Management, 33 (6): 463-468.

[7] Carroll, A. B., and Shabana, K. M. 2010. The business case for corporate social responsibility: a review of concepts, research and practice. International Journal of Management Reviews, 12 (1): 85-105.

[8] Conrad, C.C. and Hilchey, K.G. 2011. A review of citizen science and community-based environmental monitoring: issues and opportunities. 176: 273-291.

[9] Crompton, A.2013. Runaway train: Public participation and the case of HS2.Policy and Politics, http://dx.doi.org/10.1332/030557312X655963.

[10] de Luca, S. 2014. Public engagement in strategic transportation planning: An analytic hierarchy process based approach. Transport Policy, 33, 110-124.

[11] DfT (Department for Transport) 2009. Britains Transport Infrastructure High Speed Two.

Department for Transport. ISBN 978-1-906581-80-0.

[12] Dooms, M., Verbeke, A., and Haezendonck, E. 2013. Stakeholder management and path dependence in large-scale transport infrastructure development: the port of Antwerp case (1960–2010). Journal of Transport Geography, 27, 14-25.

[13] European Parliament and of the Council. 2003. Public access to environmental information. European Directive 2003/4/EC, Official Journal of the European Union, L 041, Volume 46, 14 February 2003: 26-32.

[14] Fraser, E.D.G., Dougill, A.J., Mabee, W.E, Reed, M., and McAlpine, P. 2006. Bottom up and top down: analysis of participatory processes for sustainability indicator identification as a pathway to community empowerment and sustainable environmental management. Journal of Environmental Management, 78: 114-127.

[15] Hartley, N., and Wood, C. Public participation in environmental impact assessment – implementing the Aarhus Convention. Environmental Impact Assessment Review, 25: 319-340.

[16] ICO. 2014. The Guide to the Environmental Information Regulations. Wilmslow: ICO.

[17] IPCC. 2014. Climate Change 2014: Impacts, Adaptation, and Vulnerability. Geneva: IPCC.

[18] Lee, M., Armeni, C., de Cendra, J., Chaytor, S., Lock, S., Maslin, M., Redgwell, C., and Rydin, Y. 2013. Public Participation and Climate Change Infrastructure. Journal of Environmental Law, 25 (1): 33-62.

[19] Mason, M. 2010. Information Disclosure and Environmental Rights: The Aarhus Convention. Global Environmental Politics, 10 (3): 10-31.

[20] McDonald, J.S., Hession, M., Rickard, A., Nieuwenhuijsen, M.J., and Kendall, M. 2002. Air quality management in UK local authorities: public understanding and participation. Journal of Environmental Planning and Management, 45 (4): 571-590.

[21] Mitchell, R.K., B.R. Agle, and D.J. Wood. 1997. Toward a theory of stakeholder identification and salience: Defining the principle of who and what really counts. Academy of Management Review 22 (4): 853-888.

[22] Miyoshi, C. and Givoni, M. 2013.The environmental case for the high-speed train in the UK: Examining the London–Manchester Route. International Journal of Sustainable Transportation, 8 (2): 107-126.

[23] Morgan, R.K. 2012. Environmental impact assessment: the state of the art. Impact Assessment and Project Appraisal, 30 (1): 5-14.

[24] Mulholland, H. 2009. Boris Johnson appoints Sir David King to examine Thames estuary airport plan. Guardian, 19 October 2009, 16.28 GMT, http://www.theguardian.com/politics/2009/oct/19/boris-david-king-thames-estuary-airport.

[25] Ng, Y.C. and Sheate, W.R. 1997. Environmental impact assessment of airport development proposals in the United Kingdom and Hong Kong: who should participate? Project Appraisal, 12: 11-24.

[26] Petts, J. 1995. Waste management strategy development: a case study of community involvement and consensus-building in Hampshire. Journal of Environmental Planning and Management, 38: 519-536.

[27] Sheperd, A., and Bowler, C. 1997. Beyond the Requirements: Improving Public Participation in EIA. Journal of Environmental Planning and Management, 40 (6): 725-738.

[28] Synnott, M. 2013. Reflection and double loop learning: The Case of HS2. Teaching Public Administration, 31 (1): 124-134.

[29] Tsang, E.W.K. 2013. Generalizing from research findings: the merits of case studies. International Journal of Management Review, doi: 10.1111/ijmr.12024

[30] Tomaney, J., and Marques, P. 2013. Evidence, policy, and the politics of regional development: the case of high-speed rail in the United Kingdom. Environment and Planning C: Government and Policy, 31 (3): 414-427.

[31] Topham, G. 2013. Heathrow and Gatwick shortlisted for new runways. Guardian, 17 December 2013, 12.42 GMT, http://www.theguardian.com/world/2013/dec/17/heathrow-gatwick-airports-expansion.

[32] Verbeke, A., and Tung, V. 2013. The future of stakeholder management theory: a temporal perspective. Journal of Business Ethics, 112 (3): 529-543.

[33] Wellings, R. 2013. The High-Speed Gravy Train: Special Interests, Transport Policy and Government Spending. London: Institute of Economic Affairs.

[34] YCELP (Yale Center for Environmental Law and Policy) and CIESIN (Center for International Earth Science Information Network). 2014. 2014 Environmental Performance Index. http://epi.yale.edu/.

[35] Yearly, S., Cinderby, S., Forrester, J., Bailey, P., and Rosen, P. 2003. Participatory modelling and the local governance of the politics of UK Air pollution: a three-city case study. Environmental Values, 12: 247-262.

实践中的环境民主
——芬兰环境规划及决策中的公众参与

尤哈·卡斯基宁[*]

导言

在芬兰，对于公众参与环境保护比环境问题机构化的研究历史更加悠久。一百多年前，一个小城市的居民将一家造纸厂告上法庭。在当时，造纸厂污染仅仅是局部现象，而在20世纪六七十年代则演变为全国范围内造纸产业的问题。

自20世纪60年代开始，在芬兰乃至全世界范围内环境政策均走向机构化。这种变化在政府机构中初现端倪，从七八十年代开始，政府成立了新的环保机构。这种趋势在不同的国家发展速度也有所不同。芬兰于1983年成立了环境部，成为环保机构化的先驱者。而在瑞典，环保机构化则始于1987年。

在1992年的里约峰会上，公众参与环保的原则被纳入《里约环境与发展宣言》。简而言之，这些原则包括：（1）民众应当能够获取相关信息；（2）民众能够参与环保；（3）民众应当享有公平正义。为此，我们必须构建法律框架、采集和传播信息的途径，司法必须平等、公平、值得信赖。不仅要鼓励民众积极参与，更要赋予其进入体制内工作的能力。本文分析了芬兰的环境规划和决策原则及其如何具体实现。

同时，本文将描述芬兰环境政策的框架，并把其重点放在了环境影响评估（Environmental Impact Assessment, EIA）。在芬兰，任何一项重大的建筑规划或政

---

[*] 尤哈·卡斯基宁教授，芬兰图尔库大学芬兰未来研究中心主任。本文原文为英文，中译者为宋逸群（浙江大学外语学院）。

府项目必须接受这项评估。在本文中我也讨论了一种能使公众参与决策的新工具——众包，以及众包在环境民主中到底意味着什么。

芬兰的环境政策和决策

自从 1995 年芬兰加入欧盟后，其国内环保政策的目标主要由欧盟制定。包括发展自然资本、提高资源经济效率以及保护人民健康（详见 http://europa.eu/pol/）。欧盟面临的主要任务就是出台新的能源和气候变化方面的政策。2008 年 12 月，欧盟国家领导人达成了一项关于能源和气候变化一揽子政策的共识——可再生能源的使用目标对所有成员国有效。该政策同时展望了 päästökauppa 的未来发展。欧盟希望将能源使用效率提高 20%，并减少 20% 的温室气体排放量。与此同时，欧盟希望与 1990 年相比到 2020 年可再生资源能提供 20% 的能源消耗。这些都是欧洲 2020 战略的一部分，也是所有成员国共同奋斗的目标。

芬兰的环保政策紧随欧盟制定的规则，但其环保政策的实施情况还是受到具体国情的影响。欧盟制定的规章也因没有考虑到区域特色而被公众议论。芬兰就曾抱怨过欧盟的能源政策。早在 2012—2013 年，其农村居民对于欧盟要求保护野生动物的条例就怨声载道。一方面是因为要被保护的狼和熊在 25~30 年前就已经基本灭绝了；另一方面则是因为随着野生动物数量的增长，人们担心它们会袭击居民（尤其是儿童），因此，公众希望能够相应提高捕杀量。

1. 欧盟的立法程序

欧盟立法涉及多种参与方式。在新提案出台前，欧盟委员会会组织成员国官员、公民组织、专家和民众参与听证会，广开言路。此举保证了法案建立在适当的科学基础之上。利益相关者也参与这一规划过程并且享有发言权。另一方面，听证过程也是使提案合法化的过程，欧盟对此做出了回应和承认。

在欧洲委员会会议过程中各国部长可以修改提案，欧洲议会也可以根据公民的意见调整条例。法案必须经过委员会和议会全员通过。同时，其他欧盟机构也可以提供意见。参与讨论的还包括公司、非政府组织、研究者和公民个体等。从 1970 年开始，欧盟大大小小通过了 200 多项环境法规。这些法规能否付诸实践还要看各个成员国的努力。环境法规体系十分复杂，不同成员国有不同的控制和制裁方法。如果成员国不能合理地实践法规，欧盟委员会有权将其告上欧盟法庭。因此，只有人们实行和遵守法律，法律才有意义。

有时并不需要完整的立法，即便是总体纲领和标准系统也可以达到预期目的。因此，欧盟环境政策的实现不仅通过法律，也通过法令、方针、战略和推荐等方式。1998 年通过的《奥胡斯公约》就体现了公众参与的原则（详情请见：europa.eu/legistaltion_summaries/environment）。该公约表示，政府必须确保公民能够接触到政府发布的信息和数据以提高公民在环保决策中的参与度，拓展在环保事务中公民获得正义的前提条件。欧盟清晰地遵循着 1992 年制定的《里约环境与发展宣言》中的建议和规则。

2. 芬兰环保政策的结构

当今芬兰的环保政策与欧盟的相关政策紧密相连。在芬兰，有超过 30 项法律和 10 多种法令对自然、环境、土地使用、建筑和房屋供给等做出规定。环境部是解决环保问题的主要机构，也负责房屋供给。环境部的总体战略（"共同创造可持续的未来"）确定了 2020 年将要实现的目标，它们是（详情请见：www.ymparisto.fi）：

- ➢ 减少大气中温室气体量，保持在人类能够适应的水平。
- ➢ 改善人们的生活质量，并达到建筑节能的目的。
- ➢ 房屋供给达到居民需求，市场运转良好。
- ➢ 保护生物多样性和景观价值生态系统良好循环，可持续地使用自然资源。
- ➢ 波罗的海的地表水和地下水质量至少维持在良好水平的下限。
- ➢ 防控环境风险，从根本上提高能源与材料的利用效率。

环境部的理念是勇气、专业和责任。具体包括：

- ➢ 作风开放：所有合作均保持透明。通过公开前期准备工作鼓励公众表达意见。
- ➢ 重视专业知识：从事环保工作的人员必须具备专业性。通过分享经验、提升自我以帮助他人提高专业水平。
- ➢ 负责未来的发展：决策的效应会延伸入未来。切忌牺牲长远利益换取眼前利益。

芬兰环境机构是政府的研究和服务组织，被人们称作 SYKE。它既是一个研究中心，又聚集了大量的专业环保人士。SYKE 是芬兰环境管理的一部分，主要是在环境部的支持下开展工作。然而其有关水资源的工作则隶属于芬兰农业和林业部（详情请见：http://www.syke.fi/en-US/SYKE_Info）。

经济发展、交通和环境中心（Centers for Economic Development, Transport and the Environment，ELY）负责在地区层面执行中央政策。芬兰总共有 15 个这样的中心，它们在提高地区竞争力、福利、可持续发展及减缓环境变化方面发挥了重要作用。ELY 的职责主要体现在三个领域：（1）商业和工业、劳动力、竞争力及文化活动；（2）交通和基础建设；（3）环境和自然资源。ELY 对就业和经济发展办公室的工作进行指导。不是所有的 ELY 中心都具有所有三个方面的职能，它们也可以代表彼此开展工作（详情请见：www.elykeskus.fi）。

城市、城镇和市级单位在环保方面也各尽其责。他们负责环境保护的监管和控制，同时还负责水资源和废物的管理、食品安全、颁发环保许可等。在这方面，地方政府实施的最有效的方法是土地使用规划：市政府在他们所管辖的区域具有规划垄断。总的规划方针决定了土地使用的大体方式，细致的城市/城镇规划甚至规定了每一栋建筑的用途。在土地使用规划过程中，政府必须倾听所有利益相关者的意见，后者有权对规划表示不满。

法律和法规规定了公众参与的组织形式。最常用的方法包括信息发放、组织听证会、问卷调查、采访、论坛讨论（面对面或者在线讨论）还有利益相关者和公民发表声明。法案的规模不同，参与的方式也不同。将公众参与方式与《里约环境与发展宣言》中的三个方面进行比较时，我们可以看到在大多数情况下公民和利益相关者确实有获取信息的渠道。

3. 环境影响评价

在环境影响评价的帮助下，对环境的负面效应被有效遏制。城市规划可以是一个机动车道，一个发电厂或者一个工业建筑。通过这一评价，规划的后果会在最终决策前得到评价，并且其结果会对决策产生影响。由此，环境评价就成为一种规划工具，其结果必须考虑在规划的实施过程中。

规划的负责组织必须负责评价所需要的环境研究。该组织可以是私人的也可以是政府的，例如政府办公室或者私人公司。评价的决策过程受到 ELY 中心的监督。核电站的建设计划由就业和经济部负责。所有受到规划潜在影响的人都可以参与评价。参与和信息公布是评价的基石，因为评价的目的之一就是提高公民的参与度及其影响决策的能力。环境影响评价的相关法律决定了评价可执行的规划。评价同时也可以应用于较小型的、没有包括在法律中但是有重大环境影响的规划方案中。

在环境影响评价的全过程中，其第一个环节是向 ELY 中心递交项目。项目书包括了待调查的备选方案和本方案的影响，同时还有规划区信息分发和利益相关者参与的具体方式。另外，也一定要评价规划方案对居民的影响。在专家研究备选方案和潜在影响后，所有信息都会写在评价报告中。评价报告还收录了地方居民和各个组织的意见和建议，并在此基础之上相关部门需给出自己的决议。ELY 中心最后给出自己的决议。

负责规划方案的组织负责整个过程的全部花费，包括环境研究、交流、信息传播、听证会等。同时，该组织还需要给出 ELY 中心的最终决议费用。环境影响评价法律于 1994 年生效。截止到 2011 年末，共发起了 587 个项目评价过程，其中 494 个项目完整结束。其中 1/3 针对废物管理的评价，1/5 针对自然资源利用的评价；1/2 针对其他例如动物蓄养、水资源管理、化学和金属工业等话题的评价（详情请见：www.ymparisto.fi）。

EIA 的参与方法

在过去 19 年中，环境影响评价已经成为环保政策中的一项正常程序。公众参与根据不同规章制度得到实施。个人可以发起某项方案的环境评价过程。公民有两次发表意见的机会，第一次是当项目发起时，第二次是当大多数评价研究完成时。同时，公民还在最终的评价报告中给出自己的决议。公民可以不同程度地参与听证过程。专家展示和小组讨论适用于信息发布。图像资料，例如地图、模型、模拟、3D 图像、动画和录像也可以用于听证会。参与者可以向专家提出问题并提出自己的意见。

公民意见和反馈的获取可以通过问卷调查和采访的方式。网络和社交网站也贡献出自己的力量，尤其是规划方案涉及的面积广、人数多时，"众包"方法更能为人们参与环境评价过程创造新的可能性。

我们可以说公民在环境评价过程中具有发言权。公民能够在获取信息的同时生产信息。尽管途径不是很多,公众还是能够参与决策。公民能够获得公平正义,因为 EIA 过程允许人们提出意见和不满。很重要的一点是环境评价并非是最终有效的过程。在最终决策过程中,环境评价须被考虑在内,但又不能完全决定最终决策。例如,最终决策者还是可以牺牲环境利益追求经济或者其他方面的效益。

4. 众包

网络和社交平台(Internet and social media,ICT)改变了政策、政治和决策,其中一个方面就是众包。根据 Aitamurto(2012),"众包是一种呼吁所有人网上参与任务的公开的途径",它在城市规划、产品设计、创新(公开创新)和科学问题解决(如前)中都发挥了积极作用。例如,欧盟就将其应用在新研究领域的界定和立法过程中。一个政策方面很有意思的例子就是冰岛在 2010—2011 年的宪法改革。娱乐生活中的典型例子就是 2012 年上映的电影《钢铁苍穹》。该电影的部分制作被众包给了芬兰的 Wreck-a-Movie 众包平台,任何人都可以登录设计电影特效。

众包在民主进程中的益处是很多的。参与者的多样性能够让我们在大范围内征询大量意见,因此获取信息的范围也扩大了。此外,众包还能使人们参与到决策和政策制定的过程中,起到信息传播的作用,帮助普通公民获得政治力量(Aitamurto,2012)。众包在环境政策和决策中的应用多种多样。冰岛的立法过程就是一个很好的例子。在芬兰,《公民倡议法案》于 2012 年生效。如果一项倡议能在 6 个月中获得超过 50 000 个签名,该倡议必须提上议会议程。签名可以通过用户在线获得(详情请见:www.avoinministerio.fi)。

在国家和地区的环境战略和规划设计中,收集信息的公共创新系统和 EIA 过程都能通过众包受益良多。然而,机遇也伴随着威胁。数码设备是一个严峻的问题,尽管这个问题在逐步弱化。未来人们需要更多形式面对面的交互活动,虽然在短期内众包不能取代民主和选举,但可以成为其一部分。民众也不能取代专家,但是两者都是必需的。

结论

环境决策的挑战之一是知识管理，这对决策者和公民而言都是一个问题。由于信息和数据量增长迅速，导致很多情况下研究的结果相互矛盾或者不一致。不同的利益方影响着方案的选择和实施。ICT 工具或许可以帮助处理大量的信息，但是 ICT 暂时不能取代决策过程，并且 ICT 也为虚假信息的传播提供了新的途径。

另一方面，ICT 为公众参与创造了新途径，至少那些活跃和会使用它们的公民得到了益处。ICT 缩减了立法准备和决策过程的时间（例如 EIA 过程）。网络也让意见收集更加便捷。新的 RDIF 或者 ubiq 技术让数据收集和分析更加简单（Nurmi et al., 2010）。另外一个挑战是环境规划、管理和实施的复杂性，这种复杂性是针对非专业人员而言的。当然，人们可以说，复杂的问题需要复杂的解决方式，但公众参与环境规划和决策需要一个清晰和简单的操作体系。

环境领域的规划、提案和项目实施需要时间。因为法律系统的进程缓慢，不满和申诉可以在体系内循环多年得不到解决。公民和利益相关者对此都失去耐心。与此相对应的是，利益相关者虽然能够获得公平正义，但是在很多情况下由于法律系统进程缓慢，他们的权利也不能及时得到保障。这个问题不仅体现在环保方面，也同样存在于整个芬兰法律体系中。

这些年来欧洲和芬兰的经济情况并不乐观，两者的经济结构都处于过渡时期，国有和私人经济都在试图寻找新的商机。在芬兰，开采工业因其对环境会造成影响而成为了人们争论的焦点；在城市和农村，人们对此也意见各异。因此，对于经济和环境的考量在公共辩论和政治决策中屡屡发生冲突。让我们不禁感慨，难道人们还是遵循金钱至上的原则吗？

Environmental Democracy in Practice
——Citizen Participation in Finnish Environmental Planning and Decision-Making

Juha Kaskinen *

Introduction

In Finland as well as globally, environmental policy started to institutionalize after 1960's. At first, these occurred as minor changes in governmental structures but later in 1970's and 80's, new institutions were established. Countries differ from each other in how rapidly the process takes place. Finland was one of the forerunners in this respect since the Ministry of Environment was established in 1983. For example in Sweden this was took place only in 1987.

In the Finnish case the concern of citizen participation in dealing with environmental problems has a longer history than institutionalization. The first court case against a paper and pulp factory for water pollution was over 100 years ago and the initiative came from the inhabitants of one particular small city. Back then, the problem was considered as local, the big pressure against the paper and pulp industry as a whole took place in 1960's and 70's.

At the Rio Summit in 1992, the Rio Declaration recognized the principles of citizen engagement in environmental decision-making. These principles in short are that (1) people should have access to information, (2) they should have ways to participate and (3) have access to justice. This means that there should be legal frameworks, channels for information gathering and dissemination, jurisdiction should

---

* Juha Kaskinen, Professor and director of Finland Futures Research Centre, University of Turku, Finland.

be reliable, equal and fair to all, means of participation should be effective and clear and citizens should have capacities to work within the system. In this paper I assess how these principles are achieved in the Finnish context of environmental planning and decision-making.

Before starting this assessment, we should first of all present an overview of the structure of the Finnish environmental policy. Then, I shall concentrate on the Environmental Impact Assessment (EIA). The EIA process has to be carried out for every major building plan as well as for governmental strategies and programs. On this ground, I deal with a new tool to get citizens involved in decision-making process, namely crowd sourcing. What could crowd sourcing mean for environmental democracy? A conclusion is offered in chapter three.

Finnish Environmental Policy and Decision-Making

After Finland joined the European Union in 1995, the main goals for national environmental policy have been set by the EU. The main goals of the EU environmental policy are development of natural capital, improvement of resource efficient economy and protection of people's health (See e.g. http://europa.eu/pol/). One of the main tasks of the EU is to renew its energy and climate change policy. In December 2008 the leaders of EU countries accepted a large energy and climate change package in which goals of use of renewable energy sources were set to all member countries and which outlined development of päästökauppa. The EU wants to increase its energy efficiency by 20%, decrease greenhouse gases by 20% and have 20% of energy consumption to be produced by renewable energy sources in 2020 compared to level of 1990. This all is part of Europe 2020 strategy that member countries try to achieve.

The Finnish environmental policy follows the rules set by the EU but the national context affects focus areas of national environmental policy. There can be heated discussions on EU regulation that do not take into account special local features. For example the Finnish industries have complained about the goals of the EU's energy policy. In 2012-2013 local communities in rural areas have been complaining on rules of protection of wild animals, especially wolves and bears that both were almost extinct 25-30 years ago. Now their number has increased and people are afraid of attacks on humans, especially children and want the number of killing permits to increase.

EU context of the legislation process

In preparation of EU legislation, several participatory methods were utilized. Before the EU Commission gives a proposal of a new law, the Commission organizes large hearing sessions for officials of member countries, citizen organizations, experts and a wider audience in order to take notice of their opinion. The Commission wants to make sure that the proposals are based on proper scientific sources. The stakeholders are engaged in the planning process and they have a say in it. On the other hand, the hearing process is also a mechanism to legitimize proposals and the EU clearly admits this.

In the European Council meetings the ministries of member countries can amend proposals. The European Parliament can also make changes based on suggestions from citizens. Both the Council and the Parliament have to be unanimous in the law text before it can be approved. Several EU organs also give their statement. Companies and firms, NGOs, researchers or private citizens can also give their view. Sometimes it happens that the EU legislation is not needed and that the general guidelines or benchmarking systems are enough to support the goals.

A law itself is quite useless if it is not put into force and followed. Since 1970, the EU has approved over 200 environmental regulations. It is up to the member states to see to it that the rules are followed. The environmental regulation system is really complex and in some cases not effective and member states have different control and sanction practices. The EU Commission can summon a member country before the EU Court if the member country does not put the regulations into force appropriately.

The EU conducts environmental policy not only by laws but by statues, directives, strategies and recommendations. Guidelines of citizen participation are accepted in the so called Århus general agreement/contract that member states approved in 1998 (europa.eu/legislation_summaries/environment). It states that member states must make sure citizens have access to data and information that public official poses, to improve citizen participation in environmental decision-making and extend the conditions of access to justice in environmental matters. The EU clearly follows the recommendations the Rio Declaration set in 1992.

Structure of Finnish Environmental Policy

The Finnish environmental policy and decision-making is nowadays connected to the EU's policy. In Finland, over 30 laws and tens of statutes and orders directly regulate nature, environment, land use, building and housing. The Ministry of Environment is the main governmental organization that deals with environmental issues. It is also responsible for housing. The strategy of the Ministry of Environment ("Together to Sustainable Future") sets the goals towards 2020 (www.ymparisto.fi). These are:

- ➢ The amount of the greenhouse gases in the atmosphere is limited to the level that hinders dangerous changes and enables adaptation.
- ➢ The build environment is energy efficient, viable and it improves human well-being.
- ➢ The housing conditions answer people's requirements and the markets are functional.
- ➢ Biodiversity and landscape values are preserved, the ecosystem services work well and the use of natural resources is sustainable.
- ➢ The quality of surface and groundwater in the Baltic Sea is at a good minimum.
- ➢ The environmental risks are well known and in control. Energy and material efficiency improves essentially.

The values the Ministry of Environment follows are courage, expertise and responsibility. The values that guide the activities of the ministry are (ibid):

> Boldly open: all cooperation is done openly, by open preparation of the actions the ministry encourages others to express their views and even hard decisions are taken by the ministry with good reason in an open process.
> Valued expert organization: the ministry is a well known and recognized expert, the personnel supports each other's development of skills by sharing expertise and by developing their own skills.
> Responsible future creation: the effects of the decisions will reach far into the future, the ministry does not seek quick-drawn prices at the expense of the future.

The Finnish Environment Institute is a governmental research and service organization. The Finnish Environment Institute (also known as SYKE, after the Institute's Finnish acronym) is both a research institute, and a centre for environmental expertise. SYKE forms a part of Finland's national environmental administration, and it operates mainly under the auspices of the Ministry of Environment, although the Institute's work related to water resources is supervised by the Ministry of Agriculture and Forestry (http://www.syke.fi/en-US/SYKE_Info).

The Centers for Economic Development, Transport and the Environment (ELY Centers) are responsible for the regional implementation and development tasks of the central government. Finland has a total of 15 ELY Centers, dedicated to promoting regional competitiveness, well-being and sustainable development and curbing climate change. ELY Centers have three areas of responsibility: (1)business and industry, labor force, competence and cultural activities, (2) transport and infrastructure and (3) environment and natural resources. The Centers for Economic Development, Transport and the Environment steer and supervise the activities of the Employment and Economic Development Offices (TE Offices). Not all the ELY Centers deal with all three areas of responsibility as they can also manage duties on each others' behalf (www.elykeskus.fi).

Cities, towns and municipalities have also obligations regarding the environmental policy. They are responsible on supervision and control of the environmental regulations, they have to organize water and waste management, control food safety, grant environmental permits etc. The most effective tool the local authorities pose, is land use planning; the municipalities have planning monopoly in their area. A general plan dictates the overall land use modes in a municipality and a detailed city/town plan regulates land use on the level of even one building. In the land use planning, all stakeholders must be heard during the process and they have an opportunity to make a complaint against the plan.

Laws and statutes regulate how citizen participation should be organized. Most common methods to include citizens in planning and decision-making are information dissemination and hearing sessions, questionnaires, interviews, discussion forums(face to face and online) and statements given by stakeholders and citizens. Methods vary according to size of plan or proposal. When we compare the citizen participation possibilities to the three assessed points of the Rio Declaration it is quite clear that citizens and stakeholders have (in most cases) access to information. The question is whether this information is understandable to all stakeholders.

Environmental Impact Assessment

With the help of the EIA, the potentially harmful effects of a plan, eg. a motor way, power plant or an industrial production building are minimized or totally hindered. In the EIA, the effects of a plan are assessed before the decision-making, and by doing so influence future decisions. The EIA is a planning tool and the results of an EIA must be taken into account when permission for the realization of a plan is under consideration.

The organization responsible of the planning must take care of the environmental studies needed for the EIA. The organization can be public or private, like a governmental office or a private company. The EIA process is supervised by an ELY Centre. In planning of a nuclear plant, the supervising organization is the Ministry of Employment and the Economy. All those who can be affected by the plan can participate in the EIA. Participation and dissemination are the cornerstones of an EIA.

One aim of the EIA is to increase the possibilities for the citizens to participate in and effect the planning. The law and statues on the EIA define the plans in which an EIA must always be carried out. The EIA can also be applied to smaller plans that are not included in the law but which may have significant environmental impacts.

An EIA process starts when a responsible organization delivers an EIA program to an ELY Centre. The alternatives and impacts to be investigated during the process are described in the EIA program. Also described is how dissemination and participation of the stakeholders in the area should be organized. The impact to humans must also be assessed. When the alternatives and their impacts are studied, all the information is gathered up in an assessment report. The supervising ELY Centre then notifies the area the plan concerns and has impact on. It also collects the opinions of the inhabitants and different organizations, statements of different authorities and gives its own statement based on feedback and its own expertise. The EIA process ends when the supervising authority gives its statement on the EIA report.

The organization responsible of the plan covers all costs of the process like environmental studies, communications, dissemination, hearings etc. It must also pay a fee to the supervising authority on the statements the authority has given on the EIA program and report. The law of Environmental Impact Assessment came into force in Finland in 1994. By the end of 2011, 587 EIA processes were started and 494 were completed. One third of the processes dealt with waste management, one fifth dealt with the use of natural resources. 50% of the cases were on various topics like animal farming, water management, chemical and metal industry, energy production etc. (www.ymparisto.fi).

Participative methods in the EIA

During the last 19 years, the EIA has become a "normal" procedure within environmental policy. Citizen participation is carried out according to regulations. A single person is allowed to propose an EIA process of a plan. Citizens are heard during a normal process at least twice, first when the program is published and the second time when most of the assessment studies are finished. Citizens can give their own statement on the final EIA report as well. In the hearing sessions several different methods are applied. Expert presentations and panels are used for information dissemination. Visual material like maps, models, simulations, 3D-images and animations and videos might be presented in these hearings. Experts can be consulted and bring new ideas into planning.

Questionnaires, surveys and interviews are one way of gathering citizen opinion and feedback. The internet and social media increase possibilities to include citizens in the planning, especially if the plan has impacts on wide area and on a large number of inhabitants. The methods of crowdsourcing offer new possibilities to the EIA (more on this in the next chapter).

It can be stated that the citizens have a say in the EIA process. They have access to information and are partly producing it. Methods of participation are available even if it can be argued whether they are sufficient. People also have access to justice because the EIA procedure is an established system with a possibility to make statements with complaints. One critical point concerning the EIA is that it is not a binding process. In the decision-making the EIA has to be taken into account but it does not bind the hands of a decision maker; he/she can e.g. choose an environmentally harmful alternative on economic or other grounds.

Crowd sourcing

ICT, internet and social media transforms policy, politics and decision making. One aspect of this change is crowdsourcing. According to Aitamurto (2012): "Crowdsourcing is an open call for anybody to participate in a task open online". It has been applied for example in urban planning, product design, innovation processes (open innovation) and in solving scientific problems (ibid). The EU has applied it e.g. in defining new research areas and legislation processes. An interesting example from policy arena is the constitution reform in Iceland in 2010-2011, in which crowdsourcing was applied. A nice example of crowdsourcing in entertainment is the movie Iron Sky, published in 2012. Part of the production was crowd sourced by the Finnish Wreck-a-Movie crowdsourcing platform. So anyone could go and design e.g. a special effect for the movie (a science fiction comedy).

The benefits of crowdsourcing in a democratic process are many. It makes it possible to gather a lot of information from a wide audience. It widens the sphere of new information because the diversity of participants increases. It helps to include and engage people in the decision and policy-making. It spreads information on the decision-making process itself. It might also enhance citizen empowerment (Aitamurto, 2012). If we think how crowd sourcing could be applied in the environmental planning and decision making, the possibilities are many. The preparation of the legislation as done in Finand is a good example. In Finland, the Citizens Initiative Act came into force in 2012. If a petition gets at least 50 000 signatures in six months, it must be discussed in the parliament. The signatures can also be gathered online with bank user identification (www.avoinministerio.fi).

When national, regional or local environmental strategies and plans are prepared, crowdsourcing could easily be applied. A public open innovation system for gathering and developing innovations could benefit from crowdsourcing as well as the EIA processes. The possibilities and threats go hand in hand. The digital devices are still a crucial question, even if it is diminishing all the time. So, some form of face to face interaction will be needed in the future also. Crowdsourcing does not replace - at least in the near future- representative democracy and elections but it can be part of it. The citizens cannot replace experts either but both groups are needed. Encouraging people to participate in crowdsourcing takes effort and resources as all methods of participation do. Then there is the question of overactive participants, offensive comments, insults or even hate speech.

Conclusion

One of the main challenges in environmental decision-making is knowledge management, which is an issue for both decision makers and citizens alike. The sheer amount of data and information is increasing fast and research results are in many cases contradictory and inconsistent. Different interests effect on how arguments for and against a plan or a proposal are chosen. The ICT tools may help with the amount of the information but they are not making the decisions for us at the moment. ICT opens new ways for the dissemination of mis- or disinformation.

On the other hand, the ICT tools create possibilities for citizen participation, at least for those who are active and use them. The ICT tools can shorten usually time taking processes like the EIA, or the preparation of the environmental legislation. The opinion polls are fast to make in the web. New RDIF or ubiq technologies help data gathering and analysis(Nurmi et al., 2010). Another challenge is the complexity of the environmental regulations, management and administration from the point of view of a non-expert. Of course one can say that complex problems need complex solutions and structures but engaging citizens better to the environmental planning and decision making requires a transparent, clear and understandable system for the "customers" to use.

Furthermore, as stated above, different plans, projects, proposals etc. in the environmental sector take time to be completed. One aspect of this challenge is that the legal system is quite slow. Complaints and appeals can circle in the systems for years. It irritates not only the citizens but also other stakeholders. The stakeholders have access to justice so to speak, but the legal system works too slowly in too many cases. This is not a problem only in the environmental issues, it is a characteristic of the Finnish legal apparatus.

The economic situation in Europe and Finland has been difficult for several years. The economic structure is under serious transition both in Finland and in Europe. National economies and private companies try to find new business possibilities. In the Finnish case, the most controversial issue during the last couple of years has been the mining industry which has major environmental impacts. The opinions on mining vary a lot both in the urban and rural areas. The economic and environmental arguments clash in the public debate and well as in the political decision-making. Is it still so that only money talks?

References

[1] Aitamurto, T. 2012. Crowdsourcing for Democracy: A New Era in Policy-Making. Parliament of Finland. Publication of the Committee for the Future. 1/2012.

[2] Add dates of acces of the following:

[3] Euroopan komissio (2013): Valokeilassa Euroopan unionin politiikka: Ympäristö. Euroopan unionin julkaisutoimisto. Luxemburg.

[4] Nurmi, Timo - Vähätalo, Mikko - Saarimaa, Riikka and Heinonen, Sirkka (2010) Ubitrendit 2020: Tulevaisuuden ubiteknologiat. Kehityskulkuja, sovelluksia, trendejä sekä heikkoja signaaleja. (Ubitrends 2020. Futures Ubitechnologies) Tulevaisuuden tutkimuskeskus, Turun yliopisto. 99 s. ISBN 978-952-249-042-1.

[5] Websites:

http://eu.europa/pol/

http://www.avoinministeriö.fi

http://www.syke.fi

http://www.ym.fi

http://www.ymparisto.fi

水资源管理中的公众参与
——欧盟水框架的实践与经验

斯德哥尔摩国际水研究院*

1 导言

  健康且循环良好的水环境是这个星球上所有生命赖以生存的基础。良好的水环境能够满足生物生存所需的基本条件并且服务于生命体，它能满足生活用水、防止洪水泛滥、储存水资源、帮助营养物质的循环和重复利用等。近几十年，保护水环境这一议题引起了世界各国的广泛关注、努力，并且从社会各界吸引了广泛的资源投入。中国与欧盟各国在内的越来越多的政府组织意识到在政策实施与其他不同层面的决策过程中提高公众参与度的重要性，这是作为各国政府可持续水资源管理进程推进的积极成果之一。

  在公众参与水资源管理的方法和理论研究方面，欧盟已经取得了许多经验。欧盟范围内广泛认可的公众参与水资源管理这一方法，特别是《欧盟水框架指令》（Water Framework Directive，WFD）与其他一些指令（如环境影响评价、战略环境影响评价）都提出了各利益相关者和公众在不同层面上参与到水资源管理中去的要求，并把这一要求合法化。现在这一要求已经为公众广泛接受，并且理解这种参与不仅仅能够提升水资源管理决策的效率与质量，也能够保障当地社会的水资源买入与当地人民对于水资源管理决策支持的权利。WFD 自身就是一个利益相关者广泛参与水资源管理不同层面决策制定的结果。欧盟的一些案例已经展示了公众参与水

---

*斯德哥尔摩国际水研究院（Stockholm International Water Institute，SIWI）是一个致力于水资源研究和实现水资源可持续性发展的政策组织。本文原文为英文，中译者为朱梦蝶（香港教育学院）。

资源管理的途径的不同机制，如公众直接参与、网上咨询和召开利益相关人士会议。这些案例发生在欧洲大大小小的跨国流域（波罗的海、莱茵河等）。

对于这些案例的研究可以获得以下经验：

——拥有公众参与政府管理经验或有着公众参与传统的政府可以更好地推行实施政策（如 WFD）。

——利益相关者较早的参与往往会形成一个更加积极的公众参与氛围，营造较强的主人翁意识（这要取决于政府对于公众参与的态度是仅仅咨询还是共同决策）。

——对利益相关者的分析是政府进行关于公众参与水资源管理的框架制定时的第一步，也是最重要的一步。

——对于参与管理的各方利益相关者的授权必须要明确，内外部的沟通都要到位。

——对于公众参与程序和机制的管理必须基于足够的透明度之上。

——政府与各方利益相关者需要对推行公众参与所必需的资源（时间、资金、教育等）进行必要的投入。

——政府需要了解不同利益相关者与市民由于个体与资金上的差异，参与管理的程度也不尽相同。这种差距与不平衡应该被看作是对能力建设或其他类型支持的需求指标。

——由于信息透明与公民意识对于公众参与途径的实施至关重要，政府对于交流方式、渠道与信息量需要有多样性的选择，这样才能保证不同的个体与利益相关者都能参与其中（如在年龄与性别方面的选择）。

——公众参与应该从政府整体规划的开始就被作为整体工作中的一个组成部分来看待。

2 利益相关者与公众参与的概念

利益相关者与公众参与

"利益相关者"一词往往指在规划、项目或是措施的实施过程中会受到各方面影响的那部分人群。在《欧盟水框架指令》（WFD）中，参与水资源管理主要是指公众会受到关于水资源各方面的影响以及随之产生的对水资源管理的兴趣。更确切地说，公众参与即公民（包括社会边缘和缺乏资源群体）能够接触各类信息，提供信息，及时有效地参与管理并从各方面影响政府决策。利益相关者在水资源管理中的参与度也需要视情况而定，这需要考虑到即将开展的项目类型，并且要在受到影响的群众与国家经济发展的利益之间寻求一个平衡点。所有将利益相关者包括到管理中的进程都有一个相似点，即充分利用各种可参与的元素。所有公众参与管理的进程中都有以下几点基本元素：

——参与管理进程的透明性是利益相关者参与管理进程中最基本的指导主题；

——参与管理过程的灵活性主要指公众参与管理流程结构的灵活性，使得其可以适应各种特定的体制和政治环境、利益相关者的建议以及其他社会条件；

——公众参与度在不同参与方式下会有很大变化。如单向信息传递、全面的交流合作与利益相关者自决之间的公众参与度就有很大变化（Grunig et al.，2008）。

利益相关者参与管理的核心因素有二，一是决策过程的参与，二是这种参与对决策本身的影响。这也就是说公众参与不应该只是简单地作为自上而下的决策（Grunig et al.，2008）。如果我们只是采用政府自上而下的决策，那就不必去倡导利益相关者的作用。由于 WFD 的法律要求公众应参与该政策的实施，我们就要强调决策过程的重要性、为什么重要以及怎么做。

水资源管理中的公众参与

公众参与可以体现在政策实施的各个阶段。在水资源重组的早期阶段，利益相关者的参与非常重要，这应该作为政府消除政策执行差距与实现公民对政府履行监督义务的首要步骤。相反，公众参与水平是衡量不同的水治理机制表现的一个重要标准。公众参与决策进程的途径有很多种，如在市政大厅发表个人观点，积极向咨询委员会提出建议以及在投票决议中投票等。在全球背景中，越来越多的发展中国家已经开始通过制定新的法律和政策来提高公众在各项事务中的参与度。随着人们在提高公众参与度上付出越来越多的努力，将水环境问题纳入人权已经成为水资源管理改革中的另一种趋势。2010年的联合国代表大会将水与环境卫生列为"人类共享"[①]的至关重要的人权。除了使用水资源及其带来的各项福利之外，这项有关水资源的人权还包括水资源信息的公开与公众对水资源管理的参与。

但是，大部分政府发现这些新政策法规在实际开展的过程中仍存在许多难题。为了提升公民享有水资源的权利，各级政府在向公众提供水资源以及与其相关的各项服务的同时也亟需提高其信息公开度并开发公众参与的机制，以此达到公众在水资源利用方面的利益最大化。一般来说，公众参与机制难以推行主要由以下几点造成：

——在公众参与政策和机制上的激进的改革；
——政府的活动与相关资讯缺乏透明度与公开度；
——各地的政策执行缺乏执行能力和政策制定能力；
——地方上缺乏执行该机制的财政能力；
——一些政府或个人试图保护自己的权威，抵触并干扰机制的运行；
——公众参与改革的过程和内容都过于依赖于捐助者，容易被捐助者的兴趣所左右；
——机制改革的内容与流程缺乏地方自主权；
——公民缺乏对其所希望影响到的事件的认识；
——公民缺乏维持其参与管理效率与水平的能力。

遗憾的是，在多样的社会经济条件、社会制度和环境问题之下，没有一种公众参与方式可以放之四海而皆准。因此，为了探求一种有效的方式来推广广泛而有意义的公众参与，人们必须先对公众参与机制的本质进行提问与探索，如：

——公众参与是否存在法律保障？
——是否存在其他一些可以保障公众充分参与管理的机制？如有，分别是哪些机制？
——我们对于参与管理人选是否有充分的了解（可通过收集关于参与者性别、年龄以及其他方面的资料来达到）？

[①]联合国大会，2010年7月28日，64/292决议/A文件，联合国，纽约。

——参与管理的代表是否能代表广大人民群众的利益？

关于以上问题的解决方案与要素将会在接下来的章节中从宏观上与具体案例分析两方面详细阐述。

将公众纳入水资源管理层面的目的

近年来，欧盟已经将公众参与推广到公共干预与决策制定的各个领域。为什么公众参与被政策制定者与决策者看得如此重要，其主要原因在于公众参与管理可以对决策者产生直接与间接效益。将利益相关者纳入政策制定范畴会对政策制定者产生如下直接效益：

——提高决策效率，增加决策接受度（增加合法性）；

——提高决策实施效率，减少决策实施过程中的拖延状况；

——增加决策过程的透明度；

——得到利益相关者一些经验与知识上的帮助。

同时，该举措带来的间接效益包括：

——调动公民积极性；

——鼓励社会各界互相学习（关于不同人士的节水意识）；

——增强社会上对于环境问题的认识，提高全社会保护环境的意识（Grunig et al.，2008）。

将公众纳入政策规划范畴的方法

关于公众应该如何参与管理这个问题，不同的人有不同的认识。成功的公众参与没有一个特定方式，各种各样的方法都可以用来帮助提高管理中的公众参与程度，这其中的关键就是公众要能够对直接影响其利益的问题发声。各种社会体制对于实现这一点都有不同的做法。公众直接参与决策这一过程适用于人数较少的社会群体，而水资源管理恰恰涉及较大的区域，如多瑙河与莱茵河流域。这些河流往往流经多个国家的多个省区，这些地区的法律与文化都存在差异，因此在对于水资源管理的政策上也必然存在分歧。这就导致了人们在公众参与水资源管理方面存在不同的需求。

将利益相关者纳入管理并不意味着所有涉及的利益团体都要积极地参与到管

理中。利益相关者的参与方式与程度都需要依不同情况做出相应的改变。总体上，将利益相关者纳入管理层面的途径包括对市民及时通知、将决策过程公开化、给予公众决策或推行决策（如WFD）的权利。这会使得利益相关者的如下各项权利与利益得到保障，这些权益广泛涵盖了从利益相关者利益的追求到社会的控制（Piet van Poel）：

——社区控制；
——代议制的权利；
——伙伴关系；
——咨询；
——处理社会事务；
——知情；
——获得有效劝导；
——执行。

事实上，许多项目与活动都能达到中度的公共参与——或许通过各种具有不同参与程度的项目的综合而达到。

图1（DEAT，2002）展示了邀请利益相关者参与不同目的时可以采取的不同措施以及利益相关者参与管理的各个步骤。这些措施可以被用在一个决策的不同阶段中，也可以单独使用（同前）。

	持续的公共参与技巧			
劝说	告知	咨询	合作	赋权
没有参与	开始被告知决定	影响决策	为负责任而分享决策	为负责任而假设决策
寻求操控决策的态度	提升意识和支持	决策前的提议 双方谈话 高级决策	共同决策和实施	代表责任
公益广告、新闻特写、会议、编辑工作	小册子、新闻稿、插页、展览、简报	正式听证人，会共座谈会、焦点小组、会议、研讨会、公告	解决问题的研讨会、咨询和协商、共同管理委员会	公共和私人的合作关系

图1 利益相关者参与管理过程的发展（DEAT，2002）

如图 1 所述，利益相关者参与管理的措施有一个从对其进行劝说到真正赋予其参与决策制定的权力的发展过程。到底采取哪种措施要依靠邀请公众参与管理的目的来决定。假如邀请公众参与的目的是就一个特定的决策咨询利益相关者的意见，那么就可采用听证会、公开会议和研讨会等形式。如果这个目的是把利益相关者包括到决策过程中以期望他们在决策推行时能担当部分责任，那么组成一个联合管理委员会则更加合适。

然而需要注意的一点是，虽然利益相关者参与管理的程度可以通过以上措施来进行人为调节，但是管理的结果很大程度上取决于这些参与者自身（DEAT，2002）。因此，在预计公众参与管理的可能结果时，政府不仅仅要考虑其邀请公众参与的目的，还要仔细分析利益相关者的个体条件。而利益相关者个体之间的差异（如时间、金钱和教育程度上的差异）使得对他们的分析更为复杂。这些潜在的不同也对政府提出了为这些不同能力的群众创造"公平竞争环境"的要求。对于能力相对较弱群体，政府可以运用其行政能力来帮助他们更加平等地参与管理（Earle et al.，2010）。本文中的"能力"一词主要指知识和技能，它们可以是个体所拥有的知识和技能，也可以是体现在各政府部门的程序与规则之中的知识和技能。

创建有利公众参与环境的关键因素

每一个决策的制定与执行过程都需要获得不同的反应，公众参与管理的方式也需要随环境而变化（Blokland et al.，2009）。但是，我们在 WFD 与其他相似政策的执行过程中发现了以下几点可以为公众参与创建有利环境的关键因素：

——所有涉及的利益相关者都清楚地知道参与管理的各项规定；

——相关信息与数据得到及时公布，决策与管理的透明度得到保障；

——公众参与的过程中有明确合理的角色分工；

——其他有利于增加决策管理透明度，增强参与人员相互之间以及对政府的信任的举措。

所有涉及的利益相关者在参与初期都清楚地知道参与管理的各项规定对于管理的成功至关重要。政府在邀请公众参与管理的初期就应该明确说明公众参与管理的框架、目标和一些谈判细节（以及不需要谈判决定的部分）。这可以有效防止可能产生的误会与矛盾，有助于顺利决策，同时也降低了决策运行过程中可能遇到的风险（Lange，2008）。

相关的信息与数据得到及时公布，决策与管理的透明度得到保障这也有助于公众对政府信心的建立。害怕过早发布一些"未经证实的消息"是很多行政机关常犯的错误。信息共享是建立政府与公众之间信任的基础，政府与公众互信也是公众参与政府管理的前提（Lange，2008）。

公众参与的过程中有明确合理的角色分工也是为公众参与创造有利环境的关键因素之一。图 2 阐述了公众参与过程中可能出现的一些阶段（Lange，2008）。除了建立清楚的规则、提供必要的信息，政府还应该对参与管理的利益相关者建立明确的角色分工，决定管理中的领导者及合适的公众参与方式。在角色分工的过程中，对参与者数量与类型和对组织者的资源与文化背景的准确预测十分重要（Earle et al.，2010）。这显然也是公众参与管理的一项重要前提条件。

其他有力的举措都有一个相同点，即它们都增加管理决策的透明度，增强政府与公众之间的互信，减少矛盾与分歧。

通过欧洲 WFD 的实践经验，我们还可以总结出：政府可以通过对受影响人群的调查、听证和提供的大量信息来使得公众接受一项政策。在政策的早期计划阶段就引入公众管理就目前来说比较少见，但在未来有被大量使用的可能性。综上所述，公众参与政府管理仍然需要时间探索和实践，尤其是在需要合作的公众之间存在利益摩擦的情况下。公众参与的关键就是不能排斥任何相关的群体。在公众参与管理中各派利益集团达成共识的情况是非常罕见的。尽管如此，公众参与还是可以令政府决策的透明度最大化，从而有效提高公众对决策的接受度（Lange，2008）。

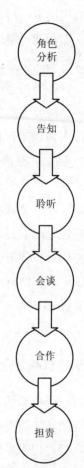

图 2　参与过程所经历的可能的阶段

3 欧盟在利益相关者与公众参与方面的相关政策

政策及其实施与参与

公众参与和公民导向的政策需求已经不是新问题。许多与公众参与相关的措施在20世纪70年代就已经出台并被人们所熟知。这一章将主要介绍一些关于欧洲支持公众参与环境保护政策的框架。利益相关者在环保上的参与并不都是自发的,许多欧洲国家的法律都对这方面进行了规定。这些法律法规可以被看做政府的大体意向,然而部分决策者仍然会在特定情况下对它们持有不同的看法(Grunig et al.,2008)。

1985—2001年制定的环境评价政策的法令致力于对环境进行高水平的保护,力图将环境保护融入各项工程项目的准备工作中,以此减少这些项目对环境的危害。他们希望通过保证环境管理中的公众参与度来提高决策的质量[①]。

——欧盟环境影响评价(Environmental Impact Assessment,EIA)指令(85/337/EEC)已被广泛应用于公共的和私营的项目,这一评价过程要求项目的实施者必须提供有关其项目对环境造成影响的相关资料,必须通知环保部门与公众召开咨询会议。公众可以在法庭上质疑主管机关对工程进程做出的决定。环境影响评价指令在2003年进行修订,使其与《奥胡斯公约》(见下文)中关于公众管理的内容相一致[②]。

——欧盟将战略环境评价(Strategic Environmental Assessment,SEA)指令(2001/42/EC)广泛应用于公共计划与项目。SEA与EIA的评价过程十分相似,但是SEA的要求更加广泛[③]。

对于强调公众参与政策制定有关键性意义的一份文件就是《在环境问题上获得信息、公众参与决策和诉诸司法公约》,也就是前文提到的《奥胡斯公约》。该《公约》中提到公众(包括公民与公民组织)在处理环境问题方面有以下权利:

---

① http://ec.europa.eu/environment/eia/home.htm, accessed 2011-09-15.
② http://ec.europa.eu/environment/eia/eia-legalcontext.htm, accessed 2011-09-15.
③ http://ec.europa.eu/environment/eia/sea-legalcontext.htm, accessed 2011-09-15.

——政府必须赋予每位公民发布其接触到的环境信息的权利,包括环境信息、相关政策措施以及可能被环境因素影响的健康咨询。在实践方面,公民可以在提出信息咨询要求之后的一个月内得到信息反馈。

——《奥胡斯公约》赋予了公众参与环境问题决策的权利。政府必须采取措施使得受影响公众以及非政府组织可以对一些政府决策进行评价(如影响环境的项目企划)。这些评价被纳入政府考虑范围,政府将在做出最终决定时提供相关信息以及说明决定的理由。

——《奥胡斯公约》提出,公民有权利对政府所做出的未经过上文提出的公众参与程序或是不符环境保护法的决策提出质疑(UNECE,1998)。在《奥胡斯公约》之后,欧盟委员会已经开始逐步将其中提到的原则纳入立法范围中去,如欧盟的《欧盟水框架指令》。

欧盟委员会与欧洲议会在2003年5月26日的决议2003/35/EC中,也有关于环境问题起草相关计划与项目开启公共参与,并对有关公共参与以及与之相关的公平性问题进行了修订。此项决议旨在促进《奥胡斯公约》提出的义务的落实。

欧盟法令的执行与环境教育

欧盟27个成员国之间的立法关系由条约规定,该条约的前身是1957年在罗马修订的《欧洲共同体条约》。目前所使用的条约则是2009年修订的《里斯本条约》。所有的欧盟成员国都必须遵守欧盟基本法律、法规和原则,包括欧盟环境法的一些条款。总体来说,欧盟可以在其范围内直接进行活动,并且可以对有"专属管辖权"范围内的成员提出要求。

在过去几年中,可持续发展的教育发展迅速,欧盟与国际社会都认识到了青年参与和对于建设可持续发展社会的重要性。同时,联合国发起了教育促进可持续发展的十年计划(2005—2014),该计划引发了一场关于青年参与可持续发展的活动热潮。

2006年修订的欧洲可持续发展战略进一步确定了教育在达成可持续发展目标中的重要作用。战略提出对青少年的教育应该包括社会、经济与环境三方面的可持续发展,从而达到自然资源的可持续利用。2008年,哥德堡会议建议以教育促进可持续发展,以及在随后2009年的《波恩宣言》都指导了该战略的实施。

一些欧盟成员国已经在国内将社会可持续发展教育作为对青年学生教育的一项目标进行强调。以瑞典为例，瑞典已经在其小学生课程目标中增加了"为学生成为可持续发展先行者而提供的知识储备，教育学生了解自己的生活方式对环境所产生的影响"一项要求（Skolverket，2011）。

除了强调教育的重要性之外，欧盟也提出就欧盟一些核心问题如可持续利用自然资源等与青年进行对话。"方案 2002/C 168/02 关于欧洲青年领域的合作框架"提出了全欧洲合作带动青年参与公民与社会生活的目标，并且确立了"参与，信息交流，青年志愿活动与更深入了解青年"这四个主题（Council of the European Union，2002）。

很显然，合作学习和进行知识储备对于水资源管理与实施欧盟各项指令有决定性作用。没有知识，利益相关者就无法有效地参与各项管理。当然，知识的传播不应仅仅局限在教室与欧盟的各项活动中，对公众各项能力的培养形式更应该要多样化（Blokland et al.，2009）。协作能力的建设是培养公民各项能力及传播知识的最有效途径，它可以在教育体制如针对水资源管理的硕士课程中进行，也可以在其他教学活动中使用（Earle et al.，2010）。

4　《欧盟水框架指令》

背景与目的

《欧盟水框架指令》（WFD）的发展主要基于日益增长的对于巩固水资源管理政策、整合水资源管理重要性的认识，它包括认识水的质量与储量，区分地表水与地下水，以及保护水生生态系统与流域管理之间相互关系的需要（McNally，2009）。《欧盟水框架指令》需要协调不同的欧盟政策，并设定精确的行动时间表，依据该时间表，所有欧洲水域将在 2015 年进入良好状态（European Commission，2010a）。《欧盟水框架指令》尤其强调了水资源管理政策中需要保证利益相关者参与这一点。

《欧盟水框架指令》（WFD）生效于 2000 年。它集中整合了现有的一些欧盟法规，如硝酸盐指令、地下水指令与城市污水指令。同时它也提供了一个对欧洲所有水资源管理政策的监管框架。这一法律致力于达成以下目的（European Parliament and Council，2000）：

——保护和改善水资源状况，防止水资源状况进一步恶化；

——在水资源保护的前提下促进水资源的可持续利用；

——要通过逐步减少排放，保护并提高水生环境的质量；

——逐步减少地下水污染，并防止其进一步污染；

——将洪水和干旱的影响降低。

《欧盟水框架指令》中有以下主要原则[①]：

——以确保所有的水域在2015年达到"良好状态"为目标；

——对水资源的保护涉及多种类型的水域；

——通过在流域层面制定流域管理计划来管理水资源。在跨国河流的情况下，国家间的合作十分重要；

——确保利益相关者、非政府组织与各地群众团体在水资源管理活动中的积极参与；

——水价政策要以"用者自付"的原则为基础；

——平衡环境与其开发者之间的关系。

《欧盟水框架指令》中对利益相关者参与的规定

利益相关者的参与是《欧盟水框架指令》的基本要素，这可以从很多方面来进行说明。首先，欧洲的政策制定者认为利益相关者参与既有可能增加水资源管理政策的工作效率，也可能更好地响应利益相关者的需求（Grunig et al.，2008）。利益相关者参与可能导致制定的环境标准水平较低，但它能够提高这些环境标准的达到率（Newig，2007）。因为若是没有利益相关者的参与，制定出的环境政策有很大的可能不被公众所接受（Kastens and Newig，2007）。同时，利益相关者的参与也可以加强他们之间的合作（Wolters et al.，2006），并提供让利益相关者及早反映涉及水资源的潜在问题或隐患的机会。

虽然WFD非常注重利益相关者的参与，但是它的实际执行很大程度上依靠

---

① http://www.eea.europa.eu/themes/water/water-management/the-water-framework-directive-structureand-key-principles, accessed 2011-09-09.

成员国和河流流域内的组织。因此，对社会各界的能力建设进行创新与对利益相关者进行必要的教育培训是现阶段的首要任务（Grunig et al.，2008）。

《欧洲水框架指令》描述了三种渐进的公众参与形式：提供信息；咨询；积极参与。实施中应该确保完成前两种方式的情况下鼓励第三种方式发挥作用，见图3（European Communities，2003c）。成员国有责任实现公众参与的过程，且要因地制宜，使其符合不同国家、地区和地方的实际情况。

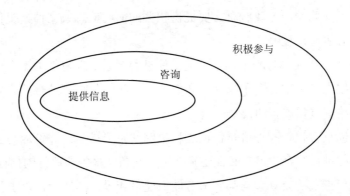

图3　《欧盟水框架指令》中的公众参与方式（European Communities，2003c）

《欧盟水框架指令》中规定，政府在进行公众管理或是召开咨询会议的时候，至少要将利益相关者包括在内。有关信息任何时候都必须对公众开放。相关的指导性文件同时建议在确定参加管理的利益相关者人选之前，政府应该对其素质进行评价，也应该对政府需要重视的一些公众反映的情况进行规定（European Communities，2003c）。

此外，《欧盟水框架指令》还规定了在流域管理规划过程其书面咨询中的三个阶段。到2006年年底，生产计划时间表和工作项目书必须被发布，其中必须包括一份关于将会展开的咨询方式的声明。到2007年年底，水资源管理中一些重要问题的中期综述必须公布。到2008年年底，流域管理计划的草案也要予以公布。成员国必须给其民众至少6个月的时间对这些文件进行评价。《欧盟水框架指令》没有对其他公众参与管理的方式进行规定，但是为了达成其宏大的环境目标，确保该指令的顺利实施，政府还是需要辅助以多样化的公众参与措施（Ridder et al.，2005）。

其他一些对于公众参与管理的建议还包括（European Communities，2003c）：
——进行流域规划时应尽早启动公众参与。
——积极的公众参与对《欧盟水框架指令》实施的各个阶段都十分有益，但是又因为公众参与能提高政策实施的效率，所以在项目的规划期进行公众管理尤为重要。
——在与利益相关者进行咨询时，明确谁在咨询、是关于什么的意见十分重要。在此期间，政府有必要提供相关的简明信息或文件。提供反馈给与会者也非常重要。
——在一个公众参与的过程中，提供信息和背景文件也很关键。一般可以通过线上（网站，电子邮件）和线下（会议）信息相结合的方式将信息传达给利益相关者和公众。
——不断进行报告和评价是保证过程透明的重要措施。评价应与公众参与的过程进行结合。指导文件还确立了一些帮助报告和评价的指标。
——任何形式的公众参与都要求有能力建设和投入的环节，以加强不同利益相关者之间的相互理解。

共同实施战略（Common Implementation Strategy，CIS）

《水框架指令》的实施对其成员国、委员会和欧盟各国以及利益相关者和非政府组织也提出一些技术挑战。为了解决在合作和协调的方式所面临的问题，成员国和挪威与委员会在该指令生效5个月后达成协议，制定了《关于水框架指令的共同战略》（CIS）。根据这一战略，《水框架指令》的实施战略的主要目标是"尽可能地使得《水框架指令》连贯而和谐地实现"，成员国需要的关于有效实施《水框架指令》的能力建设，以及在其实施中需要涉及的利益相关者和社会群体。

共同实施战略一号文件：所面临的挑战

该文件侧重于《欧盟水框架指令》在流域综合管理规划发展的大背景下一些经济要素的实施。它虽然没有最终对公众参与进行阐述，但是其中指出，利益相关者可以通过他们的专业知识进行经济分析，以及通过经济分析的结果，提高政策接受度（European Communities，2003b）。该文件还提供了过去一些利益相关者和公众运用知识进行经济分析的例子。

共同实施战略八号文件：公众参与

该文件重点介绍《欧盟水框架指令》中关于公众参与的第14条的实施。该文件将参与定义为"人们影响计划和工作的过程和结果"。为了保证《水框架指令》的实施，公众参与在规划过程中的任何阶段最好都能够体现。本文件就如何在管理过程的不同阶段落实公众参与提供了具体的建议（European Communities，2003c）。

共同实施战略十一号文件：计划过程

此文件提供了整个计划周期的总体概述，并提供了成功实施建议。它描绘了《水框架指令》对项目规划过程的具体要求，并将"公众信息咨询、各方的积极参与"列为关键，并特别指出这并不仅仅是过程中的重要步骤，而且还是每一个环节的组成部分。报告还强调，在管理过程中需要进行管理协调之间的分离，公众参与应作为规划过程的一个组成部分，认识到进行能力建设的必要，以使得各环节中《欧盟水框架指令》的实际执行最大化（European Communities，2003d）。

非正式的欧盟国家指导性文件

几乎每一个成员国通过本国的努力，在地方和区域执行机构对《水框架指令》的实施中对共同实施战略进行了补充。例子包括英国的技术咨询小组（WFD UK TAG），它为《水框架指令》的各个方面提供了广泛的指导材料，从而帮助负责实施的地方当局更好地契合欧洲的要求。同时针对一些河流流域的特别指导文件也已经出台（见5.3莱茵河的案例）。

和谐协作规划（HarmoniCOP）

在区域层面，和谐协作规划（HarmoniCOP）项目旨在集合有关流域管理规划公众参与的有用信息。该项目得到欧盟和欧洲委员会支持，在2002—2005年运营[①]。和谐协作规划制定了名为《一起学习管理——增加在水资源管理中的参与》的手册，它提供了流域范围内公众参与水资源管理的实际指导。其中涵盖了利益相关者参与过程中的一些实际问题，比如如何组织会议、如何促进社会学习和如何看待信息管理的作用（Ridder et al.，2005）。

---

① http://www.harmonicop.uos.de/index.php, accessed 2011-09-13.

对于利益相关者能力建设的论述

在本文中"能力"一词主要指知识和技能，它们可以是个体所拥有的，也可以是体现在各政府部门的程序与规则之中的。要想公民和其他利益相关者有效参与管理，他们必须需要对相关问题有足够的知识储备，这样他们才能够在必要时对技术问题提出意见，调整自己的行为。对利益相关者在水资源管理知识方面的能力建设和教育方式包括：

——基础是对水体问题形成的原因和及其理解——管理中的因果关系。

——在技术层面上提高利益相关者的自信心以加深其相互理解。

——在政治层面上为共同决策提供健全科学的基础与保障。

公民的能力建设可以在儿童和学生的正规教育体系中进行。与此同时，它还可以通过其他方式开展，如提高认识、研讨会、培训课程和其他各种方法。通过提高这些利益相关者的能力建设及其个人意识与知识，公民与政府间的不平衡都会减小，从而有助于促进有效公众参与的发展。

根据《欧盟水框架指令》对于计划过程的指导文件，这一框架要求"需要最大限度地提高所有利益相关者对《欧盟水框架指令》在实际执行中的能力"。它特别说明："一个能力建设计划总要素可以包括提高公众意识（比如大力支持保护了流域管理目标）、使公民'知其所以然'（例如通过流域管理者之间的经验交流）和各种正规的培训（例如专门的监测技术培训）。但是，各国、各流域由于社会经济条件、水资源管理问题的不同，其确切的需求会有所不同。"相关需求有以下几个方面：

——对于经济部门和非政府组织以及官员、规划者和管理者能力建设的需要（从提高认识开始）；

——对于同一流域内的国家和区域分享信息与经验的需求，互联网为这一需求提供了解决契机；

——在《欧盟水框架指令》的实施过程中，需要为每个 RBD 的能力建设活动拨出足够的人力和财政资源。

如何最大限度地发挥能力、提高认识和分享国家和流域之间的信息主要取决于各成员国的决定。不过，为了实践指导文件，已经建立了试点[①]。其中一个试点流域是英国里布尔盆地，该试点的活动为利益相关者参与管理提供了许多宝贵的参照。

那么，如何让利益相关者进行有效的参与？以下建议来自英国里布尔盆地试点流域（European Commission，2003b）：

——将关注的注意力放在利益相关者为什么、什么时候、在哪里以及怎样咨询和参与；

——将磋商进程与决策的具体条件和《欧盟水框架指令》中的过程联系在一起（无论是国家、区域或地方）；

——考虑各种基于咨询基础上的不同层面的决策之间的边界；

——考虑政府和利益相关者的资源约束力，从而开展协商进程。

5 水资源处理框架的法令实施的波罗的海国家案例

5.1 波罗的海

波罗的海是一个半封闭半咸水（大小仅次于黑海，为世界第二）的海域。它的表面积为 415 000 km^2，流域面积又是其四倍——超过 200 条河流汇入波罗的海。它的平均深度比较浅，只有 50 m 左右。此外，波罗的海的特点是它的水层垂直分层，且水循环时间久（循环一次大约需要 30 年的时间）。这些特点使得污染物很容易在波罗的海内累积（Thulin，2009）。同时，波罗的海集水面积的 93% 属于 9 个沿岸国家：丹麦、爱沙尼亚、芬兰、德国、拉脱维亚、立陶宛、波兰、俄罗斯和瑞典（UNEP，2005）。20 世纪 60 年代起，波罗的海生态系统状况的恶化就已经引起了人们注意，人们开始为改善这一情况付出努力。图 4 界定了波罗的海地区的区域范围（http：//www.helcom.fi，2011）。

欧盟实施的波罗的海地区发展战略是波罗的海水域管理的基本依据之一。该战略的创新之处在于运用了欧盟的框架来达成区域间的合作。这种合作能够综合地体现各成员国的意愿，也将各国间的历史因素的影响和所需承担的国际义务纳入考虑

---

①http://ec.europa.eu/environment/water/water-framework/prbs.htm, accessed 2011-09-13.

中来。这突出体现了实现水资源管理不仅仅需要国家之间的参与合作，也需要代表地区中不同利益团体的利益相关者参与和合作。

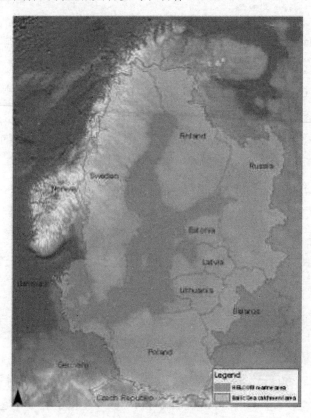

图4　波罗的海地区

利益相关者参与的机制

赫尔辛基委员会波罗的海行动计划自2006年确立以来，在其发展中组织了多个利益相关者会议①。同样地，欧盟的波罗的海地区战略也开展了一些利益相关者的活动。利益相关者会议在不同的国家都已经有所开展，它们的主题往往与战略相关，如专业知识的传递②。

---

①http://www.helcom.fi/BSAP/5thConf2010/en_GB/Fifth_Stakeholder_Conference/, accessed 2011-09-12.
②http://ec.europa.eu/regional_policy/cooperate/baltic/st_events_en.cfm#1, accessed 2011-09-14.

加强利益相关者之间知识共享的一个例子就是瑞典锡姆里斯港的海洋生物中心（Marine Biological Centre，MBC）。该中心的主要活动包括：通知并邀请公众加入波罗的海正在进行的环境保护工作，以及促进知识在该地区国家之间的交流。

意识的影响——波罗的海鳕鱼渔业案例

这是一个提高公众意识的潜在积极影响的很好证明，该案例中，公民通过非官方途径积极参与波罗的海渔业的管理并对其产生影响。

在大多数波罗的海沿岸国家，越来越多的公众认识到了波罗的海存在过度捕捞的现象。媒体在提高公众意识方面作出了很大贡献，这最终促使了公众抵制购买鳕鱼，餐馆停止对鳕鱼的供应以及非政府组织将波罗的海其他一些大量减少的鱼类列入红皮书从而引起人们关注。在波兰，渔民和渔业官员承认其存在严重的过度捕捞现象和频繁发生的非法捕捞现象。波罗的海的管理者们已经认识到了这一现象，并且制定了对波罗的海鳕鱼的保护计划。

如果将重点放在《水框架指令》实施中的利益相关者参与这一方面，该案例主要体现了利益相关者参与的三个进程：认证、网络建立和达成共识。

认证

从波罗的海地区的经验来看，对于利益相关者进行评价应该作为公众参与水资源管理的开始。这种评价包含了一系列基于对利益相关者在流域方面现有知识的提问和互动的步骤。它提供了在公众参与的初始阶段对利益相关者组成深入的分析和重新分析著名水域利益相关者结构的机会。同时，评价也可以对流域管理规划方面相对关键的利益相关者进行认证。BERNET工程在实施《水框架指令》中的认证阶段主要是通过在流域管理中准备一个关于如何处理不同利益相关者需求的大纲（Pedersen，2006）。

建立网络

该进程在波罗的海区域主要通过在流域管理过程中建立经过认证的利益相关者之间的合作来实现。BERNET领导的这一进程主要包括一系列宣传《水框架指令》的会议。这一阶段还包括建立与已认证但还没参加管理的利益相关者的联系。这一进程的经验告诉人们是明确的任务和工作方法使得利益相关者网络的建立更加顺利，这对管理初期的公众参与来说非常必要（Pedersen，2006）。

共识

由于这一进程在波罗的海区域的主要目标与环境问题的共识相关，因此它将利益相关者的参与推入了一个新的阶段。进程中包括了一些流域管理目标的讨论，但是它的中心仍然在于取得利益相关者在达成环境目标应采取方式方面的共识。这一进程的结论是否开放对话是能否达成共识的先决条件，较为实际的目标实施的共识更易达成（Pedersen，2006）。

青年参与

无论是政策文件还是针对波罗的海地区的实践指令都强调了青年参与《水框架指令》实施的重要性，但是很少有项目真正关注这方面。好在教育系统已经认识到了对青年开展参与水资源管理教育的必要性。例如在瑞典，教育大纲规定一些科目如历史中必须要有涵盖波罗的海水域问题的课题（Skolverket，2011）。关于知识共享的教育也在其他一些教学项目中得以体现。ARTWEI项目（Action for the Reinforcement of the Transitional Waters' Environmental Strategy）就是这样的一个项目，该项目旨在信息共享，根据欧盟的要求提高人们对于《水框架指令》《海洋空间规划框架》和《海岸带综合管理》的认识。该目标将会通过在重点机构之间建立伙伴关系网来实现，政府、公民和不同利益集团之间的跨境合作在其中十分重要。ARTWEI已经进行了一些宣传活动（如摄影比赛）来鼓励市民参与其中。

该案例提供的确保公众参与的经验

——尽管没有什么具体手段保证公众参与，但在地区之间开展加强公众参与举措的经验交流非常重要。

——公众和利益相关者在管理早期的参与是成功的关键，如流域管理计划的成功制定。

——信息公开、咨询和利益相关者参与（包括土地所有者、农业协会、非政府组织和其他人民群众）都是《水资源管理框架法令》在波罗的海水域能够成功实施的要素。

——不同政府和机构之间的合作进一步帮助了该地区人民发现和了解其存在的环境和社会经济问题。

5.2 北海地区

北海是一个物产丰富的地区，它周围密集地环绕着高度工业化、人口稠密的国家。海水的富营养化问题是其陆基污染的主要问题（OSPAR Commission，2000）。除挪威以外，所有在北海沿岸的国家都是欧盟的成员国，而且挪威也是欧洲经济区（EEA）的协议国之一。该组织在2009年融入《欧盟水框架指令》。1989年成立的北海委员会致力于加

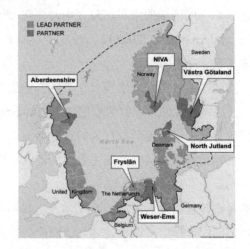

图5　欧盟水框架指令北海区域项目参加者所处地域图

强该地区间的合作，并鼓励在欧盟范围内联合出台公约计划。其他类似的组织也涉及《保护东北大西洋海洋环境公约》。

利益相关者的参与机制

按照《欧盟水框架指令》的要求，利益相关者的参与和教育是过程中很重要的。从事北海地区和当地项目实施的合作伙伴也以不同的方式遵循着该水框架指令（NOLIMP—WFD）项目。

为公众提供信息

在这一框架中整个地区通过交流活动来增加利益相关者获取信息的渠道，例如网站的发展、发送消息通讯以及组织社区展开夜间活动。网站涵盖整个地区，尽管直到项目结束后三年（2009年）这些网络主要是在线通讯。目前NOLIMP—WFD网站只在Aberdeenshire地区工作（www.3deevision.org）。

一些地区也总结出针对特定群体的具体经验，如引导私人森林所有者在河流和湖泊使用更多的森林生态管理方法（Västra Götaland，Sweden）和一本防止农

业用地的流失小册子（Nordjylland，Denmark）（NOLIMP）。

在瑞典，另一个提高公众环境意识的工具是采用 GIS 地图来显示 Gullmarn 峡湾（瑞典）不同区域的水质信息。由此便可知道水质不好的原因并提供使其缓解的方案（Interreg IIIB，2007）。

教育和青年参与

一些地区通过与当地学校合作来加强学生对参与当地环境治理问题重要性的理解，并将这种观念传递给他们的父母（Interreg IIIB，2007）。在苏格兰，广泛的教育资源分散在与当地学校有关的教师培训里。这些资源与当地的实际情况相关，因此它可以很容易地被教师应用到不同的课程中，非正式青年组织（童子军和类似）也同样是其目标之一。

教育资源包括覆盖河流、景观/水系、自然历史和生物多样性以及表达艺术和实地研究的用户指南与课程计划。他们把教材书籍、传单、视频和 cd、海报、地图、识别工具、工作表、河浸渍设备和工艺品等作为补充教材。这些资源被分发到社区里的许多学校，他们可以借给其他青年组织用于教学目的。在德国，当焦点集中在农业污染所引起的地下水污染时，一系列的研讨会便在一个有机农场中举行（Interreg IIIb，2007）。

利益相关者协商和参与

所有参与 NOLIMP-WFD 项目的地区都与涉及的利益相关者有着密切的联系。例如在瑞典，这种联系就促成了 Gullmarn 区域建成水板（Vattenråd）工程，所有的主要利益相关者都会参与到项目处理中来（Interreg IIIB，2007）①。

下面将详细描述两个具体案例（见 Hopmayer and Krozer，2008），用于说明参与本身是如何加强利益相关者的参与能力，以及如何提高他们想要参与管理活动的愿望的。

案例1：社区参与污水处理（苏格兰阿伯丁郡）

第一个是苏格兰阿伯丁郡迪河流域的问题，它需要为 Tarland Bum 亚流域中名为 Tarland 的一个小村庄更换和升级污水处理系统。因为村里的经济增长设置了现有的处理系统，但也导致了流域水质的退化。苏格兰水资源作为负责机构要求取代现有系统以满足监管的需要，它举行了一次公开的社区会议以讨论这些问题。

---

① http://www.vattenorganisationer.se/gullmarnvro/modules.php?name=Content&op=showcontent&id=257.

苏格兰水资源的提议是构造一个强化的污水处理系统（Wastewater treatment works，WWTW）并将污水排入河流。然而，当地社区需要一个继续支持鸟类栖息地并扩大草坪的系统解决方案。当地社区也需要一个比苏格兰提出的污水处理系统容量更大的水系统供村庄使用。由于这些地区性问题的影响，苏格兰水资源项目重新考虑其计划，社区代表、麦克罗伯特房地产（主要的本地地主）、当地议会、英国皇家鸟类保护协会和苏格兰环境保护局参与不断的讨论。所有这些利益相关者合作的基础都建立在分享"启发性的文件"的基础上。为了满足更广泛的利益，修改的原计划与进行的讨论都将会作为会议的背景。

让社区参与其中使得机构理解当地的考虑。接受这些考虑能够有助于对问题形成更好的定义。最终，它促进了一种既满足监管要求又能继续支持当地需求的解决方案。污水处理系统的建设开始于2004年9月，并在2005年8月完成[①]。

关于污水处理系统的讨论也有其他积极的间接影响。举办上述会议的同时，社区夜间活动需要在所有项目的执行地举行，包括苏格兰水资源新的污水处理系统。为了促进农业发展的相关措施，会议也邀请了农民参加。使农民了解到关于村里废水问题的讨论能让当地农民意识到他们不是水质辩论的焦点，而是更广泛过程中的一部分，每个人都必须承担责任。同时，农民自己也认为这个过程是公平的并愿意参与相关的管理活动。

在这种情况下，公众参与本身就能够加强利益相关者的参与能力。它可以提高公众对复杂的水资源问题的理解，理解它是如何影响各环节以及人们共同努力改善这种情况的重要性。这反过来又增强了坚持管理措施的意愿。

案例2：利益相关者参与地下水保护，德国Weser-Ems地区

Weser-Ems地区是一个农业活动占主导地位的大规模农村地区。该地区同样遭受地下水的质量问题。因为农业是主要的污染来源，所以该地区的水资源保护目标是与农民合作来实现地下水的保护。一些必要的措施也已通过法律和自愿协议的形式确定下来。

合作项目的参与者共同努力设立了一个关于测量的项目。在此项目中，每个农民能够选择在基本措施外他们偏好的额外措施来获得额外补偿。以这样的实施农民的协议是必要的，这也是确保项目成功的关键。

---

① http://www.3deevision.org:78/tarland_deliverables.asp.

在开始的时候，供水企业和农民表达了极大的互不信任和激烈的冲突。供水公司指责农业水污染并认为农民态度消极。他们认为保护地下水的最好方式是在水资源保护地区购买土地并停止农业活动。但是，这样会对当地生活和生活方式产生巨大的影响。同时农民们很排斥他们描绘的这种方式。

最后，双方同意合作。但这个过程的沟通主要是通过中介完成的。此外，所有的利益相关者每年都会碰几次面。经过一段时间后，在中介的努力下成功促成了双方都满意的解决方案。供水公司意识到除了停止农业活动以外可以通过其他途径保护地下水，而农民的利益得到尊重，他们就愿意与之合作。工作关系不断改善，不信任的程度也在逐渐降低。会议变得更有效率，讨论气氛也越来越好。

经验教训

——有时中介可以能够改善两方的关系，尤其是当利益相关者处于很不信任的状态时。

——通过更好地定义问题和目标、收集当地的知识和调查替代解决方案等途径，利益相关者能够协商改进规划的质量和有效性。这样可以识别出许多由农民自己采取的措施。

——公众参与取得了规划的广泛的社会认可和支持，也增强了利益相关者的参与意识，让他们能感到自身对于解决问题的重要性。这也有助于加强计划的实施（Hopmayer and Krozer，2008）。

——公众参与是处理农村地区非点源污染（non-point source pollution）的一个关键因素。

——即使相关措施已由法律规定下来，利益相关者也需要接受它们才能使其发挥作用。

——个人的努力和公平是项目实施的重要因素（Hopmayer and Krozer，2008）。

——建立由主要利益相关者成立的正式或非正式的"董事会"有利于在意见不同时达成一致并找到适合的解决方法。

5.3 莱茵河流域

从经济的角度来看莱茵河是西欧最重要的河流。莱茵河是世界上第72条主要河流，它也是唯一连接阿尔卑斯山和北海的河流，其长度为1 230 km。莱茵河分布在 9 个国家中，其流域面积约 185 000 km²[①]。其超过 95%的流域位于瑞士、德国、法国和荷兰。剩下的 5%的盆地位于比利时、卢森堡、奥地利、列支敦士登和意大利。有超过 5 000 万人生活在莱茵河流域。[②]

莱茵河是世界上最密集的内河，它是重要的水资源供应地区。政权的变化会对水资源的安全使用造成严重后果，它包括水运行业、国内使用、农业、自然环境和景观用水的可获得性。[③]

利益相关者的参与机制

ICPR 强调利益相关者参与在莱茵河流域管理中的重要性。"水资源的保护只能在公众知情和参与的条件下才能取得成功。在这一过程中，利益集团发挥着关键作用。无论是由于经济还是自然因素，他们代表着的都是整个水体的压力。"[④]因此，ICPR 试图使用一些利益相关者参与的机制（包括信息共享、听证、会议讨论、角色扮演、研讨会、反馈管理计划的起草以及比赛、展览等）来促使利益相关者的参与。

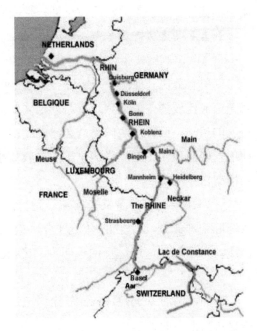

图 6　莱茵河流域

向公众提供信息

ICPR 所使用的信息渠道主要是他们的网站。其数据包括莱茵河的水体数据和相应的盆地地区网络、鱼类迁移的问题、前鲑鱼水域、野生动物走廊及保护区的情况（如 2000 年《Nature》里自然保护区和保护景观、国家公园等），还包括《欧

---

① http://www.newater.info/index.php?pid=1006, accessed 2011-09-07.
② http://www.riverbasin.org/index.cfm?&menuid=104&parentid=87, accessed 2011-09-08.
③ http://www.newater.info/index.php?pid=1006, accessed 2011-09-07 Map source:http://www.rivernet.org/rhin/imgs/maprhin.jpg.
④ http://www.iksr.org/index.php?id=148&L=3&cHash=455fdab52c&no_cache=1&sword_list[0]= public& sword list[1]=participation, accessed 2011-09-09.

盟水框架指令》在莱茵河流域早期已使用的措施（Lange，2008）。

利益相关者协商和参与

自《欧盟水框架指令》规定公众可以参与管理莱茵河流域的过程后，ICPR安排了向所有公民开放的公众听证会。自2006年以来，监管也要求通过公众参与起草莱茵河流域的管理计划。莱茵河地区欢迎各种组织和团体参与国际合作并表达自己的想法。而且，顾问委员会已经在很多地方有过成功先例。现在市民还可以使用互联网和ICPR提供的电子工具收集水体状态的信息并直接参与实现《欧盟水框架指令》。

管理计划的草案是莱茵河流域地区国际协调的结果。所有国家最初都是同意国际管理计划。在2009年上半年，公众参与起草莱茵河流域的管理计划成为可能。为了增加这个过程的透明度，ICPR收到关于管理计划草案的意见时会回复。[1]

此外，例如"Our Dreisam"这样通过照片竞赛和展览等机制提高公众意识的项目还是很多的。

教育和青年参与

在莱茵河流域，一些项目集中在涉及学校和增加青年人对莱茵河的兴趣。水资源管理青年议会已经对莱茵河流域对"团结的欧洲水域"组织提供支持。2008年，在波恩举行了主题为"青年—为莱茵河"[2]的水资源管理青年议会。45个代表不同的国家和地区的个人聚集在一起讨论莱茵河和它的未来。探讨水资源问题的目的是提高青年的问题意识从而加强地方、区域、国家和国际民主和公民权利共同治理的决心。与会者包括35名15~22岁之间的青少年，以及教师、专家、学者和莱茵河流域地区（德国、法国、荷兰和卢森堡）所有居民的代表。议会的目标是给年轻人一个参与国际共同治理莱茵河的机会，以及提供一个在国际层面上交流、学习和发展项目的平台。实现了不同程度的参与，包括信息的交流、讨论和合作以及建立责任感（Lange，2008）。

"图片之流和话语之河"是RhineNet教育项目的一个例子，该项目关注重新发现作为生存环境的河流。该项目始于2005年，整个盆地区域里的学校课程都能

---

[1] http://www.iksr.org/index.php?id=171&L=3, accessed 2011-09-09.
[2] http://www.see-swe.org/youth/eypw# , accessed 2011-09-20.

够参与该项目。

经验教训

——莱茵河流域《欧盟水框架指令》的实施并没有"蓝图",特别应该选择积极参与的公众。这是由于不同文化、地区和机构的情况所导致(Lange,2008)。

——跨界的公众参与可能是最困难组织的。RhineNet 和其他类似的举措表明文化和语言差异是妨碍跨国家公众参与的主要问题。

——大量的机构和不同的管理结构可能会进一步阻碍跨国家参与。

——能够发生影响的水平问题也在实现公众参与上发挥决定性作用。

——将地方利益整合到区域层面是一个困难问题,需要增加参与机构的联合。

Public Participation in Water Management
—EU Practices and Experiences

Stockholm International Water Institute[*]

1 Introduction

A healthy and well-functioning aquatic environment is vital for all life on this planet as it provides us with most fundamental goods and services such as supply of water, flood prevention, water storage and nutrient recycling etc. In the last decades, protection and conservation of our aquatic environment has been drawing an unprecedented level of attention, efforts and resources from a wide range of stakeholders across the world. As one of such positive trends within sustainable water resource management, a growing number of governments, including China and EU, have recognized the importance of encouraging and fostering participation of civil society in policy implementation and other decision making processes at different levels.

The EU has a long experience of developing different mechanisms and concepts for public participation in its water management work. More specifically, the EU Water Framework Directive (WFD) and other directives such as EIA, SEA etc. demonstrate the achievement of such development by making stakeholder and public participation a legal requirement in water governance at different levels. There is now a wide acceptance and shared understanding that such measures not only improve the quality and effectiveness of decision-making in water resources management, but also ensures

---

[*] The Stockholm International Water Institute (SIWI) is a policy institute that generates knowledge and informs decision-making towards water wise policy and sustainable development.

the buy-in and support from local communities. Even the WFD itself is a result of a long multi-stakeholder engagement process in which a wide range of stakeholders were given access to influence in different processes. A number of specific cases from EU demonstrate different mechanisms for public participation, such as local direct involvement, online consultation and stakeholder conferences, from both smaller and larger transboundary river basins (Baltic, Rhine, etc.) .Here is a summary of some of the main lessons learned from the public participation components of the implementation of the WFD:

——Authorities' earlier experience or tradition of promoting public participation makes it easier to initiate a participatory process when implementing policies such as the WFD.

——Early identification and involvement of stakeholders tends to create a more active participation and higher sense of ownership (depending on the authorities' intention of involving the public: consultation or joint decision making).

——Stakeholder analysis is an important first step for the authorities when establishing a framework for implementation of public participation.

——Mandates of different stakeholders involved should be clearly defined and communicated, both internally and externally.

——Management of public participation processes and mechanisms needs to be based on a sufficient level of transparency.

——The resources (time, money, education etc.) that are needed for a sufficient implementation of public participation activities should be invested by both authorities and stakeholders.

——Authorities need to understand that different stakeholders and citizens have very different possibilities to participate due to differences in human and financial resources. Such gaps or imbalances should be seen as indicators of needs for capacity building or other types of supports.

——Considering that information and awareness is crucial for the performance of any public participation mechanism, authorities' selection of communication methods, channels and information packages needs to be diversified in order to reach out to different types of individuals or stakeholders (e.g. youth or adults? Male or female?).

——Public participation should be planned as an integral part of the overall work, already from the start of the overall planning process.

2   The concept of stakeholder and public participation

Stakeholders and Public Participation

The term "stakeholder" usually refers to the part of the population that is affected by a plan, programme or measure, i.e. has a "stake" in it. In the context of the EU Water Framework Directive (WFD), participation refers to stakeholder participation in the aspect of being affected and interested in the issue. More specificically, public participation refers to the possibility for citizens (including marginalized and resource-poor groups) to access information, provide informed, timely and meaningful input and influence decisions at various levels. The degree of stakeholder involvement and participation will be different in various situations and needs to be appropriate to the type of activity planned and should seek to balance the interests of the locally affected population with the possible national economic development benefits. All processes involving stakeholders have one thing in common, and that is that they make use of various participative elements. Some essential characteristics are relevant in all participatory processes and these include:

——Transparency of the participation process as this is the underlying guiding theme for all stakeholder participation processes;

——Flexibility of process which refers to flexibility of process structures to adapt to the particular institutional and political setting, stakeholder suggestions or other input;

——Degree of participative involvement, which can range from purely one-way information to a fully integrated co‐operation or even self‐determination of stakeholders (Grunig et al., 2008).

The core elements of stakeholder participation are both that **the affected public is involved in the decision‐making process** and also important that this **involvement is later on reflected in the decision** itself. That is, participation shall not be used as a disguise for top‐down decision‐making (Grunig et al., 2008). If top-down decision‐making is needed (as indeed is sometimes the case) then this should not be promoted as a stakeholder participation process. As the WFD legal requirement states that the public should be able to participate in its implementation, it is important to motivate why this is important and how it can be achieved.

Public participation in water resources management

Public participation can occur at different stages of a decision-making or policy cycle. However, being able to optimize stakeholder participation from early stages is crucial in water reforms and should be included in governance priorities for closing the implementation gaps and also enabling the citizens to hold the authorities and officials accountable. In turn, level of public participation is an important principle that indicates the performance of different water governance mechanisms. Some of the examples of public participation mechanisms in decision making processes could be such as attending and expressing personal views at town hall meetings, actively contributing to and shaping advisory committees, voting, protesting or carrying out a referendum. In the globle context, growing number of developing countries have started to encourage and support public participation by adopting new laws and policies. Hand in hand with such increasing efforts on improving public participation, recognition of water as a human right is another trend in water governance reforms. In 2010, the United Nations General Assembly adopted water and sanitation as a human right that is essential "for the full enjoyment by all human beings"[①]. In addition to use

---

① United Nations General Assembly, 28 July 2010, Resolution 64/292, UN document A/RES 64/292, United Nations, New York.

of water and its services, access to information and public participation were also critical components included in the human right to water.

However, most government authorities have been struggling with the implementation and integration of these new laws and policies in their daily water management work. In order to maximize citizens' enjoyment of their right to water, government authorities at different levels are in urgent need to enhance their capacity to improve information accessibility and public participation mechanisms at the same time as their delivery of water and its services to the people. In general, the most typical causes of weak implementation of public participation are:

——Overambitious reforms on public participation policies and mechanisms;
——Low level of transparency and openness of government activities (processes) and relevant information sources;
——Local implementers with low competences and low institutional capacity;
——Constrained financial resources at local implementation level;
——Unsupportive authorities or individuals trying to maintain their authority and power to influence;
——The processes and contents of public participation reforms are overly dependent on the donors' interest;
——Absence of local ownership for reform's contents and processes;
——The civil society's lack of sufficient awareness and knowledge about the issues that they would like influence;
——The civil society's lack of capacities to build up and/or maintain sufficient level of efficiency in their joint engagement.

Unfortunately, different socio-economic conditions, institutional system and environmental challenges make it unrealistic to have a single best public participation strategy or mechanism that would fit all contexts. In order to find the most efficient ways to promote a more meaningful and broader public participation, the nature of the participation mechanism needs to be thoroughly analysed and assessed by answering questions such as:

——Are there any legal provisions that guarantee citizen's participation?

——Are there any other mechanisms that enable a sufficient level of participation? Which are they?

——Do we have a comprehensive understanding of who the participants are (By collecting information and data that are disaggregated by gender, age and other characteristics)?

——Is the participation representative for the wider community?

Strategies and key factors that are proven to be important for tackling the aforementioned challenges related to public participation will be further elaborated and discussed in the following chapters, both in more general terms but also based on case studies from the implementation of WFD.

Motives for involving the public in water resources management

In the recent past, the use of participatory elements has been promoted in all areas of public intervention and decision making within the EU. This has not always been the case and one might ask why this has become so important to policy-and decision makers, and to what end. The overall reasons for this emphasis on public participation can be grouped into motives directly linked to the decision makers' direct utility and more altruistic motives generating only indirect utility, as outlined below. Involving stakeholders in policy processes can generate **direct utility** for decision makers as it:

——Encourages acceptance instead of rejection of the decision (increase legitimacy) as a means to improve decision making itself;

——Improves the quality of the implementation and reduce delays;

——Increases transparency of decision making processes;

——Takes advantage of experience and knowledge of stakeholders.

Meanwhile, the **indirect utility** from involving stakeholders can be summarized as it:

——Promotes active citizenship;

——Encourages social learning processes (learn about each other's water awareness);

——Increases public awareness of environmental issues (Grunig et al., 2008).

Strategy for involving public in policy planning

Various views exist in regards to what public participation should be and the answer to this will depend upon who you ask. There is no such thing as a generic "to-do-list" for successful public participation. Oppositely, there is a wide range of methods to be used. The key element of active participation is to enable citizens to voice their opinions on issues that affect them directly. The institutional arrangements to achieve this may be very different. In smaller communities it is easier to involve participants more directly in decision-making as compared to larger entities. However, in the case of water management large areas may be covered, such as in the Danube or Rhine river basins. Such river basin areas typically cut across various administrative boundaries; grouping municipalities, provinces and sovereign states all of which may have different laws and cultures around water resource management. Such a scale creates very different demands on the public participation process.

Stakeholder involvement does not necessarily mean active involvement of all affected parties. The way and level which stakeholders are involved will differ depending on the situation. Overall, the strategies to involve stakeholders range from analysing and informing citizens, to opening up the decision-making process and enabling the public to take on responsibility of the actual decision and/or implementation of e.g. the WFD. As such, different levels of stakeholder participation can be envisaged as shown in the list below, ranging from the manipulation of stakeholder interest up to full community control (Piet van Poel):

——Community control;
——Delegated power;
——Partnership;
——Consultation;
——Conciliation;
——Informing;

——Persuasion;

——Manipulation.

In reality most projects and activities would fall somewhere in the mid-levels of participation—possibly combining several different levels in various parts of the overall programme.

The wide range of techniques to be used, depending on the purpose for involving stakeholders, as well as different possible steps during a paticipatory process are illustrated in Figure 1 below (DEAT, 2002). These can either be used as stages that build on each other in a process or as separate strategies (ibid).

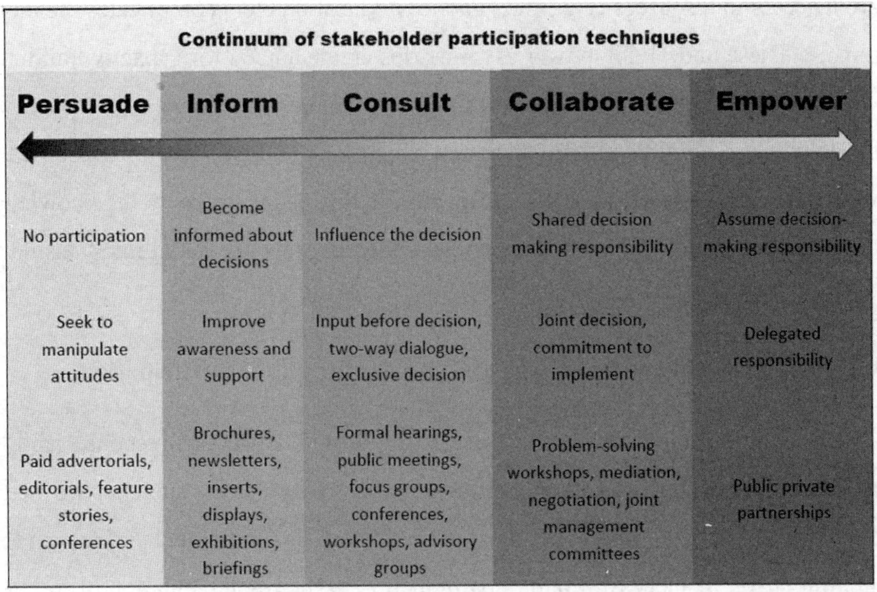

Continuum of stakeholder participation techniques				
Persuade	**Inform**	**Consult**	**Collaborate**	**Empower**
No participation	Become informed about decisions	Influence the decision	Shared decision making responsibility	Assume decision-making responsibility
Seek to manipulate attitudes	Improve awareness and support	Input before decision, two-way dialogue, exclusive decision	Joint decision, commitment to implement	Delegated responsibility
Paid advertorials, editorials, feature stories, conferences	Brochures, newsletters, inserts, displays, exhibitions, briefings	Formal hearings, public meetings, focus groups, conferences, workshops, advisory groups	Problem-solving workshops, mediation, negotiation, joint management committees	Public private partnerships

Figure 1　Continuum of stakeholder participation (DEAT, 2002)

As illustrated in Figure 1, the continuum of stakeholder participation and related techniques range from a more strict persuasive focus to an actual empowerment of stakeholders that are active in decision-making processes. When selecting a technique to support public participation, it is important to consider the purpose of the exercise. For instance, if the aim is to consult with key stakeholders on a certain decision then appropriate techniques of getting people involved would be formal hearings, public meetings, workshops etc. If the aim is to engage stakeholders in the decisionmaking process for them to assume responsibility and commit to the implementation, the establishment of a joint management committee could be suitable.

However, it is important to note that even though stakeholder participation can be planned with a certain purpose, the outcome is highly dependent upon the participants themselves (DEAT, 2002). It is therefore important to consider that the achieved level of participation can depend just as much on the individual stakeholder's resources as on the initial intention from the authority. The fact that different stakeholders have different access to resources (e.g. time, money, education etc.) complicates the picture even more. These underlying power asymmetries create a need for capacity building to "level the playing field". Capacity building programmes for the weaker counterparts could include e.g. administrative capacities which help stakeholders to participate on more equal terms (Earle et al., 2010). Capacity, in this paper, refers to the knowledge, skills and other faculties in individuals or embedded in procedures and rules, inside and around sector organizations and institutions.

Key factors in creating an enabling environment for public participation

Each decision/implementation process requires a different response and the strategies for involving stakeholders ought to be adapted to the specific setting (Blokland et al., 2009). However, a couple of factors shown below have proven to be essential in enabling successful public participation in the implementation of the WFD and similar policies:

——A clear understanding of the "game rules" from all stakeholders involved;
——An early release of data and information to support transparency of the process;
——A suitable set up of the participatory processes with clarified roles;
——Other crucial factors in creating an enabling environment all create transparency and with it trust amongst the participants of the process.

Clear understanding of the "game rules" by the parties at the beginning of a participation process, which is set by the administration or policy makers, is very important to ensure success. The framework in which the public participation would take place, the goals and the details which would, be negotiable (and those which wouldn't be negotiable), must all be clearly stated. This helps to avoid misunderstandings, frustrations and conflicts by increasing the predictability of decision-making and thus reduces risk to the implementing agency (Lange, 2008).

Early release of data and information also belongs to the necessary confidence-building foundations. To fear an early release of "unverified information" is a common mistake made by political and administrative authorities. It is important to start building a relationship of trust from the very beginning of a participatory process and sharing information is a fundamental part of this (Lange, 2008).

Actual set up of the participatory processes is also fundamental in creating an enabling environment. Figure 2 on the left elaborates on possible stages during a participation process (Lange, 2008). In addition to setting the game rules and providing the necessary background information, it is also crucial to define the roles of the various stakeholders, to decide who will lead the process and what the appropriate participatory techniques are. When doing so it is necessary to consider the expected number and type of participants, the cultural context and the resources of the organizers (Earle et al., 2010). This is of course an essential condition for public participation.

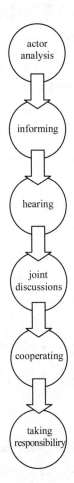

Figure2 Possible stages during a participation process(Lange,2008)

Other crucial factors mentioned have one thing in common-they create transparency and with it trust and an understanding of the arguments and concerns of the respective counterparts.

Other major conclusions that can be drawn, based on experiences from implementing the WFD in Europe, is that public acceptance often can be achieved already by involving the affected population through the means of surveys, hearings and provision of extensive information. More active involvement of the public as early as during the planning phase is still very rare, but will probably become more wide-spread in the future. To sum up, public participation needs time, especially when various players with colliding needs have to be incorporated into the process. In a participation process it is crucial not to exclude any of the relevant parties. Also, one must take into account that participation processes which lead to a consensus of all the parties involved are very rare. Still, participation can produce a maximal transparency of the decision - making process, a fact that significantly affects the acceptance levels for the decided measures (Lange, 2008).

3 Relevant EU policies on stakeholder involvement and public participation

Overview of key policies

Public participation and the demand for more citizen-oriented policies and is nothing new. Many forms of participation are already known and have been tested since the 1970's. This chapter will introduce the legal framework that supports the involvement of the public in environmental policy processes in Europe. Stakeholder participation is not always self - motivated, but it's implemented throughout Europe as it is often legally required. These legal requirements themselves can be seen as expressions of a general political will, but can sometimes also be in contradiction with the decision maker's preferences in a specific situation (Grunig et al., 2008).

The Directives on Environmental Assessment from 1985 and 2001 aimed to provide a high level of protection of the environment and to contribute to the integration of environmental considerations into the preparation of projects, plans and programmes with a view to reduce their environmental impact. They ensure public participation in decision-making and thereby strengthen the quality of decisions[①].

——The Directive (85/337/EEC) on Environmental Impact Assessment (EIA) applies to a wide range of public and private projects. The EIA procedure means that the developer must provide information on the environmental impact of its project activities, the environmental authorities and the public must be informed and consulted, and the decision taken by the competent authority regarding the progress of activities can be challenged by the public before the courts. The EIA Directive was amended in 2003 to align the provisions on public participation with the Aarhus Convention (see below)[②].

——The Directive (2001/42/EC) on Strategic Environmental Assessment (SEA) applies to a wide range of public plans and programmes. The SEA and EIA procedures are quite similar, but the SEA requirements are more extensive[③].

A key document that has contributed to the emphasis on granting the public access to policy processes is the Convention on Access to Information, Public Participation in Decision-Making and Access to Justice in Environmental Matters, often referred to as the "**Aarhus Convention**". The Aarhus Convention establishes a number of rights of the public (individuals and their associations) with regard to the environment as following:

——Everyone should be granted access to environmental information held by government authorities. This can include information about the state of the environment, but also policies or measures taken, or the state of human health and safety where this is affected by the state of the environment. In practice, this means that citizens can obtain this information within one month upon request.

---

① http://ec.europa.eu/environment/eia/home.htm, accessed 2011-09-15.
② http://ec.europa.eu/environment/eia/eia-legalcontext.htm, accessed 2011-09-15.
③ http://ec.europa.eu/environment/eia/sea-legalcontext.htm, accessed 2011-09-15.

——The Aarhus Convention enables citizens to participate in environmental decision-making. Arrangements are to be made by government authorities to enable the public affected and environmental non-governmental organisations to comment on, for example, proposals for projects affecting the environment. The comments from the public are to be taken into due account in decision-making, and information to be provided on the final decisions and the reasons for it.

——The Aarhus Convention establishes that citizens are welcome to review procedures to challenge public decisions that have been made without respecting the two aforementioned rights or environmental law in general (UNECE, 1998). Following the Aarhus Convention, the European Community has begun applying Aarhus-type principles in its legislation, e.g. the EU WFD Directive.

The "Directive 2003/35/EC of the European Parliament and of the Council of 26 May 2003 providing for public participation in respect of the drawing up of certain plans and programmes relating to the environment and amending with regard to public participation and access to justice" (Council Directives 85/337/EEC and 96/61/EC), aims to contribute to the implementation of the obligations arising under the Aarhus Convention.

The implementation of EU legislation and the environment education

The legal relationship between the EU and its 27 member states is defined by treaty, beginning with the European Community Treaty of Rome in 1957. The current treaty is the recently adopted Treaty of Lisbon (2009). All EU nations must adopt the acquis Communitaire, a body of EU laws, principles, and objectives, including certain provisions of EU environmental legislation. As a general rule, the EU may directly regulate an activity throughout the EU and impose binding requirements on member states in areas where the EU has "exclusive competence".

During the last couple of years, the concept of Education for Sustainable

Development has developed rapidly and the importance of youth involvement and education in building a sustainable society has been widely recognized both at the European as well as the global level.

Simultaneously, the United Nations launched the decade of Education for Sustainable Development (2005 - 2014), which triggers an important number of activities in this field aiming to build capacity amongst youth on this issue.

The European Strategy for Sustainable Development, adopted in 2006, further recognizes the important role that formal education and training systems should play in order to achieve the objectives of sustainable development. Education and training should contribute to a sustainable use of natural resources and it ought to address all three axes of sustainable development, namely the social, economic and environmental dimensions. The 2008 Gothenburg Recommendations on Education for Sustainable Development and the subsequent 2009 Bonn Declaration have guided the implementation of this Strategy.

Several member states have emphasized that it is part of the education system's mission to educate students on how society needs to be adapted to contribute to a sustainable development. Sweden, for instance, has added a target to the general curriculum stating that elementary school education should provide pupils with knowledge on the preconditions for a sustainable development as well as an understanding of their own lifestyles' impact on the societal use of natural resources (Skolverket, 2011).

In addition to its emphasis on the importance of education, the EU has also illuminated the benefits of involving youth in a dialogue on core issues of the union including e.g. planning for a sustainable use of natural resources. The "Resolution 2002/C 168/02 regarding the framework of European Cooperation in the Youth Field" aimed to establish a framework of European cooperation to promote the participation of young people in civil life and civil society and endorsed four thematic priorities including participation, information, voluntary activities among young people as well as a greater understanding and knowledge of youth (Council of the European Union,

2002).

Naturally, collective learning and knowledge is critical when deciding how to manage water resources as well as in applying EU directives. Without knowledge, stakeholders would struggle with getting involved in participatory processes in any meaningful way. Of course, knowledge sharing takes place also outside of formal classrooms and EU programmes and capacity building efforts can take on many different shapes (Blokland et al., 2009). Collaborative capacity building exercises are the most efficient ones and these can have a role to play both in the formal educational system, through master programmes focused on the governance of water resources, as well as in other types of training exercises (Earle et al., 2010).

4 EU Water Framework Directive

Background and the aim

The development of the WFD responded to the need to consolidate water policy and to integrate water resources management, recognizing the interrelationships between water quality and quantity, groundwater and surface water, an aquatic ecosystem approach and catchment-based management (McNally, 2009). The WFD requires coordination of different EU policies, and sets out a precise timetable for action, with 2015 as the target date for getting all European waters into good condition (European Commission, 2010a). One of the aspects that the WFD aims to strengthen in water management policies in Europe is the concept of stakeholder participation.

The EU Water Framework Directive (WFD) entered into force in 2000. It integrates several existing pieces of EC regulation, such as the Nitrates Directive, the Groundwater Directive and the Urban Waste Water Directive, and it provides a **regulatory framework for all water policies in Europe**. The WFD aims at following targets (European Parliament and Council, 2000):

——Preventing further deterioration and protecting and enhancing the status of water resources;

——Promoting sustainable water use based on long-term protection of water resources;
——Enhancing protection and improving the aquatic environment through specific measures for the progressive reduction of discharges, emissions and losses of priority substances and the cessation or phasing-out of discharges, emissions and losses of the priority hazardous substances;
——Ensuring the progressive reduction of pollution of groundwater and preventing its further pollution;
——Contributing to mitigating the effects of floods and droughts.

The WFD is based on the following key principles[①]:
——The setting of ambitious objectives to ensure that all waters meet "good status" by 2015;
——The protection of all categories of waters;
——The requirement for waters to be managed at river basin level by formulating a River Basin Management Plan. In the case of trans-boundary water bodies, this needs cooperation between countries;
——Ensuring the active participation of all stakeholders, including NGOs and local communities, in water management activities;
——Requiring water pricing policies based on the "user pays" principle;
——Balancing the interests of the environment with those who depend on it.

Stakeholder participation requirements

Stakeholder participation is an essential element of the WFD, for several reasons. Generally, this is due to the fact that European policy makers expect that stakeholder participation has the potential to both increase water management policies' efficiency as well as allow a higher responsiveness to stakeholders needs (Grunig et al., 2008). Stakeholder participation can result in a lower level of environmental standards, but on

---

① http://www.eea.europa.eu/themes/water/water-management/the-water-framework-directive-structureand-key-principles, accessed 2011-09-09.

the other hand it improves the implementation(Newig, 2007). This is related to the fact that without participation there is a greater risk of a lower acceptance of new policies (Kastens and Newig, 2007). It also tends to improve collaboration between different stakeholders (Wolters et al., 2006) and provide an opportunity for stakeholders to warn earlier of potential problems or hazards related to a water resource.

Though the WFD places emphasis on stakeholder participation, the actual implementation is left to member states and river basin organizations. As a result, the need for innovative approaches to capacity building and education of stakeholders are found primarily at that level (Grunig et al., 2008).

In the context of the WFD, three forms of public participation with an increasing level of involvement are described: **information supply**; **consultation**; and **active involvement**. The first two forms are to be ensured, while the latter should be encouraged as illustrated in Figure 3 (European Communities, 2003c). Member States are responsible for achieving public participation process, and no blueprint exists for public participation. The public participation process should be organized and adapted to national, regional and local circumstances.

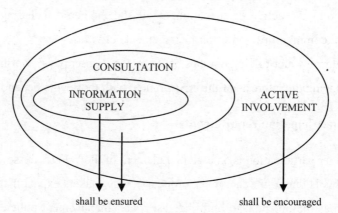

Figure 3 Forms of public participation in the context of the WFD
(European Communities, 2003).

According to the WFD, the involvement of at least stakeholders (i.e. interested parties) is required when dealing with active involvement and also the public when dealing with consultation. Background information should be available at any time for anyone. The guidance document recommends the undertaking of a stakeholder analysis to identify who should be involved in which process and stipulates that reactions of the public need to be collected and considered seriously (European Communities, 2003c).

The WFD moreover requires three rounds of written consultation in the river basin management planning process. By December 2006 a timetable and work programme for the production of the plan had to be published, including a statement of the consultation measures that would be taken. By December 2007 an interim overview of the significant water management issues had to be published. By December 2008 the draft river basin management plan had to be published. Member States needed to allow the public at least six months to comment in writing on those documents. Additional forms of public participation are not required by the WFD but may be needed for reaching its ambitious environmental goals and ensuring its success (Ridder et al., 2005).

Some other key **recommendations for public participation** include (European Communities, 2003c):

——Public participation should start early in the river basin planning.

——Active involvement can be beneficial in all stages of WFD implementation, but it is particularly important when developing the Programme of Measures, since it will most likely improve the effectiveness of the implementation and contribute to delivery in the long term.

——When consulting with stakeholders, it is important to be clear about who is being consulted and about what issues. There is a need to provide concise information or documents, which are subject to the consultation. It is also very important to give feedback to the participants.

——In a public participation process, it is important to make information and background documents available. Usually on-line (website, email) and off-line (meetings) information are combined to inform stakeholders and public.

——Iterative reporting and evaluation are important tools to make the process transparent for participants. Evaluation should be integrated with the public participation process. The guidance document also identifies a number of indicators to help reporting and evaluation.

——Any form of public participation requires capacity building and investment in order to build relations and understanding between different stakeholders.

The Common Implementation Strategy (CIS)

The implementation of the Water Framework Directive raises a number of shared technical challenges for the Member States, the Commission, EEA Countries as well as stakeholders and NGOs. In order to address the challenges in a co-operative and coordinated way, the Member States, Norway and the Commission agreed on a Common Implementation Strategy (CIS) for the Water Framework Directive only five months after the entry into force of the Directive. According to CIS, the aim of the Water Framework Directive was developed to "allow, as far as possible, a coherent and harmonious implementation of the framework directive", recognizing need for capacity building of Member States for an effective implementation of the WFD, as well as the need to involve stakeholders and civil society in the implementation of the WFD.

CIS Guidance Document no. 1: The Implementation Challenge

This document focuses on the implementation of the economic elements of the WFD in the broader context of the development of integrated river basin management plans. Although not primarily concerned with participation, this document points out that the integration of stakeholders into the economic analysis can bring substantial benefits by including their expertise as well as by increasing the acceptance later on of the outcome of the economic analysis (European Communities, 2003b). The document also provides several examples for how the stakeholders' and public concerns and knowledge have been integrated into economic analysis in the past (by making use of questionnaire surveys, public forums and stakeholder analysis).

CIS Guidance Document no. 8: Public Participation in Relation to the Water Framework Directive

This document focuses on the implementation of the Article 14 of the WFD about public participation. The document defines participation as "allowing people to influence the outcome of plans and working processes". Public participation for the implementation of the WFD is recommended at any stage in the planning process. This document gives specific help on how to implement public participation in the different steps of the management process (European Communities, 2003c).

CIS Guidance Document no. 11: Planning Processes

This document provides a general overview of the whole planning cycle and provides recommendations for the successful implementation. It maps out the specific requirements in the WFD with regards to planning process and lists "Information and consultation of the public, active involvement of interested parties" as one of the key components, specifying that this is not only another step in the process but that it should be taken in mind in every component. It also highlights the need to separate between administrative co-ordination (a managerial process) and public participation as an integral part of the planning process and recognizes the need to build capacity to maximize actual implementation of the WFD among all relevant sectors (European Communities, 2003d).

Informal European guidance and national guidance documents

Almost every Member State complements the CIS findings by its own national efforts to local and regional implementing authority's guidance in the WFD implementation. Examples include the United Kingdom Technical Advisory Group (WFD UK TAG) which provides extensive guidance materials on different aspects of the WFD, thus enabling local authorities in charge of implementation to better comply with European requirements. In some cases, separate guidance documents have also been produced for individual river basins (see case study on Rhine River, section 5.3).

Harmonising Collaborative Planning (HarmoniCOP)

At the regional level, the project Harmonising Collaborative Planning (HarmoniCOP) was designed to generate useful information about public participation in river basin management planning. The project was supported by the European Union and the European Commission and operated during the years of 2002 to 2005[①]. HarmoniCOP produced a handbook called "Learning together to manage together – improving participation in water management" and it provides practical guidance regarding public participation in a river basin context. It also covers practical issues of the stakeholder participation process, such as how to organize meetings, how to foster social learning and the role of information management (Ridder et al., 2005).

Capacity building to support stakeholder involvement within the Water Framework Directive

The cooption of capacity is defined here as the knowledge, skills and other faculties in individuals or embedded in procedures and rules, inside and around sector organizations and institutions. For citizens and other stakeholders to participate effectively, they need to have sufficient knowledge about the issues so that they may have an informed opinion on technical issues as well as to modify their behaviour when needed. The capacity building and education of stakeholders on water management knowledge forms:

——A basis for the understanding of the water body in question and for establishing cause-effect relationships in its management.

——An essential element to build confidence and enhance trust among the stakeholders concerned at the technical level.

——A sound and scientific basis for joint decision making at the political level.

Capacity building of citizens can be through the formal education system targeting children and students. In parallel, there can also be capacity building of other stakeholders in society through awareness raising, workshops, training courses and

---

① http://www.harmonicop.uos.de/index.php, accessed 2011-09-13.

other mechanisms. By building capacity among these stakeholders, the imbalances in the level of awareness and knowledge amongst them, and between them and government, will be reduced which forms a foundation for a more effective public participation in the future.

In its WFD Guidance document on planning processes, the Water Framework Directive recognizes *"the need to maximize capacity among all relevant stakeholders for actual implementation of the EU WFD"*. It specifies that: "General elements of a capacity-building programme might include raising public awareness (*e.g. to help secure broad support for the river basin management objectives*), informal transfer of 'know how' (*e.g. through the exchange of experience between river basin managers*), and formal training (*e.g. in specialised monitoring techniques*), both internal and external. However, the exact needs will vary from country to country and from river basin to river basin, inter alia according to different socioeconomic conditions, or the concrete water management issues identified. The relevant aspects are:

——The need to build capacity (starting with awareness raising) among economic sectors and NGOs, as well as among officials, planners and administrators;

——The need to enhance sharing of information and experience between countries and regions sharing river basins, with the internet providing valuable new opportunities;

——The need to allocate adequate human and financial resources for capacity building activities in each RBD as part of overall WFD implementation.

It is up to Member States to determine how to best maximize capacities, raise awareness and share information between countries and river basins. However, a number of pilot river basins were established to provide input in the development of Guidance and other policy documents, by testing the implementation of the WFD[①]. One of these pilot river basins was the Ribble basin in the United Kingdom. A number of recommendations regarding stakeholder involvement emerged from the activities here.

---

① http://ec.europa.eu/environment/water/water-framework/prbs.htm, accessed 2011-09-13.

How to make stakeholder involvement effective? The following recommendations emerged from the pilot river basin activities in the Ribble basin, United Kingdom (European Commission, 2003b):

——Focus on why, when, where and how stakeholders should be consulted and involved;
——Relate the consultation process to the specific decision-making contexts and processes in the WFD (be it national, regional or local);
——Take account of the boundaries these different decision making levels place on the consultation;
——Take account of resource constraints, both for the authorities and the stakeholders to carrying out the consultation process.

As a result of the need to implement the Water Framework Directive more or less simultaneously all across Europe, a number of regional knowledge-sharing partnerships and projects emerged, partly or completely funded by the Interreg initiative under the European Commission's Regional Development Fund. Many of these had a special focus on stakeholder involvement as part of the implementation of the WFD. Some examples of such projects include the "North Sea Regional and Local Implementation of the Water Framework Directive" (NOLIMP-WFD), "RhineNet", and a number of projects in the Baltic Sea region. These projects will be covered in more detail in section 5: Implementation of the WFD-a case compilation.

5 Implementation of stakeholder participation under EU Water Framework Directive – the cases of the Baltic Sea

5.1 The Baltic Sea

The Baltic Sea is a semi-enclosed brackish water area (the second largest in the world after the Black Sea). It has a surface area of about 415 000 km^2, but a catchment area about four times larger-more than 200 rivers drain into the Baltic Sea. It is relatively shallow with an average depth around 50 m. In addition, the Baltic is characterized by a persistent vertical stratification of its water layers and long turn-over times (it takes around 30 years for the Sea to replenish itself). These features are all factors that increase the susceptibility of the Baltic to accumulate pollutants (Thulin, 2009). 93% of the catchment area belongs to nine riparian countries: Denmark, Estonia, Finland, Germany, Latvia, Lithuania, Poland, Russia and Sweden (UNEP, 2005). A deterioration of the state of the Baltic Sea ecosystem was noted in the 1960s and efforts have been made to improve the situation ever since.

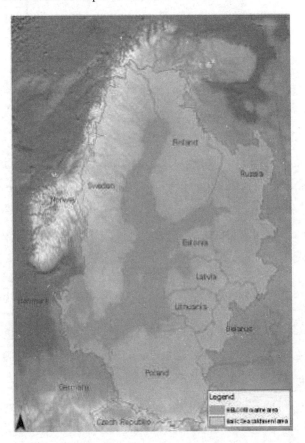

Figure 4　The Baltic Sea region (http://www.helcom.fi, 2011)

The EU Strategy for the Baltic Sea Region is one of the essential elements in management of the Baltic Sea. The Strategy is innovative in its approach of using the

EU structure to unite a "macro‐region" across multiple sectors reflecting the priorities of the member states, the history of the region and international obligations. This accentuates the need for enabling collaboration and participation, not only of different countries but of different stakeholder groups, representing the diverse interests in the region.

Mechanisms for stakeholder involvement

A number of Stakeholder Conferences have been organized during the development and implementation of the HELCOM Baltic Sea Action Plan from 2006 and onwards[①]. Similarly, several stakeholder events have been held under the EU Strategy for the Baltic Sea Region. The stakeholder sessions have been organized in different countries and they have often focused on different themes relevant to the strategy e.g. transfer of knowledge.[②]

One example of how knowledge sharing between stakeholders can be strengthened is the Marine Biological Centre (MBC) in Simrishamn, Sweden. Main activities of the center include: informing and engaging the public in the on-going environmental work in the Baltic as well as promoting the exchange of knowledge between countries in the region.

The impact of public awareness – the case of the Baltic Sea cod fisheries

A good example of the potential positive effects of an increased public awareness, and that citizens can take an active role in influencing processes through informal means is the case of the Baltic Sea fisheries.

Public awareness of the Baltic and its fish and fisheries has grown considerably in most of the Baltic riparian countries during the last years. The media has played an essential role in informing about the situation, resulting in people boycotting cod, restaurants stopping serving it and NGOs red-listing many Baltic Sea fish species. In

---

① http://www.helcom.fi/BSAP/5thConf2010/en_GB/Fifth_Stakeholder_Conference/, accessed 2011-09-12.
② http://ec.europa.eu/regional_policy/cooperate/baltic/st_events_en.cfm#1, accessed 2011-09-14.

Poland, fishermen and fisheries officials admitted to the heavy over-fishing of the total allowable catch and high frequency of illegal fishing. Baltic Sea managers had long been aware of the situation and had already prepared a recovery plan for the Baltic cod.

However, in focusing on stakeholder involvement that has taken place within the frames of the WFD, three overall mechanisms of involving stakeholders have formed during the implementation and these include: identification, networking and acceptance.

Identification

To initiate public participation in the planning process a stakeholder analysis is essential judging from the experiences of the Baltic Sea Region. This analysis entails a series of steps of questioning and interaction based on the available knowledge about the stakeholders at river basin level. It provides an opportunity to conduct a thorough analysis of the stakeholder structure in the initial phase, but also to re‐analyse stakeholder structure in a quite well‐known catchment and identify key stakeholders as regards river basin management planning. The identification phase used in BERNET's work on implementing the WFD in the region was completed by preparing a participation plan outlining how to get through to the different categories and groups of stakeholders during the river basin management planning process (Pedersen, 2006).

Networking

The networking phase of stakeholder cooperation was characterized by the involvement of selected stakeholders in the river basin management planning process for the Baltic Sea Region. The process, led by BERNET, included meetings to promote the content of the WFD e.g. on Article 5 on Analysis. This phase also included the establishment of contact with the identified stakeholders who were not already participating in the planning process. The conclusions of this phase were that clear mandates and work methods make the networking phase easier for all parties and that it is necessary to get stakeholders involved from the beginning of the process (Pedersen, 2006).

Acceptance

The acceptance phase moved stakeholder involvement to the next phase in the case of the Baltic Sea as it aimed at achieving acceptance of the environmental problems. The environmental objectives in the river basin management plans were defined during this process, but the main issue was to gain stakeholder acceptance of the programme of measures needed to attain the environmental objectives. The conclusions of this phase were that an open dialogue is a prerequisite for an acceptance process and that attainable aims make the acceptance process easier (Pedersen, 2006).

Youth involvement

Involving youth in the implementation of the WFD is stressed as important both in policy documents as well as in practical guidance notes for the Baltic Sea region, but few actual projects have been focusing on this aspect. The formal educational system, however, recognizes the importance of teaching on the subject. In Sweden, for instance, the Baltic Sea Region is a mandatory theme that must be included in teaching on e.g. history, according to the general school curriculum (Skolverket, 2011). The educational dimension of knowledge sharing is also evident in several programmes. ARTWEI (Action for the Reinforcement of the Transitional Waters' Environmental Strategy) is one such project, aiming to share information and raise awareness on EU requirements for the Water Framework Directive, Maritime Spatial Planning framework and Integrated Coastal Zone Management. This aim will be achieved by establishing partnership networks of the key institutions which means that cross-border cooperation of local and regional interest groups, citizens and politicians is crucial. ARTWEI has for instance arranged several public awareness campaigns striving to reach and activate the citizens e.g. through arranging photo competitions.

Key lessons learnt for ensuring public participation

——Exchange of experiences among different initiatives throughout the region has been important, despite there is no one specific means of ensuring public participation.

——Early involvement of the public and the stakeholders in the process is vital to ensure its success e.g. in the development of river basin management plans.

——Information, consultation and involvement of all stakeholders, including the landowners, agricultural associations, NGOs and the general public have all been vital aspect of WFD implementation in the Baltic Sea Region.

——Cooperation between different authorities and associations etc. has been necessary for understanding the environmental and socioeconomic problems in the region.

5.2 North Sea region

The North Sea is a biologically rich and productive region. It is surrounded by densely populated and highly industrialized countries. Nutrient-related problems are widespread and occur mainly from landbased pollution (OSPAR Commission, 2000). All countries within the North Sea catchment are members of the European Union, with the exception of Norway. Norway is however a party to the Agreement on the European Economic Area(EEA), which has integrated the WFD since 2009. The North Sea Commission, established in 1989 to enhance partnerships in the region and encourage joint development initiatives and political lobbying at European Union level. Other governing instruments are provided as part of the Convention for the Protection of the Marine Environment of the North-East Atlantic (the "OSPAR" Convention) adopted in 1992。

Mechanisms for stakeholder involvement

The involvement and education of stakeholders were, in accordance with the WFD, important parts of the process. The partners engaged in the North Sea regional and Local Implementation of the Water Framework Directive (NOLIMP-WFD) project approached this in various ways.

Providing information to the general public

All regions conducted communication activities to increase stakeholder access to information through, for example, development of websites, and distribution of newsletters and organization of community evenings. Websites were developed in all regions, but most of them were only kept online until three years after the conclusion of the project (until 2009). Currently, the only accessible NOLIMP-WFD website is the one set up in the Aberdeenshire region (www.3deevision.org).

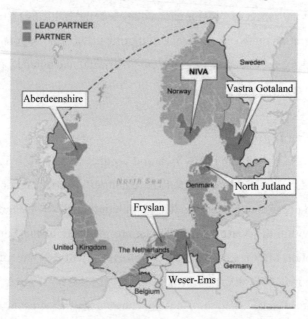

Figure 5   Map of NOLIMP-WFD project partners (North Sea Region, 2011)

Some regions also developed information material targeting specific groups, such as a campaign to induce the private forest owners to use more ecological forest management around rivers and lakes (Västra Götaland, Sweden) and a booklet describing a range of options for preventing the erosion of agricultural land (Nordjylland, Denmark) (NOLIMP).

Another public awareness raising measure tried in Sweden was to present water quality information in an understandable way through producing a GIS map displaying the water quality in the different areas around the Gullmarn fjord (Sweden), the reasons for the lower water quality and what possible solutions that could be offered (Interreg IIIB, 2007).

Education and youth involvement

Several regions worked together with local schools in order to strengthen school children's understanding of and engagement in of the local environment, and through them also reaching their parents(Interreg IIIB, 2007). In Scotland, extensive education resource boxes were developed and distributed to local schools in connection to a number of teacher trainings. The boxes were based on the local curriculum and could easily be applied by teachers in different curriculum, but targeted also more informal youth organizations (scouts and similar).

The education resource boxes contained user's guides with lesson plans covering river heritage, landscape/river systems, natural history/biodiversity, expressive arts and field studies. They were complemented by teaching materials such as books, leaflets, videos and CDs, posters, maps, identification keys, worksheets, river dipping equipment, artefacts and much more. The boxes were distributed to a number of schools in the community that use them in their education and where they can be borrowed by other youth organizations for teaching purposes. In Germany, where the focus lay on the problem of diffuse pollution from agriculture affecting the groundwater sources, a teaching trail was constructed on an organic farm, and a series of seminars were held (Interreg IIIB, 2007).

Stakeholder consultation and involvement

All regions involved in the NOLIM - WFD project had extensive activities to consult and involve stakeholders. In Sweden, for example, this resulted in the establishment of a water board (Vattenråd) for the Gullmarn area, consisting of all major stakeholders to continue working with the catchment even after the conclusion of the project (Interreg IIIB, 2007) [①].

Two specific cases will be described here in more detail (summarized from Hopmayer and Krozer, 2008), illustrating how involvement itself can lead o strengthened capacities of the participating stakeholders and how raised awareness and understanding increased their willingness to participate in management activities.

---

①http://www.vattenorganisationer.se/gullmarnvro/modules.php?name=Content&op=showcontent&id=257.

Case 1: Community involvement in wastewater treatment, Aberdeenshire, Scotland

One of the issues in the Dee river catchment, Aberdeenshire, Scotland, was the need to replace and upgrade a wastewater treatment works (WWTW) serving the village of Tarland in the small subcatchment of the Tarland Bum. The village's growth caused the existing treatment system and this contributed to the degradation of the water quality in the catchment. Scottish Water, the responsible agency, demanded that the existing system be replaced to meet regulatory requirements. It held a public meeting on the issue and presented its proposal to the local community.

Scottish Water's proposal was to construct an intensive WWTW and discharge the effluent into the stream. However, the local community wanted a solution that continued to support a bird habitat, which had been created by the existing grass plots system. The local community also wanted a WWTW with a bigger capacity than the one proposed by Scottish Water, to allow the village to expand. As a result of those local concerns, Scottish Water agreed to rethink its plan, and community representatives, MacRobert Estate (a major local landowner), the local council, the Royal Society for the Protection of Birds and the Scottish Environment Protection Agency, were involved in the ongoing discussions. In order to cooperate, all of these stakeholders were consulted and a shared "document of aspirations" was prepared. This was used as background for later meetings where modifications to the original plan were discussed, in order to satisfy the wider range of interests.

Involving the community enabled the agency to understand local concerns. Accepting these concerns assisted in formulating a better definition of the problem. Ultimately, it promoted a solution that satisfies the regulatory requirements while it continued to support local needs. The construction of the WWTW began in September 2004 and was completed in August 2005[①].

There was another positive but indirect effect of the discussions regarding the WWTW. In parallel to the above-mentioned meetings, community evenings were held on all the projects in the area, including Scottish Water's new WWTW. In order to

---

① http://www.3deevision.org:78/tarland_deliverables.asp.

promote farming-related measures, farmers were also invited to these meetings. Listening to the discussions about the village's waste water problem made local farmers realize they were not the focus of the water quality debate, but rather part of a broader process in which everyone had to take responsibility. As a result, farmers perceived the process as fair and became willing to engage in relevant management activities.

In this case, the public involvement itself strengthened the capacities of the participating stakeholders. It improved their understanding of the complexity of the water quality issue, how it is impacted by various sectors and the importance of everyone working together to improve the situation. This in turn increased the willingness to adhere to management measures.

Case 2: Stakeholders' participation in groundwater protection, Weser-Ems region, Germany

The Weser-Ems region is a large rural area dominated by agricultural activities. The region also suffers from groundwater quality issues. Because agricultural practices are the main polluters, the goal was to work with farmers to promote and implement measures to protect the groundwater. The necessary measures exceeded those already required by law and voluntary agreements to implement additional measures were initiated.

Those involved in the cooperative effort worked together to develop a programme of measures. Each farmer could choose the preferred extra measures in addition to the basic ones and be compensated for the extra costs. The farmers' agreement to implement such measures was essential to ensure the programme's success.

In the beginning of the process, the water supply companies and the farmers expressed a great deal of mutual mistrust and the level of conflict was high. The water companies blamed farming for water pollution and portrayed farmers in a very negative light. They held the opinion that the best way to protect groundwater was to buy land in the water protection area and stop agricultural activities. This, of course, would have had a great impact on local livelihoods and way of life. At the same time, farmers resented the way they were portrayed.

In the end, both sides agreed to cooperate, but communicated mainly through a mediator during the process. In addition, all of the stakeholders met a few times a year. As time passed, and the mediator was succeeding in promoting solutions that satisfied both parties, the water companies realised that there were other options to protect the groundwater instead of stopping agricultural activities, and the farmers agreed to cooperate as long as their interests were respected. The level of mistrust gradually decreased and working relationships improved. The meetings became more productive and the better atmosphere allowed a real discussion.

Key lessons learnt

——A mediator can, at times, lead to improved relationships even when stakeholders share initial high levels of mistrust and are engaged in conflict.

——Stakeholder consultations improved the quality and effectiveness of the plans through better definition of problems and goals, collection of local knowledge, and investigation of alternative solutions. Many of the implemented measures could be identified by the farmers themselves.

——Public participation achieved broader acceptance and support for the plans, increased awareness from stakeholders regarding their contributions to the problem, and a sense of ownership of problems and solutions. This in turn, led to strengthened implementation of the plans (Hopmayer and Krozer, 2008).

——Public participation is a key factor when addressing non‐point source pollution in rural areas.

——Measures need to be accepted by the stakeholders that will be influenced even if they are required by law.

——A personal approach and fairness are important factors in successful implementation (Hopmayer and Krozer, 2008).

——The establishment of "boards" of a more or less formal nature, but consisting of the main stakeholder representatives, proved successful in reaching agreement on difficult decisions and finding a suitable course of action.

5.3 The Rhine Basin

From an economic point of view, the Rhine is the most important river of Western Europe. The Rhine basin covers an area of about 185 000 km^2, distributed between 9 countries. The Rhine is 72nd on the list of the world's principal rivers with a length of 1 230 km. It is the only river connecting the Alps with the North Sea[①]. Just over 95% of the Rhine basin lies in Switzerland, Germany, France and the Netherlands. The remaining 5% of the basin lies in Belgium, Luxembourg, Austria, Liechtenstein and Italy. More than 50 million people live in the Rhine basin[②].

Figure 6　The map of Rhine riverbasin

The river is one of the world's most intensively navigated inland waterways and of major significance for the supply of water to large socioeconomically important areas. Changes in the discharge regime can have severe consequences for safety as well as for the water availability for shipping, industry, domestic use, agriculture, the natural environment and recreational purposes[③].

---

① http://www.newater.info/index.php?pid=1006, accessed 2011-09-07.
② http://www.riverbasin.org/index.cfm?&menuid=104&parentid=87, accessed 2011-09-08.
③ http://www.newater.info/index.php?pid=1006, accessed 2011-09-07 Map source:http://www.rivernet.org/rhin/imgs/maprhin.jpg.

Mechanisms for stakeholder involvement

ICPR emphasizes the importance of stakeholder involvement in the management of the Rhine basin and states that *"Water protection can only be successful, if the public is informed and involved. Interest groups play a key role, since they represent the entire range of pressures on water bodies, no matter whether economic uses ... or nature protection is concerned"* [①]. Accordingly, ICPR attempts to make use of several mechanisms for stakeholder involvement including information sharing, hearings, joint discussions through meetings, role plays and workshops, feedback loops in the drafting of management plans as well as competitions and exhibitions.

Providing information to the public

The main information channel used by ICPR is their website where data regarding the Rhine waterbodies and the corresponding basin district network, fish-migration obstacles, former salmon waters, information about wildlife corridors, protected areas (such as Nature 2000, nature reserves and protected landscapes, national parks etc.) and measurement points in the Rhine were made available early on in the WFD implementation process (Lange, 2008).

Stakeholder consultation and involvement

As the WFD requires that the public can participate in management processes related to the Rhine river basin, ICPR arranges public hearings opened for all citizens. Since 2006, regulations also require public participation in the drafting of management plans for the Rhine Basin. In the Rhine area, associations and organizations are welcome to participate in international working and expert groups and to express their ideas. In addition, advisory boards and fore were created in many places. Today, citizens can also use internet and electronic tools provided by ICPR to gather information on the status of the water bodies and participate directly in the implementation of the WFD.

The draft of the management plan for the international river basin district Rhine is

---

① http://www.iksr.org/index.php?id=148&L=3&cHash=455fdab52c&no_cache=1&sword_list[0]=public&sword_list[1]=participation, accessed 2011-09-09.

the result of international coordination in the basin district. All states initially agree to the international part of the management plan. During the first half of 2009, the public had the possibility of participating in drafting the management plan for the Rhine basin. In order to increase the transparency of this process, replies were provided to the statements that ICPR received on the draft management plan[①].

Moreover, initiatives such as the "Our Dreisam" focusing on raising the general public's awareness through photo competitions and exhibitions are popular mechanisms for increasing stakeholders' interest in the river basin at large.

Education and youth involvement

Within the Rhine basin, several projects have focused on involving schools and increasing the interest amongst youth for the Rhine. The Rhine Basin has arranged Youth Parliaments for Water with the support of the organization "Solidarity Water Europe". In 2008, a Youth Parliament for Water was held in Bonn focusing on the Rhine basin under the theme "The Young – for the Rhine".[②] All in all, 45 individuals representing different countries in the region came together to discuss the Rhine and its future. The aim was to raise youth awareness of water issues and to strengthen local, regional, national and international democracy and citizens' right to co-determination. The participants of the youth parliament consisted of 35 teens between 15 and 22 years of age, as well as teachers, experts and delegates, all residents of the Rhine basin area (Germans, French, Dutch, and Luxembourgians). The goal of the parliament was to give the youngsters an opportunity to get involved in activities associated with the Rhine, to exchange, learn and to develop projects at an international level. Various levels of participation were applied: information, exchange and discussion, cooperation and creating a sense of responsibility (Lange, 2008).

"The Rivers of pictures and streams of words" is one example of a RhineNet educational project which dealt with rediscovering rivers as a living environment. The project took place in 2005 and school classes in the entire basin area were able to

---

① http://www.iksr.org/index.php?id=171&L=3, accessed 2011-09-09.
② http://www.see-swe.org/youth/eypw#, accessed 2011-09-20.

participate.

Key lessons learnt

——There is no "blue print" for the implementation of the WFD in the Rhine basin, especially as far as the way one should choose to actively involve the public. This is due to the diverse cultural, regional and institutional circumstances (Lange, 2008).

——Cross-border public participation may be the most difficult to organize. RhineNet and other similar initiatives show that cultural and primarily language differences pose obstacles for trans - national public participation.

——The large number of institutions and different management structures can further impede trans - national participation.

——The issue of the level at which influence could be exercised, can also play a decisive role during the implementation of public participation.

——Integrating local interests into the regional basin planning level, has proven to be a difficult task in the Rhine basin and requires an increased conjunction of the participating institutions.

Endnote

This report is based upon the 2011 publication Guidebook on Stakeholder Participation – EU Practices and Experiences authored by Anton Earle, Birgitta Lyss Lymer and Josephine Gustafsson, and published by the EU China River Basin Management programme. It has been revised and updated by Frank Zhang in 2015 for the EU-China Environmental Governance Programme.

References

[1] Blokland, M. W., Alaerts, G.J., Kaspersma, J. M. and Hare, M. 2009. Capacity Development forImproved water Management. Taylor and Francis Group, London.

[2] Council of the European Union. 2002. Resolution of the Council and of Representatives of the

Governments of the Member States, meeting within the Council, regarding the framework of European cooperation in the youth field. 2002/C 1 68/02. Brussels. Accessed in the Official Journal of the European Communities at: http://eur-lex.europa.eu/LexUriServ/LexUriServ.do?uri=OJ:C:2002:168:0002:0005:EN:PDF.

[3] DEAT. 2002. Stakeholder Engagement. Integrated Environmental Management, Information Series 3, Department of Environmental Affairs and Tourism (DEAT), Pretoria.

[4] Earle, A., Jägerskog, A. and Öjendal, J. 2010. Transboundary Water Management, Principles and Practice. Earthscan, London.

[5] European Communities. 2003a. Common Implementation Strategy for the Water Framework Directive (2000/60/EC). Strategic Document as agreed by the water directors under Swedish presidency. 2 May 2001.

[6] European Communities. 2003b. Common Implementation Strategy for the Water Framework Directive (2000/60/EC), Guidance Document No. 1: Economics and the Environment – The Implementation Challenge of the Water Framework Directive. Produced by Working Group 2.6-WATECO. Brussels.

[7] European Communities. 2003c. Common Implementation Strategy for the Water Framework Directive (2000/60/EC), Guidance Document No. 8: Public Participation in Relation to the Water Framework Directive. Produced by Working Group 2.9 – Public Participation. Brussels.

[8] European Communities. 2003d. Common Implementation Strategy for the Water Framework Directive (2000/60/EC), Guidance Document No 11 – Planning Processes. Produced by Working Group 2.9 – Planning Processes. Brussels.

[9] European Commission, 2010a. Water Framework Directive. Publications office. Accessed on September 9, 2011 at: http://ec.europa.eu/environment/pubs/pdf/factsheets/water-frameworkdirective.pdf.

[10] European Parliament and Council. 2000. Directive 2000/60/EC of the European Parliament andCouncil of 23 October 2000 establishing a framework for Community action in the field of water policy. Official Journal of the European Communities 327, 22/12/2000.

[11] Grunig, M., Brauer, I. and Görlach, B. 2008. Stakeholder Participation in AquaMoney. AquaMoney.

[12] Publication: Evaluation of Stakeholder Participation in the DG RTD Project.

[13] Hopmayer‐Tokich, S. Krozer, Y. 2008. Public participation in rural area water management: experiences from the North Sea countries in Europe. Water International, 33:2, 243-257.

[14] Interreg IIIB North Sea Programme-NOLIMP. 2007. Final report 2003-2007. North Sea Regional and Local Implementation of the Water Framework Directive.

[15] Kastens, B. and Newig, J. 2007. The Water Framework Directive and Agricultural Nitrate Pollution: Will Great Expectations in Brussels be Dashed in Lower Saxony? European Environment. Vol. 17. pp 231‐46.

[16] Lange, J. 2008. A Guide to Public Participation according to Article 14 of the EC Water Framework Directive (WFD). A RhineNet Project Report. Lavori Druck and Verlag, Freiburg.

[17] McNally, T. 2009. Overview of the EU Water Framework Directive and its implementation in Ireland. Biology and Environment: Proceedings of the Royal Irish Academy 109B, 131–8. DOI:10.3318/BIOE.2009.109.3.131.

[18] Newig, J. 2007. Does public participation in environmental decisions lead to improved environmental quality? Towards an analytical framework. Communication, Cooperation, Participation (International Journal of Sustainability Communication) Vol. 1. Nr. 1. pp 51-71.

[19] NOLIMP. 2008. Local introduction of the EU Water Framework Directive-North Sea countries work together on the water quality. Information flyer produced as part of the project North Sea Regional and Local Implementation of the Water Framework Directive.

[20] OSPAR Commission, 2000. Quality Status Report. OSPAR Commission. London.

[21] Ridder, D. Mostert, E. Wolters, H.A. 2005. Learning together to manage together-Improving participation in water management.

[22] Skolverket. 2011. Läroplan för grundskolan, förskoleklassen och fritidshemmet. Accessed on September 12, 2011 at http: //www.skolverket.se/2.3894/publicerat/2.5006 ? _xurl_=http%3A%2F%2Fwww4.skolverket.se%3A8080%2Fwtpub%2Fws%2Fskolbok%2Fwpubext%2Ftrycksak%2FRecord%3Fk%3D2575.

[23] Thulin, J. 2009. The Recovery and Sustainability of the Baltic Sea Large Marine Ecosystem. In: Sherman, K., Aquarone, M.C. and Adams, S. (Editors). Sustaining the World's Large Marine Ecosystems. Gland, Switzerland: IUCN. Viii-142p. 63-75.

[24] UNECE. 1998. Convention on Access to Information, Public Participation in Decision-Making

and Access to Justice in Environmental Matters. United Nations Economic Commission for Europe. Aarhus.

[25] UNEP. 2005. Lääne, A. Kraav, E. and G, Titova. Baltic Sea. GIWA Regional Assessment 17. University of Kalmar, Kalmar.

[26] Wolters, H., Ridder, D., Mostert E., Otter, H. and Patel, M. 2006. Social Learning in Water Management: Lessons from the HarmoniCOP Project. E‐water 2006. Official publication of the European Water Association (EWA). Accessed on September 15, 2011 at: http://www.harmonicop.uos.de/HarmoniCOPHandbook.pdf.

海牙公众参与环境治理的"波德模式"

王恺　劳伦特·冯·米森*

导言

自第二次世界大战以来，政府机构在欧洲的城市规划和发展中的角色备受关注。除了政府机构，我们也需要更多地关注非政府机构以及各方社会行动者之间的作用。本研究从社会质量的角度对荷兰海牙城市中一个区域的发展加以研究。从这一视角出发，我们将研究聚焦在多个参与者之间的职能、潜力及彼此的关系。其研究范围涵盖从国家到地方的各级政府、地方或区域企业、非营利性组织、非政府组织、社区组织、学术机构等。

在本研究中，我们将从 Politeia、Oikos、Agora、Academia 和 Communication 这 5 个不同的"领域"，来探究各方行动主体在环境保护行动中所起的作用(van der Maesen, 2012)。其中，Politeia、Oikos 和 Agora 都是古希腊语。"Politeia"指政府制度、国家组织或政府的形式。"Oikos"在古希腊语里相当于家庭，是希腊城邦的基本组成单元。"Agora"是古希腊城邦运动、艺术和政治中心的意思，字面意思是"聚集地"。"Academia"指的是学术研究机构。"Communication"指关系、信息的连接和讨论及协商的方式。在此项研究里，这 5 个术语可详细地解释为：

Agora 是由社区、家庭、市民网络（青年、移民、女人、老人、残疾人和成人）组成的领域。

Politeia（i）是由政策制定者决定当地治理的性质的领域，而（ii）指由市政

---

* 王凯，国际社会质量协会研究员；Laurent J.G. van der Maesen，国际社会质量协会办公室主任。本文原文为英文，中译者为候百谦（浙江大学公共管理学院）。

部门（面向城市类别和城市政策领域）实施治理的结果。

Oikos 是由半公共和私人家庭或机构、非营利组织和企业组成的领域。

Academia 是由科学家组成的领域，分析在城市空间的社会趋势和及其矛盾的后果。

Communication 是以交流和技术为基础的交互信息领域。

本文探讨在 Laak North 这个区域（又名 Molenwijk）中开展的公众参与的社会实践活动。本研究运用上述 5 个"领域"来分析在环保议题中所扮演的角色。Molenwijk（官方名 Laakkwartier-Noord）是海牙 Laakkwartier 的一部分（见图 1）。这里是荷兰第三大城市，有 50 多万城区居民（截至 2012 年 11 月 1 日），加上城郊居民人口多达 100 万。Molenwijk 是族群多元和文化多元的城区，有来自摩洛哥、苏里南、土耳其、中国、波兰的移民居住于此。在这里，你可以听到 50 种不同的语言。多元族群的特性增加了社区管理的难度。Laakkwartier 曾经是海牙一个高犯罪率地区，但现在比以前安全了许多。由于本研究的主题倾向于民间团体的社会参与问题，因此实证研究将关注放在 Politeia、Agora、当地的 NGOs（Oikos 的一部分）和 communication 上，对企业和 academia 关注较少。

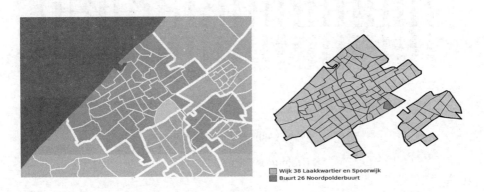

图 1　Laakkwartier 在海牙的位置（左图），Molenwijk（A）在海牙的位置（右图）

荷兰的环境问题

在工业和经济发展的过程中，欧洲各国都曾出现过严重的环境问题，荷兰亦非例外。20世纪60年代至70年代前后,荷兰人遭受着异常严重的环境污染问题。那时,民众发起了一系列行动对抗污染产业。随后,政府、企业和民众采取了更多保护环境的措施。荷兰并非只有风车和郁金香出名,它同时还拥有干净的空气和水质,以及秀美的乡村风光。"然而事实却是另一番景象。根据欧洲环境、气候、自然等指标的相关报告,荷兰的表现非常糟糕。"(译自 Dutch;Natuur en Milieu,2011)。在美国耶鲁大学发表过的一篇关于欧洲 27 个国家的环境绩效指标中,荷兰只排到了第 20 位(见图 2)。同时,结合被采访者提供的信息,空气和水质是目前荷兰面临的最重要的两个环保问题,同时节能问题与应对气候变化的问题也不容忽视。

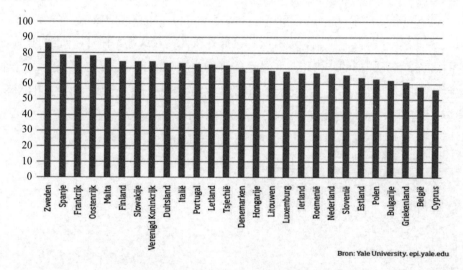

图 2　27 个欧洲国家环境效绩指数排名

(Natuur en Milieu,2011)

环境污染问题在荷兰主要由两方面的原因造成。一方面,由于密集的交通和工业分布,悬浮颗粒、氮氧化物、一氧化碳和臭氧污染着荷兰的空气。交通和工业污染是空气质量每况愈下的主要原因。另一方面,尽管已经很好地缓解了工业导致的水污染,但荷兰的地表水质量相较于欧洲其他国家仍有差距(Natuur en Milieu,2011)。荷兰人有着高效的土地使用率和出众的农耕技术。在相对较小的国土面积内,生产了大量的农业和园艺产品。正因为如此,农药、化肥的密集使用和高密度的畜牧业导致了非点源污染，影响了荷兰水资源的质量（Natuur en Milieu，2011）。

除了空气污染与水污染以外，土壤污染、噪声污染、节约能源以及可再生能源的开发等都是荷兰环境保护的重要课题。相较于荷兰其他城市，海牙显得较为幸运，因为这里没有太多的污染型工业和船舶运输业。相反，被称为"正义之城"的海牙，有许多国家政府部门和国际组织坐落于此。城市里鲜见重工业。对海牙来说，目前，主要的环保课题集中在解决建筑的节能问题以及如何开发和利用更多的可再生能源。文中将会简要介绍荷兰社会的环境治理历史。

"波德模式"与荷兰环境保护

在20世纪60年代之前，荷兰是世界上受污染最为严重的国家之一。因此，政府出台了多项与环境治理相关的法律法规。同时，民众也积极地参与到环保运动中。在荷兰，新项目只有在满足许多复杂条件的情况下，得到政府的许可，才能得以开展和实施，以此确保该项目于环境无害。在这一时期，公共部门，也就是中央和地方各级政府，在环境保护之中占据主导地位（Driessen et al.，2012）。然而，许多企业并未达到政府设定的许可条件。在某些案例中，政府也未严格履行责任，纵容了一些企业破坏规则的行为发生（Hofman，1998）。

为了应对这种情况，独立监察员应运而生，开始参与到环境治理中。独立监察员由各类专家和专业人士组成。尽管他们中有人来自于政府机构，但仍可以确保他们具有独立性，因为他们需要遵守相应的工作准则，独立监察员可根据实际情况直接采取行动。如果认定某家工厂不符合规定，他们有权力下令直接关闭工厂。并且，独立监察员法律地位极高，在没有得到他们的首肯之前，被关闭的工厂无法重新运作。倘若政府和督查员意见不统一，他们可以诉诸公堂，交由法官定夺。然而在此环境治理模式下，无论是否有独立监察员参与其中，企业（即目标群体）都只是充当着被动的接受者。一言以蔽之，这种类型的治理模式是依靠政府制定一系列技术化和可量化的准则来限制企业，以达到环境治理的效果。

20世纪80年代开始，荷兰的环境治理模式有了显著的改变。"人们慢慢意识到，公共部门并非是社会问题的唯一决定性力量。相反，政府、市场与民间团体之间的互动得到了人们越来越多的关注。利益相关方的参与是这次转变的主要特点"（Driessen et al.，2012）。1989年颁布的第一份《国家环境政策计划》（National Environmental Policy Plan，NEPP）及其后续政府文件均体现出了这一特点。1989年《国家环境政策计划》引入目标群体政策的概念："［……］对未来新的目标以及对传统模式信心的丧失，使得我们需要一种新的战略模式。这种模式不同于以往直接制定规则的权威模式，而是能够激发人的主观能动性以及'分担责任'的意愿。以往的政府单方面制定技术指标的模式将由与行业建立密切合作的新模式取而代之"（Bressers and De Bruijn，2005）。根据《国家环境政策计划》，环境治理需要目标群体间的磋商、自我约束以及合作这三者之间的有机结合（Hofman，1998）。

除了目标群体政策，NEPPs还引入了一项新的措施来确保企业与政府之间的合作——政府和企业自发签订协议。企业可以自由地制定计划和战略，已达到长久有效的可持续发展目标，如在财政和技术条件下制定二氧化碳的减排目标。

全球气候的不断变化，加之特殊的地理条件所造成的对气候变化的敏感事实，使荷兰提出的每一项环境政策都无法绕开能源问题。2013年，荷兰政府和多个社会相关机构合力提议了一项有关能源可持续发展的新协议。协议的附录简明地提到了以下共识："2013年9月6日，荷兰向一个环境友好的未来世界又迈出了重要的一步。超过40家不同的组织和机构，包括中央及地方各级政府、企业、联合会、自然及环境保护组织以及相关的民主机构和金融机构，一起通过了一项关于能源可持续发展的协议。协议的核心是一系列与节约能源、清洁技术以及气候政策相关的支撑性准则。而实施这些准则，是为了为荷兰社会提供可持续的清洁能和更多的就业机会，并使荷兰在清洁能源技术市场上抢占先机"（SER，2013）。

在荷兰，有一种非常流行的概念——"波德模式"，可以用来归纳荷兰式的合作和共识构建模式。荷兰使用这种合作和共识构建的模式的传统由来已久，可以在许多历史事件中看到（Schreuder，2001）。在中世纪，荷兰地形的某一最重要的特征便在于圩地，这是一种由低田围起的用于防范洪水的土地。在堤坝上，有一个把圩地里的水抽到运河里的风车。这种系统很难才得以建成，需要花费昂贵的金钱来维护和大量的人力和财力的投入。不论是政府还是其他团体都无法单独完成这一任务。

由于封建势力和天主教会的缺失，以区域为单位组建圩地协会代表大家行使水控制的权利。他们设立了法律和法规，并选择由民主选举选拔出来的官员来管理这些区域。而这些官员是轮岗上任的（Schereuder, 2001）。在"波德模式"下，每一个利益相关方都必须去贡献自己的力量，以实现由大家共同制定的目标。Schreduder（2001）认为，1990年后荷兰的环境治理政策也可以被称作"圩田模式"，因为其同样要求社会各方的参与。以此模式的讨论为背景，我们在下文将会关注"圩田模式"在海牙和Molenwijk的实施具体情况。

政治域：当地政府和政策

海牙环境治理的任务主要有四个，即制定政策、授予许可、提供建议和财政支持。关于政策制定，在2009年海牙市发布了名为《海牙市：成为一个可持续发展的城市》（荷兰语：*Op weg naar een duurzaam Den Haag*）的框架性政策文件。在文件中，海牙围绕六个主题提出了目标和计划。六个主题分别是能源、城市建设、公共空间、交通、国际化和可持续发展。值得注意的是，文件的前言定义了市政当局的角色："市政当局是协调者，也是监督者，监督逾越了个人视野所能了解的行为。……仅仅依靠政府是无法推动海牙的可持续发展。事实上，城市居民和企业必须要展现出自主能动性。这也解释了海牙市政府作为一个协调者，专注于推动、集中、引发和开启变革的原因。"在此，海牙市政府扮演的是一个推动者和协调者的角色。

环保目标和政策由政府和企业共同制定的。诚如前面所说，政策制定需要咨询目标群体的意见，或者将它们直接纳入讨论之中。因为专业人士以及企业在技术层面和目标的可行性方面更具有权威性。另外，如前言所说，政府是不可能单凭自身之力就实现这些目标，企业的努力不可或缺。因此，政策的制定必须参考企业的意见。关于"授予许可"，在前文已经提及。"提供建议"则体现了当地政府在环境治理上的角色转变。自下而上，而非从上至下的模式日益流行。海牙市政府在给予了企业和民众更多的自由的同时，也赋予他们更多的责任。在下文中我们将通过介绍正在实施的节能项目来详述。

在此，我们可以援引海牙的节能项目为例。海牙历史悠久，传统建筑与现代建筑并存。建筑的制冷、加热和通风能源耗费惊人。海牙市政府希望同时改建翻新老建筑，以减低能源消耗。据节能项目实施负责人 Dennis Gudden 所说，20世纪80年代前后，翻新建筑是自上而下的任务。政府告诉人们做什么、怎么做，有时候也会提供相应的补贴。那个时候，人们仅仅是等待政府有所作为。现在，政府采用相反的模式。政府通过提供专业意见和有限的财务支持来帮助民众，民众则自主采取行动进行环境治理。为了鼓励人们自主翻新房屋，政府出资改造了约70套示范房。尽管这是政府自上而下的行为，但这些示范房的主人必须要每年开放至少三天，让民众前来参观学习。在这个示范项目当中，政府不仅投入了资金，更是为民众提供了非常细致的建议。

如今，政府已经意识到自上而下的方法并不高效，它不能启发群众主动的思考。由此带来的后果是，民众依赖政府，而不愿意进行思考或者担当责任。如今，政府正尝试着鼓励民众重新思考，掌握自主权，与邻居沟通交流。民众可以齐心协力做一番事业，也可以通过分享知识和信息，达到减少生活成本的目的。然而，地方政府职能转变带来的成果在各个城市之中存在差异。有些民众非常活跃，他们会鼓励自己的邻居也一同参与到项目中去，而另一些则因为各种理由没有参与其中。

海牙市政府如今面临着以下几个问题：第一，决策出台耗时过长，需要经历谈判、讨论、修正、解释，甚至是法庭判决诸多环节，这些环节无疑都占用许多时间。然而，决策耗时过长的问题在那些重要的、由政府主导的项目之中并不常见。第二，作为协调者和推动者，市政府并不能保证政策目标最终得以贯彻。尽管政策和计划都是由政府、企业和民众共同制定，但其实施的责任却只是落在企业和民众身上。政府并没有得到授权对实施过程进行把关。企业和社区可以遵循自己的方案来完成政府的可持续发展目标，但结果是否令人满意与政府是无关的。第三，尽管市政府、企业和民主社会之间的沟通不可或缺，有时候却是困难重重。不同的工作方式和工作语言意味着有效的沟通并不容易。有效的沟通是基于透明性和公开的，但这两点对于某些参与者来说并不容易。第四，NIMBY（邻避症候群，not in my backyard）是全球性现象，在荷兰亦有存在。邻避症候群会导致某些重要项目的进度拖延，这让当地政府感到非常头痛。第五，当地市政府从上级政府得到的财务支持越来越少，因此对环保项目的支持也日益减少。第六，企业和民间团体提出了自己的想法和计划，希望能够得到政府的批准和更多信息的支持，比如说土地使用方案、土地所有权的信息等。然而，市政府目前并没有为提供这些信息做好准备。

社会组织域：非政府组织和企业

本研究并没有采访特定的 NGO 或者公司的雇员，相反，本研究引用了 Klink 之前在荷兰所做的实证研究成果和对其他组织采访来说明 NGO 在荷兰社会的作用。如前所述，《国家环境政策计划》（NEPP）需要政府机构、企业和民间组织的广泛合作。受此新环境治理政策的影响，如今荷兰政府通过资助众多非政府组织，以保证民间团体部门健康发展（Klink，2007）。荷兰 NGO 在环境议题上表现得非常强势。因此，在荷兰有超过 200 万人加入 NGO 的事实并不令人感到意外（Klink，2007）。

NGO 的两大财政来源分别是政府资金和从民间团体筹集的资金。荷兰大部分的 NGO 接受了来自政府的大量资金支持（Klink，2007）。不仅如此，有些荷兰 NGO 还接受来自欧盟，甚至欧洲以外的经济支持。此外，为了得到学术上的帮助，这些 NGO 与大学、研究机构以及国外的 NGO 保持紧密的联系（Klink，2007）。而且，雇佣全职工作人员，还让这些 NGO 拥有更强组织管理能力。在荷兰，注册成立 NGO 的流程相对简单，只需填写一些简单的文件表格即可。但对 NGO 的管理则相对严格。从 2007 年起，NGO 必须每年提供证明他们的活动和对社区贡献的报告（Klink，2007）。

尽管荷兰的 NGO 从政府手上获得了可观的资金，总的来说他们还是保持了自己的独立性。一些受了政府资助的 NGO 还会周期性地抗议政府实施的项目，因为这些资助并没有背后的政治含义（Klink，2007）。并且，NGO 也并不畏惧影响政府的决策。为了解决问题，NGO 用尽所有可行的方法，包括游说、宣传、抗议、科普和对保护区实施直接管理（Klink，2007）。然而，需要强调的是，NGO 通常更愿意选择与政府合作而非与之作对。因此，与众多利益相关者达成共识是一个常态。NGO 和政府机构通常在协商中有平等的地位。换句话说，平等的参与者通过达到共识的过程来解决分歧，这种方式深深地植根于荷兰文化。

社区域：沟通和个体

本研究对几位来自社区组织（Community-based organizations，CBOs）的工作人员进行了采访。经了解，这些组织分别都有不同的着眼点，有些着眼于环境和可持续发展问题，例如城市农业；其他则关注宽泛的主题，例如健康、体育和福利。在 Molenwijk，部分社区组织积极参与到健康、体育、护理及青年就业等当地事务之中。大部分组织都独立于政府，却可得到政府甚至是来自欧盟的持续资金支持。接下来，本文将着重介绍"健康、福利和体育项目"（即荷兰语中的 Gezondheid, Welzijn en Sport, GWS 项目）以及"微型经济活动合作社"项目（即荷兰语中的 Cooperatief Eigenwijzer）。

健康、福利和体育项目（GWS 项目）

GWS 项目由 Laakkwartier（包括 Molenwijk）的当地政府和若干社区共同组织启动，并得到欧盟提供的资金支持。该项目被认为是海牙城市可持续发展的示范项目，其目的在于从公众参与度、生活质量、居民发展机遇和城市发展新方法等方面提升和改进 Molenwijk 的可持续发展水平。之前的经验显示，尽管 Molenwijk 有着不同的社区组织，它们之间并没有对当地未来如何全面发展进行任何交流（van der Measen et al., 2012）。而 GWS 项目的其中一个任务就是给当地的社区组织、民众和专家提供一个共同为当地的现状和未来出谋划策的平台。该项目在一幢废弃的学校教学楼内成立了"交流中心"来接待当地居民。这所学校（带有一个花园）属于另一个社区组织。该组织征得了当地政府的同意后，暂时将旧的教学楼回收，为 Laakkwartier 多个创新项目提供服务。这样的"交流中心"为当地居民发表意见提供了平台。在每个月第二个星期三的晚上，交流中心会组织一个由当地居民、专家、社区组织、警方代表、政党或者政府代表共同参与的会议。这个定期举办的会议被称为"多彩 Molenwijk"（即荷兰语中的 Kleurrijk Molenwijk）。

微观经济活动合作社

微观经济活动合作社是 Molenwijk 的一个合作组织，也是一种创新的社区商业模式。合作社致力于帮助那些能够让人们发挥才学、财富和其他潜力为社区做贡献的小企业。位于学校花园中的"城市农业"项目正是微观经济活动合作社扶持的众多创新项目之一。学校占地面积 1 400 m^2，其中花园为 700 m^2，由合作社所有。

举例来说，在微型经济活动合作社的帮助下，Elemam 得以运作这个学校的花园。2005 年他搬到海牙并且加入几个合作项目，帮助失业的居民和小商人。2008 年，他被微型经济活动合作社任命管理位于学校花园的城市农业项目。他向当地政府申请了一笔津贴，但是要启动这个项目仍然是困难重重。于是他想到在学校花园举办开放日，邀请居民到花园做一些开心的农活，并将食物卖给到场的每一个人。如今，即便家附近就有大型超市，许多人还是愿意来参观他的花园并购买食物，也在这个花园里，不同族群的人在购物或讨价还价的过程中相互沟通，彼此了解。学校花园每个季节都会举行开放日（除了冬季），而每期开放日都有一个特定的主题以便于孩子们和老人们交流。

Quartier Laak 组织

Quartier Laak 是 2012 年成立的一个社区组织。该组织成立的初衷是保留、美化和发展 Laakkwartier。为了达到这个目标，Quartier Laak 启动和运行一系列对 Laakkwartier 当地居民和企业开放的项目。Quartier Laak 组织总共有 3 位员工。居民不仅要有参与地方管理的权利，也有为地方管理作贡献的义务。这一点已经被整合进了 Quartier Laak 的工作策略中。Quartier Laak 正尝试在 Laakkwartie 举办一系列能够有利于当地的活动，但人们要想参与这些活动必须有所付出。对当地的商人来说，他们可能会被要求以低廉的价格或者免费出售他们的产品，甚至可能为其他项目提供资金扶持。Quartier Laak 的策略被称为"价值创造网络"，因为 Quartier Laak 鼓励在 Laak 生活和工作的居民共享他们的知识、专业技能、产品和服务。

个人角色

不可否认，社区中某些个人对当地政府的决策能够产生一定的影响力，比如律师、医生、政治家和地方议会的成员。他们在能够代表社区与其他机构（如政府）谈判的过程中提供有意义的知识、信息和建议。此外，社区要求生活在当地的政治家在参与政策讨论时能够游说其他议员，为本社区"发声"。这对于居住着专家或者政治家的社区来说无疑是一个好消息。除了有政治影响力的人之外，学者也被动员参与到城市环境的讨论中。例如 Tycho Vermeulen 是瓦格宁根大学的一位科学研究者。他参加了托干海牙的城市农业项目。作为一个科研工作者，他能够提供理论依据，并解释城市农业对于城市和公务员、商人及公民非常重要的原因。尽管 Tycho Vermeulen 拥有专业知识，但他更多的是在商业层面上采取行动。他在人脉建设、筹款、游说议员以及与不同机构交流等方面表现得非常活跃。

以此为经验基础，Tycho 强调项目参与者之间"好的交流"和"内部独立相互依存"是非常重要的，并认为"好的交流"应该建立在"相互依存"和"开放"的基础上。换句话说，项目参与者参加这个项目的原因是他们可以从他人身上获益并且他们能够自然而然地实现信息共享。Jurienne Hollaar 也强调了在一个社区中，活动"发起者"的重要性。在某些情况下，如果当地居民的权利（例如安全和健康）受到威胁，他们可以报警或者成立团体对抗威胁者或者政府。在采取这种措施的时候，需要有具备特定知识和技能的人参与进来，以准备像论据和文书等专业材料。而在另一些情况下，一个区域的居民也许对于当地的发展并不积极，但这也可能取决于地方的社会形象的好坏。

尽管民间团体在荷兰表现得十分活跃，但其在发展过程中经历了不少困难。首先，就像当地政府发现与民间团体交流困难重重一样，民间团体发现与政府交流也不简单。很多精力和时间都被花在了无穷无尽的听证会和谈判上，这显然降低了社区组织和公民个人的参与热情。更重要的是，不同的社区组织之间的交流也并不充分。社区组织在社区的事务上十分活跃，但是它们似乎都各自为政，而不情愿与同类组织进行交流和信息共享。其次，动员居民参加区域管理亦十分困难。但这似乎取决于区域的基本情况。其中不乏所谓的"困难区"，相对来说当地的居民更加贫困，且没有接受过高等教育。他们对于区域的发展并不积极主动。

相比较而言，更加富裕和受过更好教育的居民更为主动。他们更具创造力，有时候甚至会成为发起者。再次，资金缺乏是一个存在的问题。政府机构和企业的经济支持是主要的资金来源，筹款也是一种方法。但如今，无论哪一种方法，如今获得足够的经济支持变得越来越困难。最后，地方自主权的扩大具有一定的挑战性。因为资金的缺乏，以及可能存在的利益相关者之间不充分的交流，扩大地方的自主权有一定的困难。

结论

在海牙，空气污染和水污染的现象依然存在，但因为海牙境内没有那么多重工业，污染与荷兰其他地方相比没那么严重。然而，从结果上看，我们发现当地政府更加关心节能问题。作为一个国际化城市，海牙有很多国际政府组织、非政府组织和环境组织。到2040年时，海牙希望能尽全力变成一个气候中和的城市。而且，荷兰有很大一部分国土面积在海平面以下，海平面上升对于这个国家来说是一个灾难性的威胁。如今，过度消耗能源会导致过量排放温室气体，引起全球暖化和海平面上升。因此，海牙人更关心节能策略。

此外，荷兰的城市的加热设施并不是很先进，且因为气候较冷，它们面临着家庭和办公建筑的节能问题。20世纪60年代和70年代严重污染催生了针对环境保护的政府机构。在环保议题上，政府最终意识到让企业（"目标群体"）和社会民众参与进来的重要性。在21世纪，能源问题和全球暖化成为荷兰首要的环境问题。

同时，随着政府的去集权化，以及中央政府和地方政府减少支持力度，市政府发现自己无法解决当地所有环境问题。所以，企业、NGO和社区要分担更多的责任。另一方面，随着民间团体在荷兰社会的权力不断增大，他们需要在环境保护领域发挥更大的作用。因此，正如很多被访对象所说，一个协同的工作方式是荷兰社会的本质（nature），这可以有术语"波德模式"来解释。

当地政府更像一个协调者或者服务者，它鼓励公民和团体承担自己的责任，用自己的计划和方式来保护当地的环境。当地政府提供有用的咨询和有限的资金给团体，但这并不是团体管理中决定性的力量。在荷兰有许多强大的NGO，政府和公众是他们的资金提供者。即便如此,他们仍是政治上独立的,并不惧怕与政府站在对立面,但大部分情况下它们还是与政府处于合作的关系。

在海牙,社区组织和居民委员会在社团管理上有更多的自治权,他们通常是许多鼓舞人心的活动的发起者。他们承担起使当地变美和更好发展的责任,而非仅仅等待政府采取措施。从针对 Laakkwartier 的研究看,我们发现当地社区组织和一些活跃的发起人并不介意这些举措的大小,因为他们相信"蝴蝶效应"——再微小的变化都会在未来带来很大的改变。而且,除了科普宣传活动,Laakkwartier 的社区组织更喜欢定期与不同的当地组织和居民进行交流。整体而言,荷兰海牙的情况与中国相似,虽然有一些主动的积极分子,但调动当地民众踊跃参加(环境保护活动)还是比较困难的。

The "Polder Model" of Public Participation in Environmental Governance in The Hague

Kai Wang and Laurent J.G. van der Maesen*

Introduction

In the European context, much attention has been laid on the role of "state agencies" in urban planning and development (van der Maesen, 2012) since the Second World War. However, despite the importance of the "state agencies", more attention is required to be paid to agencies in the societal actors. This research concerns the development of a district of the city The Hague, the Netherlands from the perspective of Social Quality. In the social quality approach (van der Maesen, 2012), attention is dedicated to the differentiation between multiple actors, their specific responsibilities, possibilities and their inter-relationship. It concerns local, regional and national authorities, local (municipal) agencies, local and regional companies, non-for-profit organisations, NGOs, community-based organizations, academic agencies, etc.

This study is made on the expemes of public paticatation in Laak North, The Hague and the counties was made in five different main "worlds" of actors, ie., Politeia, Oikos, Agora, Academia and Communication. "Politeia", "Oikos" and "Agora" are three ancient Greek words. "Politeia" nowadays refers to the system of government, state organizations or form of government. "Oikos" is also an ancient Greek word equivalent to household, house or family. It was the basic unit of Greek

---

* Kai Wang, the researcher of the International Association of Social Quality. Laurent J.G. van der Maesen, the director of the International Association of Social Quality.

city-states. "Oikos" is contemporarily used to describe social groups. "Agora" was the central spot of ancient Greek city-states, the centre of athletic, artistic and political life of the city. It literally means "gathering place". "Agora" was the place of goods and information exchange in Greek cities where diverse actors of the society can join. "Academia" refers to the research and scientific institutes. "Communication" refers to relationships, informational connections, and the ways of discussion and negotiation. In this study, the five terms had more specific content.

• **Agora** is the world of communities, families, and networks of citizens (constituted by urban categories of daily life as youth, migrants, women, elderly, handicapped people, adults).

• **Politeia (i)** is the world of those policy-makers determining the nature of local governance and **Politeia (ii)** refers to the municipality departments (oriented on urban categories and urban policy areas), operationalising the results of governance.

• **Oikos** is the world of semi-public and private households or organizations, NGOs and companies (oriented on the manifold of urban policy areas of housing, education, health care, employment, economy, etc).

• **Academia** is the world of scientists, contributing to public and non-public urban policies, analyzing consequences of societal trends and their contradictions in the urban space.

• **Communication** is the world of communicative and informational based connections and techniques, supporting the understanding of a comprehensive and possible sustainable urban development.

As the location of this study, Molenwijk (official name is Laakkwartier-Noord), is part of Laakkwartier, The Hague (see Figure 1). With a population just over 500 000 inhabitants (as of 1 November 2012) and more than one million inhabitants including the suburbs, it is the third largest city of the Netherlands. Molenwijk is a multi-national and multi-cultural neighborhood. Migrants like Moroccan, Surinamese, Turkish, Chinese and Polish people live in this neighborhood. In Molenwijkone can hear 50 different languages. Multi-ethnic character increases the complexity of neighborhood management. Laakkwartier used to be district of The Hague with a high criminal record, but it has now become much safer than before. During the empirical study, more attention was given to the role of Politeia, Agora, local NGOs (part of Oikos) and communication. Enterprises and academia were given less concern since the main theme of the research project is public participation with more attention on civil society.

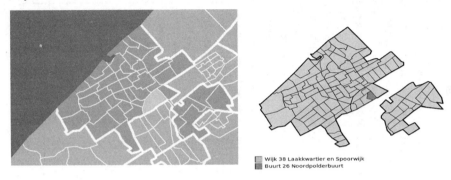

Figure 1　Laakkwartier (green) in The Hague (left) and Molenwijk (red) in The Hague (right)

Environmental problems in the Netherlands

Industrialized countries in Euorpean has experienced many severe environmental problems alongside their industrial and economic development. The Netherlands is not an exception. Dutch people experienced extreme pollutions during 1960s and 1970s, and at that moment, the public raised plenty of movements against polluting industries. Afterwards, many actions were undertaken by the governments, industries and the public to protect the environment. The Netherlands is famous of not only windmills and tulips, but also of its clean air and water, and beautiful countryside landscape. "The reality is actually another case. On the European ladder of important indicators of environment, climate and the nature, the Netherlands performs badly" (translated from Dutch: Natuur en Milieu, 2011). Yale University published an updated ranking of 27 European countries by the Environmental Performance Index. The Netherlands is only ranked 20th (see Figure 2). Combined with the information provided by interview respondents, air quality and water quality are the top two important environmental issues in the Netherlands in general. In addition, energy-saving and climate change adaptation are also important.

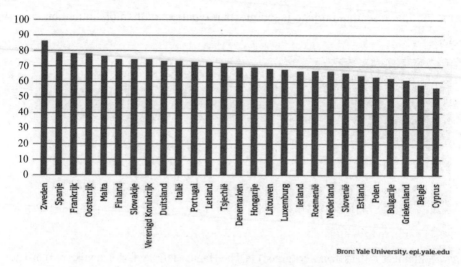

Figure 2 Ranking of 27 European countries by the Environmental Performance Index (Natuur en Milieu, 2011)

In the Netherland, the major causes of pollution due to intensive traffic and industries, the air is polluted by small particles, NO_x, CO and O_3. Traffic and industries are both the causes of air pollution. Water pollution caused by industry has been well solved in the Netherlands. However, the quality of surface water in the Netherlands is not so good in Europe (Natuur en Milieu, 2011). This is because of non-point source pollution caused by intensive usage of manure and fertilizer and high concentration of livestock breeding in Dutch agricultural sector (Natuur en Milieu, 2011). Having a smaller area, the Netherlands produces an extremely high volume of agricultural and horticultural products. This is because of the highly efficient usage of land and advanced cultivation technologies.

Next to air and water pollution, soil pollution, noise pollution, energy-saving and demand of renewable energy are also concerns in environmental issues. The Hague seems to be lucky because there is fewer polluting industries and shipping in its territory. The Hague is the City of Justice, which means there are mainly national governmental agencies and international institutes or organizations located in The Hague. Heavy industries are rare in the city. Currently, the major tasks in terms of environment in The Hague are energy-saving in buildings and discovering renewable energy. In the next section, the brief history of environmental governance of Dutch society will be introduced.

"Polder Model" in Dutch environmental protection

By 1960s, the Netherlands was one of the most polluted countries in the world. The governments issued many regulations on environment protection, and meanwhile, the public also actively involved in protection movements. The public was very enthusiastic in many fields, including environmental governance (Wouter Veening). A new project can only start after obtaining the permit from the government. The permit consists of many complicated conditions, which are aiming at guaranteeing environmental-friendly quality of the potential projects. At that time, the public sector, which refers to the central or local government, was dominant in such environmental governance (Driessen et al., 2012). However, some companies did not comply with the conditions in the permit. In some cases, the government did not strictly fulfil its duties, allowing companies to break the rules (Hofman, 1998).

In response to this situation, Independent Inspectorates were introduced in environmental governance. Independent Inspectorates consisted of experts or professionals. They might be part of governmental agencies, but they were independent because they acted based upon their own laws; they can act when they think they have to act. Independent Inspectorates can close down factories if they think they do not comply to the conditions in the permits. Independent Inspectorates have a strong legal status, if they say no, the factories cannot function. If there is disagreement between the government and Independent Inspectorates, then they may go to court and let the judge decide. No matter whether Independent Inspectorates joined or not, the companies - so-called target group - were passive acceptor in such mode of environmental governance. The norm of this kind of governance was that environmental problems can be solved by the government by setting technical and quantitative regulations (Hofman, 1998).

From 1980s onwards, there was a great shift in Dutch environmental governance. "Governance emerges as a concept that acknowledges that the public sector is not the only controlling actor when it comes to the solution of societal problems. Instead, more attention is given to interaction between actors pertaining to the state, the market and civil society. Stakeholder involvement is one of the main characteristics of this shift" (Driessen et al., 2012). From 1989, the first National Environmental Policy Plan (NEPP) and its successors showed the changes. A so-called "Target Group Policy" was introduced in NEPP-1989: *"[...] New ambitions and the lack of confidence in traditional approaches called for a strategy and style other than the authoritarian style that accompanied the use of direct regulation. The new strategy aims specifically at eliciting private initiative and 'shared responsibility'. Instead of setting technology-forcing standards unilaterally, the approach builds on close collaboration with industry"* (Bressers and De Bruijn, 2005). In the light of NEPP, the principle of environmental governance is dealing with environmental problems through consultation, self-regulation and collaboration among target groups (Hofman, 1998).

Apart from Target Group Policy, NEPPs introduced a new instrument to ensure the collaboration between industries and governments. Covenants are voluntary agreements between businesses and governments. Businesses are free to work out their own plans or strategies to reach the long-term goals of sustainability, for example emission reduction, based on their financial and technical situation.

In the light of climate changes, and the fact that the Netherlands is vulnerable to impacts of climate change, energy issues are related to every environmental policy. In the year of 2013, Dutch government and diverse sectors from society worked out a new agreement for energy-focused sustainable development. The Annex presents a concise summary of the agreement. *"6 September 2013 - The Netherlands has today taken an important step on the way to an environmentally-friendly future. More than forty organizations – including central, regional and local government, employers and unions, nature conservation and environmental organizations, and other civil-society organizations and financial institutions-have endorsed the Energy Agreement for Sustainable Growth. The core feature of the Agreement is a set of broadly supported provisions regarding energy saving, clean technology, and climate policy. Implementing these provisions is intended to result in an affordable and clean energy supply, jobs, and opportunities for the Netherlands in the market for clean technologies"* (SER, 2013).

There is a popular term to label Dutch way of cooperation and consensus-building, called "Polder Model". The tradition of cooperation and consensus building in the Netherlands has been evident in various historical situations (Schreuder, 2001). In the Middle Ages, one of the most important features in the Dutch landscape has been the polder, an area of low-lying land reclaimed from a body of water and protected by dikes against flooding. On the dike stood a windmill that pumped the water from the polder into a canal. The system was difficult to build, expensive to maintain and required sufficient manpower and capital at all times. Individual authorities or entities cannot cope with this task independently.

With the absence of Catholic Church control and feudal power, small rural communities organized territorial units in polder boards representing water control interests. They established rules and regulations and chose democratically elected officials to govern the areas on a rotating basis (Schreuder, 2001). In such "Polder Model", every party has to contribute to reach the goal and act according to the rules made by them collaboratively. Schreuder (2001) argued the environmental policy in the Netherlands since 1990s can be presented as "Polder Model" because it requires support from all segments of society.

In the next section, our attention will be paid to the specific situation in The Hague and Molenwijk.

Politeia: local government and policies

In general, four main tasks of The Hague Municipality in environmental governance are: policy-making, permit-granting, advice-providing and finance-supporting. In terms of policy-making, in 2009, the Municipality of The Hague published its framework policy document named *The Hague: Becoming a sustainable city* (in Dutch: *Op weg naar een duurzaam Den Haag*). The document presents the general objectives, achievements and future plans in terms of six main themes: energy, urban design, public space, mobility, international city, sustainable municipal organization. It is noticeable that in the foreword of this document, the role of the municipality is: *"The municipality has a natural role as coordinator and also oversees actions that surpass the individual's perspective for action. [...] we cannot boost The Hague's sustainability on our own. As a matter of fact, the city in particular - its residents and businesses - will have to take the initiative. That is why the Municipality of The Hague, in its role as coordinator, will focus on facilitating, smart grouping, inspiring and initiating change."* We can see that the Municipality of The Hague acts as a facilitator or coordinator.

Those policies or objectives were collaboratively made by government and businesses. As mentioned above, the target groups should be involved or consulted during policy-making. Professionals and businesses are more acknowledgeable to the technical aspect and feasibility of those objectives. Moreover, as the foreword says, the municipality cannot achieve those objectives only by its own efforts; the efforts from businesses are indispensible. Therefore, businesses' opinion must be enclosed in the policy-making process. In terms of permit-granting, the previous text introduced the responsibility of the government. Advice-providing can represent the shifting role of local government in environmental governance. Instead of a "top-down" process, "bottom-up" initiatives are getting more and more popular. The municipal government is giving more freedom, as well as responsibilities, to residents and businesses. A building energy-saving project was chosen as a representative case during exploration, which will be introduced below.

Here we can take Energy-saving projects as an example of illustration. The Hague is a historic city, which has lots of old-fashioned buildings, and modern buildings as well. The level of energy consumption in buildings for cooling, heating and ventilating is remarkable. The municipal government of Hague wants to renovate both old and modern buildings to reduce energy consumption. According to Dennis Gudden, who is employed by the municipality to conduct some energy-saving projects, in 1980s, the process of building renovation was top-down. Government told people what and how to do, and it sometimes offered subsidies for this. At that time, people just waited for the government to act. Now, the government arranges this in another way. In general, the government helps people by providing professional advices and limited financial support, and people take the initiatives themselves. But in order to encourage people to renovate their houses, the government invests money to around 70 houses to do the renovation as a demonstration project. The demonstration project is rather top-down way. These 70 houses have to open to the public at least three days per year to show the people what the outcome is. In the demonstration project, the government not only invests a lot of money, but also provides very detailed instructions.

The government realizes the top-down process does not work efficiently nowadays. Top-down processes do not trigger the thinking process of people. They rely on the government but do not think and take responsibility themselves. Nowadays, the government tries encourage people to think, take initiatives and communicate with neighbours. People can do something together; they can share knowledge, information and reduce the costs. However, the outcomes of those projects differ in different places. Some residents are really active, and they encourage their neighbors to join together.

The Municipality of The Hague is confronted with some problems in this regard firstly, the decision-making process is really time-consuming. Negotiation, discussion, modification, explanation and even judging in the court take much time. However, the long decision-making process is not commonly seen in some "large" projects, which are rather top-down organized. Secondly, as coordinator and facilitator, the municipality cannot make sure the policy objective would be achieved in the end.

Although the plan and policies are made collaboratively by the government, businesses and sometimes citizens, the implementation is the responsibility of businesses and citizens. The municipality has no enforcement power to control this; businesses and communities have their own plans to achieve the agreements with government of sustainable objectives. However, whether the outcomes are satisfactory is out of government's hand. Thirdly, appropriate communication between municipality, businesses and civil society is indispensible, but at the same time, it is difficult. Diverse working styles and working languages make good communication hard to maintain. Good communication is based on transparency and openness, but it is not always easy to some involved actors. Fourthly, the NIMBY (not in my backyard) attitude is a global phenomenon. The Netherlands is not an exception. NIMBY can delay the necessary developments from a certain point of view, which is a headache to local government. Fifthly, municipal governments get fewer and fewer financial support from upper governments, which make it more difficult to municipalities to invest in local development projects. Sixthly, many enterprises and civil society organizations have worked out their plans and asked the municipality for permits and further information, such as land-use plans, land ownership information and so forth. However, the municipality is not prepared yet to provide this information.

Oikos: NGOs and enterprises

In the exploration, the researcher did not interview person from specific NGO's or companies. Instead, information from Klink's empirical studies in the Netherlands, and interview respondents from other organizations were quoted to present the role of NGOs in Dutch society. As mentioned before, the National Environmental Policy Plans (NEPP) calls for extensive collaboration and cooperation between governmental agencies, businesses and civil society. Influenced by this new approach of environmental governance, Dutch government now ensures a healthy civil society sector by subsidizing and giving grants to numerous NGOs (Klink, 2007). NGOs in the Netherlands are very robust in environmental issues. It is no surprise that more than two million Dutch people are members of NGOs (Klink, 2007).

Two main financial support resources to NGOs are governmental funds and fundraising from civil society. The majority of NGOs in the Netherlands receive a substantial amount of funding from the government (Klink, 2007). Moreover, some Dutch NGOs can get financial support from European Union agencies or even outside Europe (Klink, 2007). In terms of knowledge support, Dutch NGOs are in close contact with universities, institutes and international NGOs (Klink, 2007; Wouter Veening; Henk Heijkers). Furthermore, they have full-time workers, which make them have stronger capacity in organizational management. The registration process is relatively easy in the Netherlands. Only a few paperwork is needed. Since 2007, NGOs must submit reports documenting their activities and contribution to the community (Klink, 2007).

Even though getting considerable funds from government, Dutch NGOs maintain their independence in general. Some NGOs regularly protest governmental projects even though they receive funding from the government. This is because the fund is not politically motivated (Klink, 2007). NGOs are not afraid to influence the government. They use all methods available to them to address issues, including lobbying, awareness activities, protested activities, educational outreach and direct management of protected areas (Klink, 2007). However, it should be stressed that more often than not, NGOs favour collaboration with government rather than acts of protest. Hence, consensus-building with numerous stakeholders is the norm. NGOs and governmental agencies usually have equal status in negotiation. In other words, disagreements are dealt with in the process of consensus among equal participants, which is deeply rooted in Dutch culture.

Agora: Communities and individuals

During the empirical exploration, several persons from community-based organizations (CBOs) were interviewed. These organizations have diverse profiles. Some of them focus on environmental or sustainability issues, such as urban farming; others concern broader and general topics, such as health, sports and well-being. In

Molenwijk, there are some CBOs that are taking care of many other issues in the neighborhood, such as health, sports, nursery, youth employment, etc. Most of these CBOs are independent from governmental bodies, but they can receive substantial funding from the government, and even from European Union. Here, the paper will introduce the Health, Welfare and Sport Programme (in Dutch: Gezondheid, Welzijn en Sport, GWS programme and Cooperation for Micro-economic Activities (its Dutch name is Cooperatief Eigenwijzer).

Health, Welfare and Sport Programme (GWS programme)

The GWS program is undertaken by several CBOs in Laakkwartier (including Molenwijk) and the municipality of The Hague. This program is funded by the European Union. The GWS program is regarded as "demonstration project" of sustainable urban development in The Hague. It aims to improve the sustainability of Molenwijk in terms of participation, life quality, and development opportunities for residents and new knowledge of urban development. The previous experience shows that although there are diverse CBOs in Molenwijk, they do not have forms of communication with which to develop a comprehensive vision about the future of the neighborhood (van der Measen et al., 2012).

One of the tasks of the GWS program is to provide a platform to brace local CBOs, residents and professional to join together to discuss the current situation and the future of their neighborhood. The program set up a "communication centre", which locates in a reused school building, to welcome local people. The school (with a school garden) is owned by another CBO, which asked the approval from the municipality to temporarily reuse the old school building for different innovative projects in Laakkwartier; and the municipality approved. The meeting place provides a stage to local people to raise their voice. In the evening of every second Wednesday of the month, the communication centre organizes a meeting that welcomes local residents, professionals, CBOs and representatives from police, political parties or government. The fix-planned meeting is called "Colourful Molenwijk" (in Dutch: Kleurrijk Molenwijk).

Cooperation for Micro-economic Activities

The cooperation for Micro-economic Activities is a cooperation organization in Molenwijk and a new type of neighborhood business. It helps small businesses, in which people can contribute their knowledge, money and other potential to help the development of the neighborhood. Urban farming in the school garden is one of their initiatives. The school garden is owned by the Cooperation for Micro-economic Activities. The school occupies around 1 400 m^2 of land, including a 700 m^2 garden.

Taking Elemam case as an example, with the help from the Cooperation for Micro-economic Activities, Elemam operates the school garden. In 2005, he moved to The Hague and joined some cooperative projects to help people without a job and small businessmen.In 2008, he was appointed by the Cooperation for Micro-economic Activities to undertake the urban farming project in the school garden. Elemam asked the municipality to offer some subsidies; it was difficult to get it started. The school garden organizes an open-day to invite people to the garden do whatever they want and sell food to everyone. Elemam also sold his food. It is called "mouth-to-mouth" promotion. Nowadays, many people come to visit his garden and buy some food even though there is a big supermarket nearby.

In addition to business, Elemam has another motivation to run this school garden. He wants to make this garden as one of meeting places for local residents where they can meet and talk with each other. Molenwijk is a multi-ethnic neighborhood. In the garden, people from different ethnic groups can have conversations during purchasing or bargaining and get to know each other. During the open-day, children cook the meals and invite the elderly to share the food with them. Each open-day has a specific topic, so that children and old people can talk with each other.

Quartier Laak

Quartier Laak is one CBOs founded in 2012. Quartier Laak was established for the purpose of preserving, beautifying and socio-economic strengthening

Laakkwartier. Quartier Laak realizes its goals by initiating and executing various projects, which are accessible to all residents and businesses in Laakkwartier. In total Quartier Laak has three workers. Residents have the right to participate in neighborhood management but also an obligation to contribute to neighborhood management. This is incorporated in Quartier Laak's working strategy. Quartier Laak is trying to provide a lot of activities in Laakkwartier that are supposed to benefit the district. However, they are not for free. For local businessmen, they probably are required to offer their products for free or at a reduced price, and even offer investments. The strategy of Quartier Laak is called a "value-creating-network", because Quartier Laak encourages people who live and work in Laak to leverage each other's knowledge, expertise, products and services.

Individuals

It is undeniable that some individuals in community have certain power to influence the local decisions, such as lawyers, doctors, politicians and council members. They can provide meaningful knowledge, information and advice on behalf of communities in negotiation with other agencies, such as government. Moreover, the community asked the politician living in the same community to lobby in political discussions. It is good news if the community has some influential professionals or politicians. Next to the people who have political influence, some persons from academia are also motivated to engage in urban environmental issues. Tycho Vermeulen is a scientific staff-member at Wageningen University. For instance, Tycho Vermeulen engages in several urban farming projects in The Hague. As a scientific person, he can provide some concept meaning and explain why urban farming is important to cities to civil servants, businessmen or citizens. Although he has this expertise, Tycho Vermeulen, acts more on the business-like aspects. He is active in networking, fundraising, lobby, and communication with people from different agencies.

With this experiences, Tycho underscores the importance of "good communication" and "inter-dependence" between project participants. He believes "good communication" should base on "inter-dependence" and "openness". In other words, project participants join the project because they can be benefited from others and they can share information transparently with each other. Jurienne Hollaar highlighted the importance of some "initiators" in a neighborhood. In some cases, if the local resident's rights, such as safety or health, are threatened, they might call the police or form a group to the threats-maker or government. By doing so, some people with certain knowledge and skills are necessary in such cases. They have to prepare something, like arguments and paper-works. In other cases, residents in a neighborhood might be passive in neighborhood development; however this may also depend on the social status of this neighborhood.

Although the civil society in the Netherlands is active, it experienced some difficulties as well. Firstly, like the municipality finds it difficult to communicate with civil society, the civil society finds it hard as well the other way around. Lots of energy and time is spent in endless hearings and negotiations, which remarkably decreases the enthusiasm of community-based organizations (CBOs) and individual citizens. Furthermore, insufficient communication also exists between different CBOs. CBOs are active in community's issues, but they seem to work separately and are not willing to communicate or share information with their peer organizations. Secondly, it is hard to mobilize residents to join neighborhood management. However, it seems to depend on the profile of the neighborhood. There are some so-called "difficult neighborhoods" where residents are relatively poor and do not have a high level of education. People there are not active and motivated in their neighborhood development. On the other hand, residents who are relatively wealthier or have good education are more motivated. They are more creative and sometimes it is them to take the initiative. Thirdly, lack of money is always a problem. Financial support from governmental agencies and enterprises is the major source. Fundraising is also an approach. However, nowadays it is getting more difficult to obtain sufficient financial support. Fourthly, the up-scaling of local initiatives is challengeable. Due to insufficient financial support, and probably inappropriate communication between different stakeholders, it is hard to upscale small local initiatives.

Conclusion

In The Hague, the air pollution and water pollution still exists, but the problem is less severe than in other Dutch cities since The Hague does not have much heavy industry in its territory. However, looking at the results, we find out that the local government is more concerned with energy-saving. As an international city which has many international governmental, non-governmental, environmental organizations, The Hague tries its best to be climate-neutral city by 2040. Moreover, since a large proportion of the Netherlands is under the sea level, sea level rising is a disastrous threat to this country. Nowadays, extra energy consumption means extra emission of greenhouse gas, which causes global warming and a rising sea level. Therefore, The Hague cares much about its energy-saving strategies.

In addition, due to less advanced heating facilities and a relatively colder climate, Dutch cities face difficulties in energy-saving in household and office buildings. In the past, severe pollution in 1960s and 70s testified the governmental institution in environmental protection. The government realized to incorporate companies ("target group") and civil society in environmental issues. In the 21st century, the energy crisis and global warming became the primary environmental concerns of the Netherlands.

Meanwhile, with the decentralization of governmental systems and less support from provincial and national government, municipal government finds it is getting incapable to tackle all local environmental issues. More responsibilities are distributed to companies, NGOs and communities. On the other hand, the rising empowerment of civil society in Dutch society required a role in the environmental protection arena. Furthermore, as many interview respondents said, a collaborative working style is the nature of Dutch society, which can be explained by the term "Polder Model".

The municipality is rather a coordinator or facilitator. The Municipality encourages communities and citizens to take their own responsibilities and work out their own plans and measures to protect the local environment. The municipality helps to provide useful information and limited funds while it does not have decisive power in community management. The Netherlands has more powerful NGOs. The government and the public are the fund providers of Dutch NGOs. However, even Dutch NGOs receive funding from government; they are politically independent, and not afraid of protesting against government. But more often than not, they try to cooperate with the government.

In The Hague, CBOs and Residents' Committee have more independent power in community management. They usually are the initiators of many inspiring activities. They take responsibility to make their own neighborhood more beautiful and improve local development, instead of relying on governmental measures. From the exploration in Laakkwartier, we found that local CBOs and some active initiators do not care too much whether the initiatives are small or large; they believe in the "butterfly effect", which means tiny changes can bring huge changes in the future. Moreover, despite educational activities, CBOs in Laakkwartier prefer regular communication between different local organizations and residents. Similar to the Chinese situation, there are some motivated activists, but it is still hard to mobilize local residents to join actively.

References

[1] Bressers, H. and De Bruijn, T. 2005. Conditions for the success of negotiated agreements: partnerships for environmental improvement in the Netherlands. Business Strategy and the Environment, 14, 241-254.

[2] Driessen, P.P.J. et al. 2012. Towards a conceptual framework for the study of shifts in modes of environmental governance - Experiences from the Netherlands. Environmental Policy and Governance, 22, 143-160.

[3] Hofman, P. 1998. Public participation in environmental policy in the Netherlands. Thailand Development Research Institute, 13 (1), 25-30.

[4] Klink, T.E. 2007. The role of environmental NGOs: From China to the Netherlands. Macalester International, 20, 105-128.

[5] Natuur en Milieu. 2011. Ranking the stars. The Netherlands in comparison with other European countries in the field of environment, nature and climate.

[6] Schreuder, Y. 2001. The Polder Model in Dutch economic and environmental planning. Bulletin of Science, Technology and Society, 21, 237-245.

[7] SER. Social and Economic Council of the Netherlands. 2013. Energy Agreement for Sustainable Growth. Available at "http://www.ser.nl/~/media/files/internet/talen/engels/2013/energy-agreement-sustainable-growth-summary.ashx".

[8] Van der Maesen L.J.G. and Walker, A. (ed) 2012. Social Quality: From theory to indicators, (Basingstoke: Palgrave Macmillan), see also 44-69.

[9] Van der Maesen, L.J.G. 2012. The functions of social quality indicators. In Social Quality: From theory to indicators, (ed.) Van Der Maesen, L.J.G. and Walker, A. (Basingstoke: Palgrave Macmillan), 243-248.

作为规范导引和作为政策现实的环境政策：中意环境治理的比较研究

根亭　林卡*

导言：全球发展和环境挑战

当前正处于全球化的时代，国与国之间的相互影响日益加深，尤其是来自不同地域的人们在政治、经济、社会领域的相互联系也正日益加强。这一全球性特征在环境保护领域尤其突出。正因为我们居住在同一个星球，共同分享同一个居住环境，我们的生存状况也深受形形色色的人类活动所影响。在过去的两百年间，伴随着工业化和城市化的进程，人类不断利用各类自然资源来提高自身的居住条件，并最终成为自然界的主宰。然而，现今历史已经发展到了一个决定着未来人类社会是否得以永续发展的关键阶段。从环境退化的趋势上来看，二氧化碳排放量不断增加意味着全球气候变暖的势头依旧持续。此外，研究人员发现暴露在空气污染中会使人们患癌症的概率大幅度提高（Loomis，2013）。由于空气中的 $PM_{2.5}$ 浓度快速增加，在印度、伊朗和中国等国，相关研究也表明这将使与癌症相关的非意外死亡率增加6%以上（Chen，Goldberg and Villeneuve，2008）。

因此，在21世纪，环境政策将会成为各个国家和国际组织决策讨论的核心议程。这一严峻的形势导致一系列有关环境保护方面的国际公约、国家间的战略合作协定、跨国条约的签订以及相关国家法律的颁布。这些官方文件和政策声明将会给民众在社会环境方面的理念产生深刻积极的影响，同时也能够提高人们的环

---

* 根亭，浙江大学公共管理学院博士生；林卡，浙江大学公共管理学院教授。本文原文为英文，中译者为白莉（浙江大学公共管理学院）。

保意识。诚然，与半个世纪之前的情况相比，我们相信当下的政策制定者与公众对环境的认知和保护意识显然更为强烈（Gallagher and Tacker，2008）。这就为各国在环境保护方面制定和实行相关政策措施设立了规范依据。

事实上，各国政府在环境保护方面业已达成了强烈的共识。1992年在里约热内卢召开的联合国环境与发展大会就是一个绝佳的例子。在此次大会上，超过172个国家的政府代表以及2 400多名非政府组织代表签署并通过了《里约环境发展宣言》（Levin and Bradley，2010）。这份宣言所设立的总目标就是通过在各个国家、社会相关利益部门和人民之间不同水平的合作往来以建立一种公平的新型全球伙伴关系，为签订尊重全体成员的利益和维护全球环境与发展体系完整的国际协定而努力，认识到地球的完整性和相互依存性（Rio Declaration，1992）。不同政府、机构之间不断扩大旨在保护地球免受污染的广泛共识，各国也吸收了上述目标作为本国制定相关政策的指导原则。同时，这一规范基础也是支撑民众和非政府组织开展草根环保活动的官方依据，并以"同一个世界，同一片天空"作为环境保护行动的明确策略（Rio Declaration，1992）。

尽管这些国际组织和政府机构付出了诸多努力，我们不难发现在环保方面规范指导中的理念期望与环境政策执行的现实之间存在着巨大的鸿沟。地球的生态系统持续退化，生物多样性也面临着不断减少或丧失的严峻挑战。环境污染依然是人类发展议程中的首要问题，特别是由于当今世界人口急剧增长，导致用以满足人类消费需求的日常生活生产必需品倍感紧张（WHO Outlook，2013）。仅从亚洲地区的发展历程来看，伴随着新兴经济体崛起的代价是严重的环境破坏。这类国家往往过分依靠需要大量资源消耗的制造业拉动并支撑经济增长，但却对环境产生了巨大的压力。中国、印度或其他亚洲国家虽然已逐步确立了自己世界工厂的地位，然而这些国家中的一部分企业常常在营利过程中忽略了环境保护的迫切需要。

因此，我们需要讨论在环保领域中的规范需求与政策执行之间的矛盾。这一矛盾可以看作是发展全球环境治理的主要障碍，同样符合各国内部的现实情况。为此，本文将聚焦对这一矛盾及其成因的探讨上，并对中国和意大利的环境保护方式进行比较研究。这两个国家在环保方面处于不同的发展阶段，更重要的是在环境治理的决策依据和政策执行方面也表现出截然不同的特征。我们将讨论两国的决策制定机制和在具体领域政策执行过程中面临的主要问题及表现出的不同特性，并探索可能的解决之道。

环境治理领域的几大关键议题

要理清在环境治理方面的核心议题，我们可以从有关环境政策本质属性的讨论入手。在 Bührs 看来，环境政策这一概念可以看作是"环境"和"政策"两个部分的结合。"环境"因素主要指与社会维度（如生活品质和健康状况）和经济维度（包括资源管理和生物多样性等议题）密切相关的自然生态环境。而另一因素"政策"则可以用来描述"一个由政府、党派、企业或个人接受或发起的行动或纲领的进程"（Bührs，1991）。基于这一认识，我们可以将环境政策的性质理解成专门针对因人类活动产生的环境问题而制定的一系列政策措施，这类活动所派生的影响反作用于人类社会，而且往往对人类福祉（如健康的身体与"绿色清洁"的环境）的实现造成十分消极的影响。

环境治理领域的几大核心议题也隐含在这一关于环境政策的定义之中。首先，这一概念强调了人类行为和行动的重要性，也正因为环境保护十分依赖人类活动（这里主要是指亲环保或反环保人士）。人们对于绿色环保理念的认识和对待环境的态度极大地影响着他们日常行为的选择。诚然，尽管个人的行动可能受到私利而非社会共同利益的驱使，我们依旧可以在宏观层面强调政策理念背后的价值取向之于公众行为的指导意义。这就凸显出规范性原则或规范指导对政策行为的重要性。可以理解的是，当一个国家仅仅以 GDP 的增长作为社会发展导向时，环境政策的重要性就会被忽视。因此，在环境保护方面所达成的国际协定对于提升政府和民众对环境问题的重视和关注是有其十分重要且积极的现实意义的。

从行为角度来看环境治理的问题，对个人、组织和国家的环境行为的关注与本文所谈论的议题密切相关。有些行为直接或者可能带来环境变化，例如砍伐森林和焚烧生活垃圾（Stern, Young and Druckman, 1992），而有些行为会间接地重塑可能会直接导致环境变化的活动选择的语境（如 Rosa and Dietz, 1998；Vayda, 1988）。对此，在环境保护领域建立切实有效的规范指导机制并以此来引导民众、组织乃至国家的行为行动将会有其十分重要的价值，特别是能够推动人们的绿色环保行动。例如，1992年欧盟颁布的《马斯特里赫特条约》将环境保护的目标上升到"为促进和谐发展，各成员国应当营造持久稳定、无通胀且相互尊重的社会环境"。2007年通过的《里斯本条约》更是将环境保护作为欧盟发展的根本目标，要求各成员国资助和执行相关环境政策（EEPS Europe 2020，2013）。这些政策可能在个人、组织或国家层面不能立竿见影，但它们能够为各成员国在环境保护方面的政策制定提供规范依据和价值导向。

在环境政策领域第二大焦点议题就是有关政策工具的问题。政策理念或理想的政策模式的实现需要一定的手段和策略，唯有如此才能够达到政策的预期目标。因此，当谈论环境治理时，我们不得不将政策实施的手段（或是工具）作为讨论的核心议题。这些工具手段多种多样，有的是由国家政策形成的，另一些又是从基于不同目的所设立的指标体系或评价标准中派生出来的。这些政策工具可以包括温室气体排放的方式、排污费以及环保基金系统（这类基金以现金补偿形式鼓励企业或个人承担一定的污染防治责任，并以此来达到环境政策的预期目标）（Stavins，1988，1991）。

此外，相关标准或指标体系也可以归入这类政策工具中。以旨在环境政策的ISO 14001标准为例，它的颁布是通过制定排放物的衡量标准和减少其对环境的影响等方法来有效地评估各个企业或机构组织的环境绩效。从这个意义上来说，它仍是环境政策的一个政策工具。例如ISO 14000标准体系，就可以被定义为"通过提供切实有效、可持续的信息来重视并推广富有成效的环境友好型政策工具。同时为了构建一个环境应用信息资料库，还需要建立一个收集环境保护领域成功实践经验的交流网络/平台"。另一类政策工具就是在环境保护和治理方面各国签署的国际条约和合作协定。以欧盟的环境评估程序所运用的政策工具为例。Pearce发现"自1990年代早期开始，这一官方评估机制就已经得到完善并广泛地推广开来"（Pearce，1998）。被多个发达国家所接受的《京都议定书》是另一个例子。它的主要内容是分两个阶段（2008—2012年为第一阶段，2013—2020年为第二阶段），各国承诺以法律的形式限制或减少本国的温室气体排放量。

环境保护与治理领域的第三大议题就是政策实施的有效性/成效。当相关规范指导确立了国家社会的发展导向，并辅以政策工具作为达到目标的策略手段，环境政策的执行就成为其目标能否得到切实推进及其所预期的环境治理效果能否有效实现的决定性因素。有关这些问题的讨论通常与政策行动的动因与限制相关。这类研究就会涉及国家与其他不同部门之间的制度/组织关系（如公民社会、民间组织等），甚至对于国家机器/体制内部来说，也会涉及中央与地方治理的关系论述。这类关系在不同社会背景下所表现出的不同特性对政策实行成效的大小产生非常深远的影响，而且这些特性仍会影响环境保护的总体效果。举例来说，在中国，政策利益的差异性，尤其是中央与地方政府在环境问题上的分歧是显而易见的，这也直接影响了双方在环境政策执行上的不同态度。因而，在此问题上的相关比较研究将会帮助人们理解环境治理的一整套政策体系，同样，这也有助于理解国家层面在环境保护的决策行动背后的基本原理。在下文中，我们将对中国和意大利的环境治理制度进行比较研究，从而吸取不同社会提高环境保护质量的经验教训以及为未来全球环境治理提供若干建议。这一比较研究将从上述三个维度展开，即环境保护与治理领域的规范指导、政策工具以及政策实施过程中地方与中央政府的治理关系等。

意大利的环境治理

作为欧盟的成员国，意大利在环境问题上的政策制定基本遵循欧盟所颁布的各类规范指导性条例，即将环境政策作为本国决策过程的重要部分。在这一领域，欧盟已制定并通过了第七环境行动计划以及一系列相关政策声明，并于 2013 年 11 月开始生效。基于欧盟环境政策 40 年的发展经验基础上，第七环境行动计划整合了环境领域一系列的战略行动计划和举措，其中包括《资源效率进程计划》《2020 生态多样性策略》《关于提高欧盟环境服务中收益交付的通报》《低碳经济进程计划》（EEAP，2013）。这就为欧盟各成员国在 2020 年之前推进各自的环境政策和实施行动提供了统一的规范指导。而不同成员国运用这一指导性文件的具体情况取决于中央政府的选择。然而这一指导方针不仅仅左右了国家政策的议程走向，但同时也对非政府组织的活动以及全体欧盟公民的态度和日常行为产生影响。

在政策执行的工具方面，在早期，政府主要运用成本效益分析作为分析环境问题的手段，在这个阶段，环境问题主要被看成是一种包袱，而非全社会所必须承受的道德责任。然而，这一分析仍在一定程度上唤起了人们的环保意识，并且推动了政策制定者出台相应的政策措施以回应现实问题。正如欧共体在 1992 年颁布的《第五环境行动计划》所宣称的那样，这一计划明确地是为了"在对环境与自然资源储备方面具有重要影响的若干政策措施和行动方面，构建极具意义的成本/效益分析方法论与指导规范，并推动这一政策工具的发展"所制定的（Vogel，2007）。其后，作为新时期环境领域的规范指导，《京都议定书》建议各国将排放配额和信用体系的交易作为新的政策工具利用起来（Hood，2010）。

欧盟的辅助性制度对于各个成员国来说也是推动国家政策行动的手段之一。通过辅助性制度，欧洲最为发达的几个国家都将采取严厉的污染减排手段来减轻欠发达的欧洲国家的减排限排压力，使后者得到更多的排放指标（IEEP，2012）。此外，在《马斯特里赫特条约》中的"污染者补偿"原则和预防原则（即"生产者责任延伸制"）（Papadiaki，2012）指导下实行的政策工具同样对自然环境的循环过程产生积极影响。根据这些原则，生产者和进口商都将在各自商品的整个"生命"过程中承担产品对环境所造成的影响，从原材料的获得，再到商品的生产、分配直至使用的过程（Papadiaki，2012）。

然而，意大利的环境治理面临的主要障碍存在于政策执行层面。这些障碍包括以下几个方面：(1) 落实欧盟政策规范的程度；(2) 经济成本；(3) 治理制度的碎片化。关于执行欧盟相关政策指导方针的程度问题上，意大利政府作为立法的委托机构行使着落实欧盟各项指令的权力和责任，但由于中央-地方政府的行政体制上的原因，在执行这类政策时意大利一直出现系统性的延误。这直接导致意大利绝不是环境保护方面最富有成效的欧盟成员国。在一份来自欧洲议会的官方文件里指出"欧盟各项指令在意大利的总体执行情况要么被延误，要么就是根本没有得到落实。通常情况下，当欧盟指令转化为意大利的国家法令之后，仍无法达到这些指令所设立的预期目标或效果"（DG HOME audit，2006）。

第二个问题就是经济成本。尽管意大利的环境管理部承担了执行欧洲法例的主要责任，但相关地方政府仍不得不去筹集和转移这类政策实施所需要的财政资源（Ministry for Environment，2012）。意大利的中央政府确实分配了部分预算来支持地方治理的需要，然而上文提及的环境管理部缺乏必要的资金来源，仅扮演了主要的管理角色。由于不同层级的权力机构可能追求截然不同的发展目标（尤其在中央与地方层面），有时所谓的辅助性制度也使欧盟政策的实施过程中的责任和角色越来越难以定义。因此，一旦经济危机席卷意大利甚至整个欧洲时，很容易出现政府取消欧盟委员会所要求的在环境政策上的资金支持的情况。

第三个问题就是环境治理制度上的碎片化。在意大利，国家、地区、省、市各级层面的差异化导致了这一制度的碎片化情况，这为能否切实有效地落实欧盟或意大利本国的环境政策和法律法规增加了一些困难。因此，相比其他成员国，意大利是当下欧盟委员会侵权诉讼立案最多的国家，这类案件往往与废弃物管理有关。造成意大利在环境事务的政策执行方面如此现状的一个主要原因是不同机构之间缺乏交流沟通网络。负责执行欧盟指令并依据相关国际法规对其实施情况进行监督的地方政府缺乏履行职责的驱动力。因而，中央和地方的关系，或者说是欧盟治理作为地方发展和国家政策制定过程中的协调机构成为了诸多困难中的主要矛盾。

举例而言，当我们的目光投向意大利北方自治省中的博尔扎诺市和特兰托市，这是欧盟失业率最低的地区（Eurostat，2013），同时也呈现出了高环境质量与低失业率之间的有趣联系。意大利地方政府的经济发展问题常常被归结于南北差异的扩大。这一地理经济差异确实源于一些历史和经济因素的影响。在南方地区（例如普利亚省），当地的许多政党如"重视环境和自由的左翼"是地方政府倾向实施绿色政策的一个典型例子（Ministry for Environment report，2010）。因此，普利亚省这一案例说明在南方地区同样将清洁环境作为地方发展的首要任务。

正因为这些困境的存在，意大利成为欧洲环境质量最差的国家。尽管在地方大气污染的减排总量上位居欧盟第一（OECD，2013），但那是由于欧洲污染最严重的30个城市中超过一半位于意大利。在对意大利和德国进行的一项比较研究中表明近10年间德国在清洁能源（太阳能等）的产量远远高于意大利，这还是在德国年平均日照小于意大利的情况下实现的。虽然意大利清洁能源领域的发展受到欧盟和本国政府的大力资助，然而财政上的管理不当与腐败使得这笔资金未能发挥应有的作用。这也导致地方政客出于政治目的对这批基金进行了重新配置，例如近期受到欧盟资助的竞选活动（Nihan，2008）。

因此，对意大利来说，提高环境质量是一项相当艰巨的任务。然而，有一种观点认为意大利对履行欧盟规章方面缺乏足够的关注。这一点可以从对清洁能源对国家能源供应的贡献分析上得出。因此，如何提高欧盟、国内各个地区和各个地方政府（通常情况下相对独立，缺乏互动）之间的交流与沟通就成为意大利在环境保护领域采取进一步行动前的最大挑战。作为区域治理机构的欧盟与作为地方治理主体以及成员国的意大利之间的差异将会为双方在推进环境治理过程中的角色扮演留有很大的发挥空间。

中国的环境治理

在中国，人们的日常生活和社会经济的发展正受到环境问题所带来的严峻挑战。根据世界卫生组织的相关数据，世界上在大气、水、废弃物管理、森林砍伐、荒漠化等方面污染最严重的 10 座城市中有 7 座在中国（WHO, 1998）。由煤炭燃烧所引发的大气污染中产生的二氧化硫和烟尘又进一步导致占中国陆地总面积30%的地区发生酸雨频率和数量持续增加（Gilley, 2012）。从 21 世纪初开始，环境游行活动由基于农村的抗议转变为城市运动（Xu, 2014）。2013 年，一项由《中国日报》发起的网络民意调查收到了超过 67.5 万份问卷回复，其中，环境保护是公民首要关心的议题之一（新华网，2013）。环境恶化的现状可能会导致公众的不满情绪，由此带来集体上访，甚至在某些情况下引发大规模的群体性事件（Wu, 2013）。特别在贫困地区，经济发展所付出的代价必然伴随着土地污染，但利益牺牲者往往是社会底层的农民而非富裕人群。

在此背景下，中国政府早已开始制定与环境相关的政策、法律和法规。1997年颁布的《中华人民共和国节约能源法》是第一部为提高整体能源利用效率而制定的法律，旨在打击和杜绝能源浪费的现象。2004 年，时任国务院总理的温家宝又明确提出"建设资源节约型社会"的口号。这一主张在 2005 年之后受到了各级政府的大力支持，因为当时社会的发展理念和路径导向已经由原来的单纯追求经济发展转变为社会的全面发展，其中就包括了环境质量的提高。2013 年 2 月，国家发改委联合有关部门修订并发布《产业结构调整指导目录（2011 年本）》，其中明确指出中国需要转型并升级传统优势产业，提高其竞争力，通过新技术的运用来革新产业结构。在"十二五"规划阶段，国家发改委发布了新建或翻新原有的以煤炭发电供应的建筑为清洁能源设施。为此，国家发改委设立了 138 个风险投资基金，其中 38 个基金将被用来发展节能、环保和新能源领域来取代中国以往以煤炭为主导的能源消费结构。

值得关注的是，推动中国制定环境治理议程的一个重要驱动力是与国际组织的互动联系，特别是与欧洲各个机构之间的交流合作。以意大利的贡献为例，自 2000 年起，中意政府就启动了中国-意大利环境保护合作项目（首先与环保部建立合作）（Ministry for Environment，2010）。这一项目的宗旨是为了帮助中国提高环境质量，支持其可持续发展，并推动两国企业的战略合作。在随后的几年间，意方又将官方合作的范围扩大到其他中国政府组织和学术机构（Ministry for Environment，2010）。因而，这一合作关系也包含了增进两国在环境议题上的交流对话这一目标，并通过促进学术研究的合作来深化这一联系。

在过去 10 年间，一系列政策手段、措施和工具被运用到服务国家环境目标的政策执行过程中。尽管在大气污染的问题上政府已经实行了相当严密的监管措施，近年来，中国政府仍宣布投入 2 270 亿美元用于一项环境治理的五年计划。中国政府也将继续增加对可再生资源的资金投入。其中，风力、太阳能、生物质能、海洋能和地热能等清洁能源将会是下一代能源供应领域的优先发展部门。尽管在太阳能和太阳能板生产领域居于世界领先水平，中国将会进一步实现清洁能源领域的多样化。仅在 2012 年，中国在水电站、核电站和风力发电站上的投入分别达到了 1 277 亿元、778 亿元、615 亿元人民币。正因为有了这些巨额投资，水力发电的能力才会居于全球第一位（24 900 万 kW），核能发电也达到了 1 257 万 kW。鼓励清洁能源的使用也是一种积极的环境政策措施。未来，作为主要能源消费品的煤炭将大规模、大范围地被清洁能源所取代，后者包括了可燃化石、太阳能、风能和水能等。

除了政府在环境问题上的财政拨款外，另一个基本的政策措施是在信息披露或信息公开上。这项举措作为中国环境政策的基本内容已经得到环保部副部长潘岳的官方证实，他认为披露相关环境信息可以使公众理解地方政府的环境决策和规划制定，限制官僚作风和权力滥用，充分发挥媒体舆论以及全社会对执法活动的监督作用。披露环境信息可能将强制污染企业去承担应有的经济和社会成本/代价。在政策实践过程中，中央政府设立了污染源信息公开指数（Pollution Information Transparency Index，PITI），并将中国 113 个市级政府的环境信息公开程度进行排序（Tang, 2012）。这一政策行动不仅无形中激励了地方政府，使其改进所在城市环境信息公开的实践，还加深了政府与非政府组织在环境问题上的紧密合作关系。

然而，尽管从政府的视角来看，中国在环境问题上的政策支持和投入是相当可观的，但事实上仍缺乏一套真正能够依照中央政府的理念而实施的政策机制。中国在环境方面的法律框架在某种程度上来说是完整且有效的，但中央和地方政府之间的鸿沟使得国家政策在地方层面实施起来十分艰难，这也导致地方政府有时不得不制定与中央政府相异的政策目标，并为之奋斗。缺乏政策的有效执行这一情况在当下正进一步加剧。例如，旨在治理大气污染的《大气污染防治法》早在 1987 年就已经颁布了，然而在现实中它仍未完全付诸实施，中国大气污染的情况仍面临严峻形势。

同样，由于中国的威权主义政治体制，自下而上的政策行动更是特别复杂。在中国，各个环境利益相关方通常是遍布全国，但又无组织，更缺乏统一有效、能够整合民众与非政府组织意见投诉的反馈机制。很多中国的非政府组织尚未得到官方注册的承认，因此也不享有法律保证的生存和活动权利。虽然只要他们不挑战政府权威，他们的存在就会被官方默许（Wong, 2005）。因而在许多城市，尽管国家已经建立了基于利益相关者要求的宏观环境政策体系，但依然缺乏真正的政策落实机制。其结果是尽管民众仍不断就残酷的现实环境问题进行请愿，但许多地方官员仍将实行严厉的环境政策视为社会经济增长的一个潜在威胁。

由于各个地方政府在理解环境保护重要性这一问题上持有不同态度以及地域情况差异，中国各地在环保领域的政策执行程度各异。这就在另一个方面为地方政府提供了相当大的政策创新空间。例如部分城市的污染源监管信息公开指数报告会以排序的形式出版公布，这就使哪些企业、机构做得好或做得不好一目了然，同时也为下一步采取何种措施提供依据。在嘉兴和宁波两个城市，地方政府公开了污染企业的名字，并向当地媒体提供了相关信息，这就在无形中提高了企业污染环境的社会代价。正如宁波市环境保护宣传教育信息中心主任所评价的那样，"许多企业不在乎违法排放罚款，但他们却十分在乎他们的银行贷款审批是否会受到媒体曝光的影响。"

中意环境治理的比较分析

对意大利和中国在环境治理上所表现出的不同特性进行比较，能够为我们进一步的讨论提供一些思路和研究问题。基于上述讨论，我们能够观察出双方的相似之处。就政策理念和规范来说，两国政府对于环境保护的体制运行的需求和需要等方面都持大体一致的看法。尽管这类规范性指导文件是由两国或区域不同的立法机构所制定的：在欧洲和意大利是由欧盟，在中国则是由中央政府；但整体上看双方的价值观念都是相近的，且采用的政策工具也基本相同，包括排放交易、环境信息披露制度、清洁能源的使用等。这一相似性的产生源于环境保护的需要已成为全球性议题，也是出于政策学习的考虑以及中国吸收了欧洲国家的相关政策理念和手段措施。同时，本研究也揭示了中意两国在环境保护和治理领域所面临的共同问题。在政策执行方面，中央和地方政府的关系成为左右政策实施效果的一大决定性因素，特别对两国社会中的环境保护情况产生了深远的影响。

进一步而言，由于两国截然不同的中央-地方政府关系，我们也能看到双方在政策绩效上的差异。意大利现行的环境政策指导纲要是由欧盟委员会制定的，它要求各个成员国将其转化为国家政策并付诸实施。出于这个原因，这类文件要具有很强的规范要求和政策导向功能，然而它的实施却很大程度上依赖于各国政府的决策。这些基本方针和政策提案可以分为三种类型：法律规章、政策指令和具体决策，它们在不同的成员国具有不同的现实作用。基于欧盟法律的地方化、本土化，成员国时常无法全面实施欧盟所规定的法律章程，甚至许多情况下执行程度欠佳。对于部分成员国，在将欧盟规章整合到本国环境政策体系的过程中，常常出现在地方层面的政策实施延迟的现象。但这是因为欧盟的指导方针对于每一个具体案例来说太笼统空泛，因而很难真正得到全面落实。又由于欧盟各个国家地区情况高度的多元化，这些规章制度往往缺乏具体价值。

此外，欧盟委员会或欧盟公民经常对各成员国发起侵犯诉讼程序。这是由多方面原因导致的，如政策实施程序的延误，甚至有时拒绝批准欧盟指令的本土化，而这类指令往往需要通过国家议会才会在成员国全面生效。事实上，我们的确看到了类似案例，欧盟法院负责判定国家政策与共同体政策是否协调一致，特别对于意大利来说,由于诸如废物和大气质量等环境事务管理存在不当使得欧盟法院不得不介入进行必要的政策干预(DG HOME Audit,2006)。因此,欧盟成员国从自身的角度出发在环境政策实施的过程中扮演了一个相当复杂的角色,也正因为如此,欧盟政策在各国的执行情况都大相径庭(各国国家宪法/立法模式的不同)。在本研究对意大利模式的案例分析中,欧盟政策的生效首先必须得到意大利国家议会的通过,而在其他国家却可能得以直接运用。

另一方面,地方政府同样具有自身的利益考量。一旦国家利益无法与地方治理的需要相结合,政策执行将变得毫无效用。虽然各个成员国有义务将欧盟制定的各项政策措施付诸实践,然而,相关的预算投入是一项极具挑战性的任务。他们可能不得不求助于欧盟的资助来完成环境政策所设立的预期目标。例如,意大利受到的资助水平就远高于欧盟平均水平。然而,由于一些国家政治体制的分权化,这笔资金常常被用于比环境保护更为迫切的发展需要。在意大利,直到2001年,有关环境议题的中央实权才逐渐下放到地区省份,后者就必须承担起处理与地方政府和经济体、组织的关系。

意大利政治体制的全景可以看成是一种联邦主义的实现过程,这是因为政策执行、财政拨款、行政干预的实际权力都控制在区域地方政府手中。因而,在意大利,草根环境治理的动力相对薄弱。然而,这一自上而下的行政管理模式应当与其民主国家的背景联系起来理解,例如通过核心议题的谈判以及中央(联邦政府)与地方政府之间的协商来制定和实施相关环境政策,而这一过程又没有否认地方权力机构具有相当程度的自治权。

从权力机构的组织层面来看,类似的情况也能在中国发现。尽管中国的治理体系是高度中央集权化的,但在政策行动的制定选择上地方政府仍有很大的发挥余地。中央政府能够制定环境政策的规范方针和发展目标,但如何实施这些政策很大程度上取决于当地(政府)的客观情况。Lieberthal 和 Oksenberg 形容中国这一体制为"碎片化的威权主义",或如 Lieberthal 所说,这一体制在国家层面可以起到十分积极的作用,但在地方层面就会出现政令不通不畅的问题(Lieberthal,1991)。这一碎片化的过程有多方利益集团的较量,将中国带入一个未知的图景中。在环境事务上,一旦中央政府制定颁布的法律法规应用到地方层面,地方政府的态度就成为环境政策执行效用高低的决定性因素。在此背景之下,我们应当推动草根阶层的环境治理活动，并通过建立保障公民权利的法律秩序来规范公民在环境保护的需求，以这一"先上层后草根"制度的建立来监督当地企业的生产行为，后者常常因外来投资的需要而受到地方政府的支持。

　　这一影响同样发生在地方次级层面。尽管在国家层面已经建立了一系列环境标准，地方政府仍可以通过放宽或收紧环境执法力度来实现截然不同的实际规章。这些激励政策能够部分解释环境质量在地域上的显著差异，即使在宏观层面都执行一套相同的配套环境标准。因此我们看到在重视以经济发展为导向或是亲环境友好型地方政府管辖之下，中国各地不同的环境发展趋势。

　　我们已经得到一些关于地方政府无法履行环境责任的真实案例。这些问题往往会引发大规模群体性事件，根据官方数据统计，仅在2012年就有超过5万起环境抗议活动（Xu, 2014）。因此，我们需要寻找相关领域的社会创新或者政策执行的成功案例，以回应民众在环境利益上的需求，从而找出能够容纳国家机构和各方利益相关集团介入的参与机制。

结论和讨论

　　在国际背景中，环境问题一直是当前决策讨论的焦点议题。欧洲的环境政策是从草根阶层发展起来的，而且在传统上，欧洲民众具有强烈的绿色环保意识，这也深深地影响了欧洲各国的政策制定者。然而，欧盟的环境决策制定过程在很大程度是自上而下的，但在欧盟成员国和地方政府间存在的民主协商机制又能够帮助减少官僚主义和地方化的影响。由于其在环境运动方面长期形成的传统，因此欧盟环境治理这一自上而下的模式并没有在大多数成员国中成为一个现实问题。公民社会的发展应当关注在长远目标的实现上，但又干涉或波及个体优先发展的生存需要。当然，这一情况不符合新加入欧盟的成员国的国内情况，如罗马尼亚、保加利亚和克罗地亚等，这类国家由于在国内市场缺少竞争性经济系统，所以仍在环境问题上针锋相对。因此，这一政策建议能够增加不同层面体系之间的沟通交流，并兼顾整体上的环境共识。

　　中国距离全面实现其绿色发展目标仍有很长一段路要走。在中国，针对污染企业生产活动的环境效益评估仍然取决于地方政府的态度，而这一态度又分为积极和懈怠两类。与此同时，对这些积极性的潜在影响被政策的制定者和执行者之间的巨大差距给破坏。另一个值得考虑的重要问题是在环境政策的实施过程中明确的权责划分。在等待国家政策制定者先行动的这一状况下，地方政府的主动性和积极性将成为关键性的问题。在过去，由于地方政府为了确保经济的持续增长，使得环境保护政策的实施在地方层面很难展开，导致环保政策存在实施上的困难（Tang, 2012）。但现如今，民众的环境意识正不断增加。这一因素驱使一些地方政策一线执行官员以创新的方式，即通过强化鼓励非政府组织的发展问题，与各类公民机构一道探索寻找环境问题的解决之道。因为这些非政府组织大多数更愿意与地方政府而非国家层面的官方机构进行接触，因为与后者打交道似乎不现实，需要考虑的情况相对来说也更复杂。

　　从这一角度上而言，我们可能认为中央与地方关系之于环境保护政策绩效的影响是确实存在的。从环境保护和治理的长远发展历程看，我们必须引导民众和地方决策制定者更积极参与到环境事务的政策行动上来，尤其是对地方层面的政策执行予以特别重视。本研究对中国和意大利在环境规范与政策制度的复杂体系等方面进行了相关介绍，同时我们对这些制度在政策执行方面的影响作了具体的国别分析与比较，并突出了其实时变化的特性。这一变化可能发生在几个方面，其中环境治理领域的社会创新应当用来推动更多的污染防治活动。各类污染物数据库的未来影响程度也将部分取决于是否能够获得新因素以构建出有利于强化政策执行的制度工具。

Environmental Policies as Normative Guide and as Realities of Policy Implementation: A Comparative Study of Environmental Governance in Italy and China

Gentian Qejvanaj and Ka Lin*

Introduction: global development and environmental challenge

In the current era of globalization, the mutual influence of nations and people in different parts of the world in the political, economic and social spheres is intensified, in particular in the field of environmental protection. As we are living in the same world, sharing the same environment, our living conditions are affected by human actions of different societies. In the last two hundreds years, human beings became super specie in nature, exploiting natural resources to improve their living conditions, through processes of industrialization and urbanization. This process has reached a critical point that determines the future of human society and its sustainability. Looking at the trend of environmental degradation, the global warming continues and carbon dioxide grows (NOAA ESRL). Furthermore, researchers also presented evidence that the risk of developing lung cancer significantly increased when people are exposed to air pollution (Loomis, 2013). Since the fast increase of $PM_{2.5}$, in countries like China, India or Iran, the non-accidental mortality related to cancer increased by over 6% (Chen, Goldberg and Villeneuve, 2008).

---

* Gentian Qejvanaj, the doctoral reasearcher in College of Public Administration, Zhejiang University. Ka Lin, the professor in College of Public Adminstration, Zhejiang University.

Thus, in the 21th century, environmental policies became the key area of policy debates, for national and international organizations. This challenging situation resulted in a large number of international conventions, inter-state agreements, cross-national contracts, and national laws on environmental protection. These official documents and policy statements have a strong positive impact in shaping people's view on the environment of human society, and raising people's awareness of environmental protection as well. Indeed, if we compare the situation with half a century ago, we can surely conclude that the environmental awareness is much stronger than before in the mind of policymakers and the general public (Gallagher and Tacker, 2008), and this sets down the normative basis of policy actions of different countries on environmental protection.

To give a few examples illustrating a strong consensus of policymakers on environmental protection, we can refer to the Rio conference in 1992, which made the declaration on environmental protection. It was signed by over 172 governments and 2 400 representatives of non-governmental organizations (NGOs) (Levin and Bradley, 2010). The document aims at establishing a new and equitable global partnership, through the creation of a new framework among stakeholders working towards easing the cooperation among states, in order to promote agreements in the interests of long standing environmental integrity and sustainable growth (Rio Declaration, 1992). The ideal of saving the earth from pollution set up the guiding principle for the states to design relevant policies. This normative ground also supports the civilians and NGOs to take action at the bottom line of society on green activities, with a clear idea of "one world one sky" for protection (Rio Declaration, 1992).

Despite the efforts by these international and national agents, we still see a large distance between the idea and expectation in normative guide, and the reality of policy implementation. The earth's ecosystem continues to be degraded, and biodiversity is a serious challenge. Pollution remains a top issue in the agenda of human development, in particular due to the overcrowded population in the contemporary world, demanding more and more daily consumption goods (WHO Outlook, 2013). As observed in the

Asian region, the rise of new economies is accompanied with serious environmental declination, with especially the manufacturing industries bringing huge environmental costs. Countries like China, India or other Asian countries have entered a status of world factories in the new era, but the need of environmental protection is often ignored by many companies in their profit-making efforts.

Thus, we need to discuss the contradiction between the normative demand and the policy implementation. This contradiction can be seen as the major barrier for developing environmental governance in the world, as well as in the case of national states. Accordingly, this study will explore this barrier and its underlying reasons by comparisoning environmental policies in Italy and China. These two countries have different stages of development in regard to environmental protection, and even more, the difference on the policy making rationales and their policy implementation in environmental governance. We will discuss these differences and the related problems in order to find a solution for them.

Some key issues in environmental governance

To define the key issue of environmental governance, we can start from the debate over the nature of environmental policy. In Bührs's view, the term environmental policy can be perceived by the two components of this term, i.e., environment and policy. The factor "environment" refers to the physical ecosystems in relation to two dimensions, i.e., the social dimension (such as quality of life and health) and the economic dimension (including the issue of resource management and biodiversity). Another factor "policy" can be defined as a "course of action or principle adopted or proposed by a government, party, business or individual" (Bührs, 1991). Understood on this basis, we can comprehend the nature of environmental policy as those policies focused on the problems arising from human impact on the environment, which retro acts onto human society by having a (negative) impact on human well-being such as good health or a "clean and green" environment.

This definition on environmental policy implies some key issues for the discussion of environmental governance. First, this definition stresses the significance of human behaviors and actions, as the environmental protection very much depends on the human (pro-environmentalist or anti-environmentalist) actions. These actions are very much influenced by people's view on the green values and their attitude towards the environment. Indeed, although the actions of single individuals might be driven by personal interest against common interests, we can still underscore the value of the policy ideas for guiding people's action in general. This relates to the normative basis of these actions, by which the normative guide of actions become important. Understandably, once a nation takes a GDP-oriented pursuit as development goal, the significance of environmental policies should not be overlooked. In this respect the international agreement of environmental protection can contribute to increase the attention and concern on environmental issues.

To look at the environmental behavior of individuals, organizations, and the state, some behavior directly (or approximately) causes environmental change, such as clearing forests or disposing of household waste (Stern, Young and Druckman, 1992), whereas other behavior has a more indirect effect, by shaping the context in which choices are made that directly cause environmental change (e.g. Rosa and Dietz, 1998; Vayda, 1988). In this regard, to set up a normative guide of environmental protection to influence people, organization, and the state's behaviors, would have a particular value to generate people's green activities. In Europe, the Maastricht Treaty in 1992 declaimed a goal "to promote harmonious and balanced development respectful of the environment". The Treaty of Lisbon set environment protection as a fundamental objective of the EU, demanding that the EU member states finance and implement an environmental policy (EEPS Europe 2020, 2013). These policies may not work effectively in practice at the individual, organizational and national level, but they do function as a normative orientation for guiding the member states' policy line.

The second issue of the environmental policy debate is the policy instrument. The policy idea (or ideal) needs policy means (or policy instrument) to implement or to reach the policy goals. Thus, when talking about environmental governance, we have to refer to the matter of policy instrument as an important issue for discussion. There are various kinds of instruments, some made by state policy and other by indicators systems or standards of evaluation with different purposes. These policy instruments can include the means of trade permits, pollution charges, and deposit fund systems, which encourage firms (and/or individuals) to undertake pollution control efforts for their own interests and for a collective goal (Stavins, 1988, 1991).

Still, the standard/indicators system is also a policy means.The ISO 14001 standards concerned with environmental policies are effective means to outline environmental performance of organizations such as in the business sector, so serving as a policy tool. According to some note, the ISO 14000 is generally defined as "to emphasize productive environmentally friendly tools by providing useful and sustainable information, with a communication framework which gathers the best practices available in the environmental protection field, in order to build a system of environmentally applicable information" (ISO 14000 family of International Standards p. 6). The other instrument is international agreements signed by national states. The instrument of environmental appraisal procedures in the European Union is one example. Pearce finds that "Since the early 1990s formal appraisal procedures have improved and are applied more widely" (Pearce, 1998). The Kyoto Protocol accepted by many developed countries is another example of legally binding limitations/reductions in their emissions of greenhouse gases in two commitment periods (Kyoto Protocol, 1997). The first commitment period of this contract applies to emissions between 2008-2012, and the second commitment period applies to emissions between 2013-2020.

The third issue is the effectiveness of policy implementation. Once the normative guide sets up the orientation of development, and the policy instruments find the way how to achieve the targeted goals, the policy implementation becomes the determining factor how far could to reach at the policy goal and how effective environmental governance can be achieved to implement these policies. Debate on these issues relate to dynamics and/or the restriction of policy actions. This discussion will refer to the institutional relations between the state and civil society sectors, and even for the state mechanism itself, the relation between the central and local governance. Different features of this relation in various societies have profound influence over the effectiveness of policy performance, which features still affect the overall outcome of the environmental protection. For instance, in China, dissimilarities of policy interest between central and local governments are obvious, and this influences their attitude in policy implementation. Comparative studies can be important to researchers to understand the system of environmental governance, and also, it helps to understand the rationale of the state actions on environmental protection. In the following section we will develop a comparative study on the systems of environmental governance in Italy and China, in order to draw out some lessons on how to improve the quality of environmental protection in different societies, with some general implications on global governance. The comparison will be made with regard to the three dimensions mentioned above, i.e., on the normative guide, the policy instrument, and the policy implementation in relation to the central-local relations of governance.

The Italian case of environmental governance

Italy as member state of the European Union follows its environmental policy as a guide line in its policymaking process. In this field, the EU adopted the 7th Environment Action Programme (EAP) with a number of policy statements, which entered into force in November 2013. Built on the 40-years experience of EU environment policy, the EAP integrates a number of strategic initiatives in the field of environment including the Resource Efficiency Roadmap, the 2020 Biodiversity Strategy, the Communication on improving delivery of benefits from EU environment services, and the Low Carbon Economy Roadmap (EEAP, 2013). This set up the guidelines for the member states to develop their environmental policy action in a common agenda for up to 2020. The situation of applying this guideline to different EU states depends on the national states; however this guideline influences not only the state policy agenda, but also the actions of NGOs and the attitude of general EU citizens with regard to the environment.

As to policy means, in the early days the cost-benefit analysis was the main measure to experience the impact of environmental issues as a burden rather than a moral duty to undertake by society. However, the people's awareness increased on the issue and will push policy makers to undertake policy actions. As claimed by the Fifth Environmental Action Plan promulgated in 1992 by the European Community, this plan was explicitly made for "… the development of meaningful cost/benefit analysis methodologies and guidelines in respect of policy measures and actions which impinge on the environment and natural resource stock" (Vogel, 2007). Later on, the Kyoto Protocol worked as a guideline suggesting using trade on emissions quota and crediting system as policy means (Hood, 2010).

Still, the EU subsidiary system to the national state is also a means of policy action generating state action. In this system the most developed EU countries will enforce harsher reductions on polluting emissions in order to help the developing EU member states imposing lighter limits on them (IEEP, 2012). In addition, the policy means in the principles of "The Polluter Should Pay" in the Treaty of Maastricht and the precautionary principle ("Extended Producer Responsibility Principle") (Papadiaki, 2012), will also have positive consequences on the recycling processes. According to these principles, the producers and importers of goods share responsibility for the environmental impact of their products for the entire duration of the products' life, from the acquisition of raw materials, through the manufacturing process, the distribution and use of the products (Papadiaki, 2012).

However, the major barriers for environmental governance in Italy remain in the area of policy implementation. These barriers include the follow issues, a) the extent to commit to EU policy guide; b) the economic costs; c) institutional fragmentation of governance. With regard to the first issue it is a question to what degree to commit to the EU policy guide. The Italian Government as a legislative delegate is empowered to implement EU Directives, but it suffers asystematic delay in implementing the provisions due to its State's and local authorities' administrative structure.

Consequently, Italy is surely not among the most active member states in terms of environmental protection. A document from the European Parliament states that "The implementation of EU Directives (in Italy) generally is delayed or missing. Very often the implementation of EU Directives into a national legislation does not reach the objectives set by the directives" (DG HOME audit, 2006).

The second problem is the economic costs involved. The Ministry of environmental administration takes the major responsibility to implement EU regulation, however the relevant regions have to locate and divert the needed financial resources to implement these demands (Ministry for Environment, 2012). The central government did allocate a budget to support regional action for the tasks, but its Ministry of environment lacks the financial resources and generally merely plays a managing role. A subsidiary system sometimes makes it harder to define responsibility and roles in the EU policy implementation process, since the authorities at different levels may pursue different goals of development at the national level and local level. Thus, once the economic crisis hit Italy and Europe, environmental policy funds could easily be withdrawn by the EU Commission.

The third problem is the institutional fragmentation of governance. In Italy, the differentiation at the regional, provincial and municipal levels creates an institutional fragmentation which makes effective implementation of the regional rules sometimes very difficult. Among the EU members, Italy has the highest ongoing infringement procedure opened by the European Commission, and these infringements are often related to waste managing. A main reason for policy implementation on environmental affairs is lack of communication between institutions. The Regions that implement EU directives and oversee the implementation based on international agreements of the European Unionare not very motivated to perform the procedure regulations defined by the law of the State. Accordingly, the central-local relations, or the EU governance as the coordinating bodies of regional development and the national state policymaking process become the main problem among various difficulties.

For example, if we look at the northern autonomous provinces of Bolzano/Bolzen and Trento, which have a record of the lowest unemployment in the EU (Eurostat, 2013), merging than an interesting link between high environmental quality and low unemployment. It is too often that the economical issues of local governments have contributed to increase the gap between north and south (Ministry for Environment report, 2010). Indeed, there are some historical and economical reasons for this geographical economical difference. In southern regions like Puglia, political parties like "the left for environment and freedom" are an example of local parties inclined to commit to green policies. This Puglia case shows that a cleaner environment is seen as a priority also among Southern regions.

With these difficulties, Italy locates among the last position of environment quality in Europe, despite overall reductions in emissions of local air pollutants that were among the largest in any OECD country (OECD ITALY 2013 highlights, 2013). Indeed, more than half of the 30 most polluted cities in Europe are in Italy. A comparison between Italy and Germany intelligibly shows German clean energy (or sun energy) production was higher than the Italian one in the last ten years, despite the general lower number of sun hours per year in Germany. The sector of clean energy in Italy has been highly financed by the EU and by the national government, but financial mismanagement and corruption made these funds operate ineffective. This caused local politicians to reallocate these funds for political goals, looking at closer political campaign to be financed with EU money (Nihan, 2008).

Thus, to improve the environmental quality is a demanding task for the Italian state. The Italian lack of attention to ratifying EU regulations can be seen by analyzing clean energy contribution to the national energy supply. Therefore, to improve communication between the EU, the regional and local governments (which often act in total independence from each other), becomes the challenging further step in the field of environmental protection. The contrast between the EU (as a regional governing agent) and Italy (as the member state of local undertaker) would leave a large space for both sides to play their role in promoting environmental governance.

The Chinese case of environmental governance

In China, the challenge to people's living and to the economical growth from the environmental problems was serious. According to the World Health Organization (WHO, 1998) among the ten most polluted cities in the world (on air, water, waste management, deforestation, desertification, etc.), seven out of ten were located in China. Air pollution caused by coal combustion generated sulfur dioxide and soot resulted in the increase of acid rain falling on about 30% of China's total land area (Gilley, 2012). Beginning in the late 2000s, the environmental protest movement has shifted from rural-based protests to urban movements (Xu, 2014). In 2013, an online poll of China Daily with more than 675 000 participants enlisted environmental protection on the top citizen issues (Xinhua, 2013). This may generate collective petitions or massive protests in some cases due to the rising discontent of the general public (Wu, 2013). Especially in some poor regions, it seems that land polluting is the necessary cost for economic growth, which is however paid by the poorer layer of the society rather than the richer.

Against this background, the Chinese government has been formulating environmental regulations and policies. The Energy Conservation Law of 1997 was the first one to improve overall energy efficiency by fighting energy waste, and in 2004 Wen Jiabao demanded to make China more energy efficient. This effort was especially encouraged after the mid-2000s, when the idea of development has been altered from the route of a single-minded economic pursuit to the route of an all-around development, including social development and environmental improvement. In February 2013, the National Development and Reform Commission (NDRC) cooperated with relevant administrations to amend the 2011 edition of the Guideline Catalogue for Industrial Restructuring, in which it pointed out that China needs to restructure and upgrade its traditionally-advantageous industries, enhance its competitiveness and improve its industrial structures by adopting new technologies. In the 12th Five-Year Plan period, NDRC have launched construction renovation by building in the coal electric power so far, 138 venture capital funds have been set up, in which 38 funds are designed to develop the energy-saving, environmental protection and new energy sectors to replace China's coal based energy sector (China's Policies and Actions for Addressing Climate Change, 2013).

One thing helping China to set up its agenda of environmental governance is the interaction with the international organizations and in particular the European agents. Taking the Italian contribution as an example, since 2000, Italy and China launched the Sino-Italian Cooperation Program for Environmental Protection together with the Ministry of Environmental Protection in China (Ministry for Environment, 2010). The goal of this project is to help China to improve its environment, support her sustainable development and promote cooperation between enterprises of the two countries. In the following years, the Italian side has expanded the cooperation to other Chinese governmental bodies and academic institutions (Ministry for Environment, 2010) therefore the goal of this collaboration is also to enrich the dialogue on this issue by including academic research in it.

Serving for the state's environmental goals, a number of policy measures are adopted in the last decade. The state announced a five-year US$277 billion plan to address the issue, despite an already tight regulatory on air pollution issues. China will continue to increase investments on renewable energy. Wind, solar, biomass, ocean and geothermal energies are priority sectors for the next energy supply generation, being already world leader in solar energy and solar panels production, China will further diversify its clean energy sectors. It invested 127.7 billion yuan in hydropower stations, 77.8 billion yuan in nuclear power plants and 61.5 billion yuan in wind power in 2012, by the end of 2012. Within this investment, the capacity of hydropower, which ranked first globally, reached 249 million kW, and nuclear power plants for 12.57 million kW (China's Policies and Actions for Addressing Climate Change, 2013).

To encourage the use of the cleaner energy is also a policy measure. Coal being the main energy producer will be replaced by cleaner energy as much as possible, such as using combustible fossils, solar panels, and the power of sun, wind and water.

Besides the state's financial allocation to environmental affairs, one of the basic policy measures is on information disclosure. This is confirmed by Mr. Pan Yue, the Vice Minister of Environmental Protection in China: "Disclosing environmental information enables the public [to] understand environmental decisions and plans of local governments, [...] restrains the 'will of officialdom' and abuse of power, [...] and gives full play to the role of media and society to supervise law enforcement [...] Disclosing environmental information may also force polluting enterprises to shoulder their due economic and social cost" (Tang, 2012). In policy practice, the government set up the Pollution Information Transparency Index (PITI) to rank 113 municipal governments across China on their levels of environmental disclosure (Tang, 2012). This action created incentives for municipalities to improve its disclosure practices, and generate a closer engagement between governments and NGOs on environmental affairs.

However, even though the policy commitment is serious from a government point of view, still a genuine policy implementation to put in action according to the central government will is lacking. The Chinese legal framework on environment may be seen as valid and complete according some signs, but the gap between central and local government makes the implementation of national policy arduous at local level, and also local governments sometimes strive for different goals than the national governments. The lack of implementation worsened the issue until today. For instance, the Air Pollution Prevention and Control Law was enacted against air pollution as early as 1987, however it has never been fully put into force and the air pollution situation remained serious.

Also, due to the restrictive conduct of China's authoritarian regime, a real action bottom-up is particularly complex. The stakeholders in China are often unorganized and spread all around the country without a framework able to concentrate the complaints coming from citizens' and NGO's. Many Chinese NGOs are not officially registered and have no legal right to exist, although they are tolerated by the government as stated by Wong 2005, "as long as they do not challenge the authority of the state, the officials keep 'one eye open, the other closed'." Therefore in many cities, although the state has set up an overall environmental policy in line with stakeholder claims, the will of a real implementation is still missing. Subsequently, despite the harsh environmental issues presented by citizens, many local officers see a tight environmental policy as a threat to a long term growth.

With different attitudes of understanding about the significance of environmental protection for local authorities, the municipal governments have different degrees of policy implementation in accordance to local circumstances. This yet allows a large space of policy innovation for the local governments. For example in some cities, some further forward steps are made in the field. In Jiaxing and Ningbo cities, the local government made active actions against pollution involving the general public.

A comparative analysis of environmental governance

Comparing the features of environmental governance in Italy and China leads to some considerations for further discussion. Established on the issues presented above, we can observe some commonplaces and similarities on both sides. With regard to policy ideas and guidelines, the states of both sides share many common views on the need and demand for environmental protection systems. They are made by different legal bodies, i.e., by EU Commission in Europe and by the Central State in China, but the ideas are closely and their adopted policy instruments are also similar: the emissions trading, the information transparency, the use of cleaner energy, etc. This similarity is caused by the needs for environmental protection as a global issue, but also due to the sake of policy-learning and assimilation of ideas and policy means from European states to China. Meanwhile, this study also unravels some key issues in both cases of Italy and China. With regard to policy implementation, the central and local relation is a significant determining factor strongly affecting environmental protection in these societies.

To move the issue further, we can also observe the difference between the policy performances of both sides due to their different central-local relations. In Italy, the policy guideline is set up by the EU Commission which demands member states to confirm and implement them. For this reason, these policies have a strong function of the normative demand and policy guide, whereas its ability to implement them is subject to the decision of the national states. These policy guidelines and policy proposals can be distinguished into three formats, i.e., regulations, directives and decisions, having a different effect on each member state. On the basis of local assimilation of EU legislation, member states often fail to fully implement EU regulations. In some member states, when integrating EU legislation into the national one, they delay a real implementation at local level, since EU guidelines are often too broad and generic to be fully implemented in every specific case, and due to the extreme diversified dimensions among EU local identities, those regulations often turn out to be unproductive.

Moreover, often citizens or the European Commission itself, have open infraction procedures against member states due to several reasons, like delays in the implementation procedure or sometimes even a total inadequate ratification of EU directives, which by definition have to be ratified by national Parliaments before being fully effective. Indeed, we see several cases where the European Court of Justice(ECJ) is demanded to consider whether national policy is in line with the Communitarian one, and especially in the Italian case, the managing of some issues such as waste and air quality has been cause of several interventions by the ECJ (DG HOME Audit, 2006). Thus, the member states from their side have a more complicated role in this process, as the implementation of EU policy at national level is different from state to state, based on the different national constitution models. In the Italian case, EU policy ratification first passes through the national parliament and through direct implementation in other cases.

On the other hands, local governments also have their own interests. Once the interest of national states does not coincide with the demand of regional governance, the policy implementation will become ineffective. Though the national state has a duty of bringing EU policy to life, the budgetary commitment related to the policy is a challenging issue. EU funding may be needed to accomplish the set goals. For instance, Italy is one of the member states receiving an amount of funding higher than EU average. However, this fund is often used for some more urgent needs rather than environmental protection.In Italy, only after 2001 effective power on environmental issue has been delegated to the regions, which have to deal with local entities and local governments.

Thus, the whole picture of the Italian system may be seen as an ongoing federalism, where policy implementation is mostly in regional hands which are directly financed and controlled (and interfered as well) at this level. However in Italy, the dynamics of environmental governance are relatively weak. Nevertheless, this top-down model of administration should be understood in the democratic context, i.e., through the negotiation at the top level and again, between the central (like the federation) and the local (the state) government, which does not deny a strong autonomy of the local authorities.

Looking at this organization model, similarity can be found with the Chinese model. Although the Chinese system of governance is very much centralized, there is a large space for the local governments to make their own actions. The central government can set up the guideline and goals of development on environmental policies, but how to implement these policies depends on the realities of localities. This system has been described by Kenneth Lieberthal and Michel Oksenberg as "fragmented authoritarianism", or as Lieberthal called it, this system can make certain good practices at the national level, while it may become ineffective at local level (Lieberthal, 1991). This fragmentation with several actors playing the same game, delivers an undefined picture. With regard to environmental affairs, once the central government in charge of regulating and promulgating national laws is applied by local authorities, the attitude of local governments become determining factors on how effective the policy implementation is on environmental issues.Under such circumstances, there is a need to reinforce the dynamics from the bottom line, and regulate the demands of citizens for environmental protection into a legal order by a truly enforcement of citizens, rights, by a regulation at both top first and bottom level after against investing companies, which too often enjoy backup from local governments willing to attract investment from abroad.

This effect may also occur at the subnational level. Although environmental standards are set at the national level, local governments can achieve different de facto regulations by relaxing or tightening environmental enforcement. These incentives partly explain why environmental quality has significant spatial heterogeneity, even if the same environmental standard is enacted across the country. Therefore we see different trends around the country, with either more growth oriented or more environmental friendly local governments.

We had several examples of local governments to be a failure in fulfilling their responsibilities. These problems are often incurred a large number of mass incidents, with official reporting that more than 50 000 environmental protests occurred in China in 2012(Xu, 2014). Thus, we need to look for social innovation or the "best practices" of policy implementation in response to people's demands on their environmental interests, and to find the solution engaging both the state agents and multiple stakeholders.

Conclusions and discussions

In the international context, the environmental issue is a highlighted topic in the current policy-making debate. The European environmental policy stems from a tradition of deep green awareness among the European population also influencing policy-makers. However, the policymaking process within the European Union is very much top-down, but democracy and compromise at both the Unions and the local levels can modify the influences of bureaucracy and localization. Therefore, with its long tradition of environmental movement, a top-down model of environmental governance in the EU is not a problem in the majority of the member states. The civil society can focus on long term goals, which do not interfere with the own self subsistence of the single individual priority. Of course, the countries of the recent enlargement, such as Romania, Bulgaria and Croatia, still face hard talk on environmental issues due to the fact that their economic system is not competitive enough yet to be able to compete in the internal market. Thus, the policy proposal can be to increase the interaction of institutions at different levels, while respecting general environmental agreements.

China still has a long way to come to achieve its green goals. In China, the overall cost-benefit evaluation of a real action again polluting companies is still up to the attitude of local governments, which are often split between pro active and inactive. Thus, the potential impact of these initiatives is undermined by a still wide gap among policy makers and policy implementation. An important issue is to establish a clear responsibility distinction among who is in charge to implement the policy. If compared with the Italian case, the overlapping of EU regulation on regional regulation brought an undefined set of responsibility on what to implement, delays in reaching communitarian goals, and also at times a waste of EU funding. In China, the problems are of another kind. In state of waiting the state policy-makers to act first, the local initiatives become a key issue. In the past, environmental policies remained difficult to be enforced at a local level, where officials often retain to economic incentives to ignore them(Tang, 2012). Currently, a generalized awareness on environmental issues among common people is increasing. This factor makes some policy frontrunners at local level to act in a creative way to find solutions in alliance with civil agents, through uprising or by addressing NGOs, which are mostly willing to get in touch with local authorities rather than national, which anyway seems to far and complicated to be taken in account.

Looking at this prospective, we may think about the influence of the central-local relation over the policy performance of environmental protection. From a long term process, we need to bring both citizens and local policy-makers to engage more in environmental issues actions, paying particular attention to the implementation of this policy at the local level. In this study, a complex system of rules and institutions in Italy and China is presented, and we assess the impact of these institutions on policy implementation to lead to a real change. This change might take place in several ways demanding social innovation in environmental governance to shape out in more activities against pollution. The future impact of pollution databases depends in part on whether additional input can be created in order to create an institution that is more effective to reinforce the policy implementation.

References

[1] Buhrs, T.; Bartlett, Robert V (1991): Environmental Policy in New Zealand. 'The Politics of Clean and Green' Environmental politics. Vol. 1 (2) pp 1-9.

[2] Bradsher, K.(2010): China Fears Warming Effects of Consumer Wants. The New York Times, acessed08June2013 http：//www.nytimes.com/2014/05/22/business/international/chinas-fears-warming.html？_r=0.> 2014. 06.15.

[3] Chen, H., Goldberg, M.S., and Villeneuve, P. J.(2008). A Systematic Review of the Relation Between Long-term Exposure to Ambient air Pollution and Chronic Diseases. Reviews on Environmental Health, Vol. 23 No. (4) 243-297.

[4] Chung (2010): Ningbo Tops Pollution Transparency List, China Global Times. Accessed 05 June 2014Retrieved from http：//www.globaltimes.cn/china/society/2010-12/606531.html.

[5] De Sombre E.R. (2000): Domestic Sources of International Environmental Policy: Industry, Environmentalists, and U.S. Power Global Environmental Politics. Global Environment Pollution, Vol. 1 (5): 112-114.

[6] DG ENVI, (2013): Process leading to the 7th EAP, EU official journal, : Retrieved from http://ec.europa.eu/environment/newprg/process.htm. Accessed 01 June 2014.

[7] DG Internal Policies of the Union.(2006). Policy Department: Economic and Scientific Policy. Status of Implementation of EU Environmental Laws in Italy, EU official journal (IP/A/ENVI/IC/2006-183) IPOL/A/ENVI/2006-38 PE 375.865.

[8] European Commission. (2013). European Commission Welcomes the 7th Environment Action Programme Becoming law, EU Official Journal p. 1-7 No Issue, Press document.

[9] Galeotti, M., Alessandro L., and Francesco P..(2006),"Reassessing the Environmental Kuznets Curve for CO_2 Emissions: A Robustness Exerciser," Ecological Economics, 57, 152-163.

[10] Gilley, B. (2012): Authoritarian Environmentalism and China's Response to Climate Change.

Environmental politics. Vol. 21, No. 2, March 2012, 287-307.

[11] Godement, F. (2014). China at the Gate: A Changed Relationship with Europe. Critical Issues Seminar Series. Summary of Talk, Hill Rag Magazine, February 2014.

[12] Gunter, Michael M. Jr. and Rosen, Ariane C. (2011): Two-Level Games of International Environmental NGOs in China, Wm. and Mary Pol'y Rev. Vol. 270 (2) pp.128-53.

[13] Hart M. and Cavanagh J.: (2011): Environmental Standards Give the United States an Edge Over China, Center for American Progress. Vol. 1 (3) pp. 23-47.

[14] Hood, C. (2010): Current and proposed emissions trading systems. Review Existing and Proposed Emissions Trading Systems: Information paper from International Energy Agency (IEA). Head of Publications Service, Paris, France: pp.1-34.

[15] Raustalia K., (1997): States, NGOs, and International Environmental Institutions International Studies Quarterly, Harvard Law School Vol. 41 (3) pp.719-740.

[16] Jensen T., (2009): The Democratic Deficit of the European Union. Living Reviews in Democracy, Vol. 1 (2) pp. 23-59.

[17] Jiang, K.J., Hu, X., Chuang, X., and Liu, Q., 2009. Zhong guo 2050 nian ditan qingjing heditan fazhan zhilu [China's 2050 low carbon outlook and development path]. Beijing: Energy Research Institute, National Development and Reform Commission.

[18] Kahn J. and J. Yardley "As China Roars, Pollution Reaches Deadly Extremes". The New York Times (26 August 2007).

[19] Karlsson, C., Parker, C., Hjerpe, M., and Linne'r, B.-O., 2011. Looking for leaders: perceptions of climate change leadership among climate change negotiation participants. Global Environmental Politics, 11 (1), 89-107.

[20] Levin, K., and R. Bradley (2010), Working Paper: Comparability of Annex I Emission Reduction Pledges (PDF), Washington DC, USA: World Resources Institute. pp. 129-149.

[21] LI Jinshan, YE Tuo, Fragmented government: theories and facts in China. An Interdisciplinary Journal on Greater China. The China Review. An Interdisciplinary Journal on Greater China, Vol.14 (1) pp. 23-64.

[22] Lieberthal, K. and Michel O.(1988)Policy Making in China: Leaders, Structures, and Processes. Princeton University Press, Vol. 23 (7) pp. 178-203.

[23] Li, Y.W. and Miao, B., 2011. Managing deep uncertainty: Shanghai and Yangtze river delta governments' adaptation to climate change. International Studies Association Annual Conference, 16–19 March, Montreal, Canada.

[24] Midlarsky, Manus I., (1998), "Democracy and the Environment: An Empirical Assessment, " Journal of Peace Research 35 (3): 341-361.

[25] Nihan Akyelken. Analysis of European Union Environmental and Energy Policies related to long Distance Freight. Working paper No 1037, October 2008, Oxford University Centre for the Environment.

[26] Oates W.E.and Portney P.R.(2003): The political economy of Environmental Policy Handbook of Environmental Economics, Volume 1 (3) pp. 67-96.

[27] OECD Publications, "Instrument Mixes for Environmental Policy" (Paris, 2007) "Environmental Policies and Instruments, " Accessed 10 June 2014.

[28] Ostrom, E. 2000. "Collective Action and the Evolution of Social Norms." Journal of Economic Perspectives, Vol. 14 (3): pp. 137-158.

[29] Papadaki O., (2012): Evolution of European environmental policy, Regional Science Inquiry Journal, Vol. 4 (1), pp. 152-58.

[30] Pearce, D.W. (1998), "Environmental appraisal mad environmental policy in the European Union", Environmental and Resource Economics. Vol. 23 (2) pp. 34-59.

[31] Price, L., Levine, M.D., and Zhou, N., 2011a. Assessment of China's energy-saving and emission-reduction accomplishments and opportunities during the 11th five year plan. Energy Policy, 39 (4), 2165-2178.

[32] Portney P.R.and Robert N. S., (2000): Public Policies for Environmental Protection, John F. Kennedy School of Government, Harvard University Press, Washington, D.C.: RFF Press.

[33] Rio Declaration on Environment and Development. The United Nations Conference on Environment and Development, Environmental management, Rio de Janeiro 3 -14 June 1992.

[34] Tan Y., (2012): Transparency without Democracy: The Unexpected Effects of China's Environmental Disclosure Policy.

[35] Upton John. "China to spend big to clean up its air". Grist.org. Grist Magazine, Inc.

[36] Xu Beina, China's Environmental Crisis, Council on foreign relations. Accessed February 5,

2014 http：//www.cfr.org/china/chinas-environmental-crisis/p12608>.

[37] WHO Press Release No 22 1 17 October 2013 IARC：Outdoor air pollution a leading environmental cause of cancer deaths，Geneva，Switzerland.

[38] Wu J.，Deng Y.，Huang J.，Morock R.，Yeungs B.（2012）Incentives and outcomes：China's Environmental Policy，IRES Working Paper Series No. 23 Issue 1，University Of Singapore，Singapore.

论美国的环境信息披露制度及其对公众参与的激励作用

杨泽明*

概述

  2014 年 12 月是印度博帕尔市骇人听闻的毒气泄漏惨案 30 周年祭。1984 年 12 月 2 日夜里，一场灾难性的工业事件在人们的酣睡中爆发。联合碳化物有限公司在当地生产杀虫剂的化工厂泄漏了超过 40 t 的异氰酸甲酯气体，这种物质是该工厂生产过程中的中间产物，它是一种高度致命性的有毒气体。

  泄漏气体很快形成了一团巨大的有毒烟雾并向四周扩散，造成了 2000 多人当场死亡的惨剧。死亡人数中，绝大部分是居住在工厂附近贫民窟的居民。据估计还有超过 15 000~30 000 人因毒气导致的相关病症而死亡，另有上万人因此而伤病。那些幸免遇难者也饱受化学灼伤、呕吐、呼吸道问题、肺功能下降以及失明等病症的折磨。事情一旦发生，由于没有相应的应急机制，得不到及时有效的解决，会进一步恶化得更加严重。此外，负责接收受害者的医院缺乏对该毒气关键信息的了解，因此也未提供有效的医治。

  事后，联合碳化物公司试图以蓄意破坏罪之名将罪责推卸给一位员工。不过，到 2010 年，印度法庭判定该公司 7 名 70 多岁的管理和技术人员对该毒气泄漏事件负有罪责，并对他们判处两年有期徒刑和小额罚款。然而，他们中的大部分在

---

* 杨泽明，圣克拉拉大学法学院教授，前美国国家环境保护局法律顾问。本文原文为英文，中文译者为付志宇（浙江大学公共管理学院）。

上诉期间因保释而逍遥法外。数十年来围绕此毒气泄漏案的侵权诉讼一直进行着，最终在确定联合碳化物公司对上万受害者提供 4.7 亿美元的损失赔偿方案后才停止。

博帕尔毒气泄漏惨案令人震惊的电视影像对美国民众产生了巨大的冲击，也使美国公众对环境污染和对化学物质的使用及排放的危害意识大大提高，最终促成了 1986 年美国《应急计划与社区知情权法》（Emergency Planning and Community Right-to-Know Act，EPCRA）的颁布。1990 年，EPCRA 被修改成为美国《污染预防法》（Pollution Prevention Act，PPA），它有两个具体的目标：推动化学事故应急预案的制定以及向公众提供其社区所使用的有毒和危险化学品的相关信息。

为了达到这两个目标，《应急计划与社区知情权法》和《污染预防法》建立了美国《有毒物质排放清单》（Toxic Release Inventory，TRI），这是一个有关被工业设施使用和对外排放的有毒化学品和污染物的数据库。《有毒物质排放清单》（TRI）是"污染物排放与转移登记制度"（Pollution Release and Transfer Registry，PRTR）一个更为群众所熟知的版本，也被一些人称为"污染排放数据库"。

近年来，"污染物排放与转移登记制度系统"（PRTR systems）已经成为了保护环境和吸引公众参与的重要工具。这些制度能够促进信息公开，并在信息公开的情况下更广泛地推动社会参与环境决策。而 TRI 是美国主要的污染物排放与转移登记制度，它已成为公开有毒化学物质排放和处置信息的关键机制。它现在有美国国家环境保护局管理的免费访问在线数据库。

TRI/PRTR 的污染物排放要求很容易理解，即它们要求工业设施收集并向公众公开其排放到环境中的有关有毒化学品和废弃物的信息。基于其既实用又有效的特点，越来越多的国家采纳该制度。

污染物排放与转移登记制度（PRTR）在美国的应用：《有毒物质排放清单》（TRI）

《应急计划与社区知情权法》第 313 节明确指出该法公开信息的根本宗旨是

"为公众提供有毒化学物质排放的信息,以协助政府机构、学者以及其他人员进行研究和数据搜集工作,为有关法规、指导方针和环保标准的完善提供帮助;和其他的类似用途。"

在 TRI 的要求下,美国国家环境保护局每年需要收集各个工业设施当年排放的有毒化学物质以及废弃物管理的相关数据,并通过 TRI 的数据库向公众公开。目前,TRI 数据库包含有来自美国 20 000 个工业设施的超过 600 种有毒化学品的排放和处理数据以及这些设施循环使用进行能源再生和有毒物质的最终处置的信息。

那么哪些企业需要采用有毒物质排放清单制度呢?事实上,虽然美国总共有大约 29 000 家工厂,但必须遵守 TRI 制度的企业却不超过总数的十分之一。这主要是 TRI 限制了需求提交报告的企业范围以及履行报告义务的活动区域的比率的结果。

第 313 节所规定的需要服从申报的工业部门主要集中在制造行业。但是美国国家环境保护局也拥有国会所赋予的增减需要申报的工业部门的职权。自 1986 年以来,美国国家环境保护局已多次行使此项权力,将包括金属采矿业与电力行业在内的多个工业部门划入需要提交报告的企业的范围内。单独的工业设施虽不是独立的企业单位,也可以是进行 TRI 申报的责任单位。这可以通过 TRI 为社区提供有毒化学品排放信息的基本信息。由于一个企业可以拥有许多工业设施,因此要达到这个目标需要具体到了解特定的工业设施及其排放数据(而非公司层级的对外排放数据)。与此同时,TRI 的要求仅适用于雇佣 10 个及以上全职员工规模的工厂,因而解除了小型工厂的报告负担。

除了工厂规模,TRI 还限定了生产活动的报告阈值。在一个日历年度内工厂若生产/加工/排放了超过 25 000 磅的有毒化学品或是在其他使用过程中使用了超过 10 000 磅的有毒化学品,就必须向 TRI 进行申报。美国国家环境保护局也对某些特定的化学物质降低了报告阈值,以便将某些可能引发特殊关注的物质控制在一个更低的生产、加工和使用水平。例如水银的报告阈值是每年 10 磅。

企业每年需要将上一日历年度的申报信息在当年的 7 月 1 日前上交给国家环境保护局。尽管无法做到全面仔细的检查,但美国国家环境保护局仍然会对企业上交的信息进行数据质量审查,然后再向公众公开。这也意味着 TRI 的信息资料需要等到几个月之后才会被公布于众。美国国家环境保护局维护的一个网站负责对外公布这些信息。

正是因为 TRI 数据库向社会大众提供有毒物质排放信息的目标，商业秘密保护条例对 TRI 信息只有有限的适用性。该条例确实允许企业对生产过程中所使用的特定化学物质的信息进行保密。但如果企业需要保密，则必须同时提供有关生产过程中所使用的该分类的化学物质的信息。

TRI 制度最终引入了一系列的强制性条款。美国国家环境保护局可以在企业违反报告要求时对企业处以高达每天 2.5 万美元的罚款。然而，对企业是否达到应该申报门槛的界定则是执法的挑战。国家环境保护局可能不知道法定阈值是否被跨越了，因而也就不知道报告义务是否已经形成。此外，为了减轻 TRI 规定的信息采集义务，申报允许企业使用"根据其他法律规定所收集的现成数据（包括监测数据），或在没有现成数据时使用合理估计的估计值。"该条款也不需要向 TRI 具体报告"任何排放到环境中的有毒化学物质的数量、浓度和频率等监视和测量数据"。

最后，这个法律与美国其他环境法律一样也有公民诉讼条款，它允许个体对违反 TRI 报告要求或违反国家环境保护局规定的企业合理执法。与其他联邦环境公民诉讼条款相同，公民原告需要提前 60 天通知被告才能完成公民诉讼申请。但是，案例法却成为了这类关于举报违法行为的公民行动的实际阻碍，因为案例法允许企业在公民申诉完成前通过补交相关的表格和资料，从而达到法律要求，不构成违法行为。如此一来，被告企业一旦发现有环保组织计划对其提出诉讼时便能逃脱法律诉讼。

《有毒物质排放清单》（和污染物排放与转移登记制度）的作用

具体来说，TRI/PRTR 等制度向社区提供了有毒物质排放和废弃物管理的信息，也协助政府官员、工业企业、民间组织和公民个体进行环保决策。TRI 的一个激励动机便是去告知社区和联邦政府、州政府及地方政府有毒物质排放到环境的相关信息。这一前提是为决策提供信息的知情决策对于有效的环境治理至关重要，并且环境信息披露是环境保护的核心。更确切地说，污染物排放和转移登记制度系统能够提高环境管理水平和公众的作用。

（1）污染物排放和转移登记制度为当地社区提供了可能对他们产生直接影响的污染物的重要信息。社区使用这些信息可以帮助自己了解环境中存在的危害，从而能够将环境的风险降到最低，也可以围绕共同关注的问题进行更好的组织。

（2）这些信息还可以被工业企业或污染者自身使用用以监管减少污染排放。这一做法能够减少污染的机会。企业的高层管理者会因为已经向公众披露这些信息而更加重视日常的排污报告。他们通常有权利在企业内部实施降低污染的调整计划，例如更高效的物料管理或识别可供选择的更为"绿色"的生产方法。

（3）公众使用环境排放数据可以用于了解公众的态度和消费行为，并且利用竞争压力和市场激励企业减少污染。即使这些信息没有使企业产生压力或者促使企业采取环保行动，在没有其他积极干预的情况下，仅仅环境信息披露本身也能够让污染者对公众舆论和压力做出反应。

（4）污染物排放与转移登记制度可以提示政府对环境保护的优先顺序做出决策，从而改善环境的问题。因此污染物排放与转移登记制度自然而然就成为了政府官员重要的信息补充来源。然而值得一提的是这个目标与有效的公众参与密切关联。好的环境治理和政府决策需要考虑到受污染影响的周边社区的价值观和利益诉求，而这种价值观和利益诉求正是通过社区的积极参与表现出来。

成效

TRI 信息是怎么在实践中运用并且获得了怎样的效果呢？美国国家环境保护局于 2003 年的一次调查检验了这两个问题并且得到了如下结论：

——社区使用 TRI 数据展开了与地方工业企业的对话，旨在促进企业减少污染排放并开展以改进安全措施为目的的污染预防（pollution prevention, P2）计划。

——公益组织、政府和学者以及其他人士使用 TRI 数据对公众开展有毒物质排放和潜在风险教育活动。

——工业企业利用 TRI 数据寻找污染防治（P2）机会，设定减少有毒物质排放的目标并做出降低污染排放量的承诺和进展计划。

——联邦、州和地方政府用 TRI 数据来帮助设定决策的优先顺序，将环境保护资源用于最紧要的问题上。

——管理者使用 TRI 数据设定许可范围，度量范围的遵守程度，进而为执法行动确定目标企业。

——公益组织利用 TRI 数据来向公众说明现行法规迫切需要强执行力并落到实处，以及对新的环保法规的需求。

——投资分析师用 TRI 数据向有环保投资意向的客户提供建议。

——保险公司用 TRI 数据来预测潜在的由环境引发的债务。

——政府利用 TRI 数据来评估和修改基于有毒物质排放或整体环境绩效的税收和费用。

——咨询公司及其他企业使用 TRI 数据来发掘潜在的商业机会，例如进行污染防治（P2）营销和对需要进行 TRI 报告的企业的相关技术进行控制等。

总而言之，TRI 数据被广泛应用于各个方面，联邦和州政府用它来制定应急计划；TRI 报告企业等工业企业也使用它来改进他们的操作；社区和民间环保组织用它来避免其居住社区遭到污染的危害；而学术研究人员用它提出环境监管体系的改进建议。

除了以上对政府官员、公民社会以及企业的效用，TRI 更展现出了其在环境保护上的成效。事实上，在利用 TRI 公开披露污染排放信息这一行为成为合法权利之前社会就已经强烈感受到了它的效果。一个关于 TRI 早期影响力的故事曾广为流传。孟山都公司在当时还是世界几大化工企业之一，它的领导者理查德·马奥尼曾经向他的经理们宣称他的公司将在 5 年内把有毒污染排放量减少 90%。这条备忘在当时新法规所规定的 TRI 申报日期的前一天（1988 年 6 月 30 日）对外公布。而令马奥尼讶异的是，早在几日前，他的员工已经对外公布了孟山都公司"在 1987 年向空气、水域和陆地排放了 3.74 亿磅的有毒物质"。为什么在根本不知道公司的污染排放量时，马奥尼会做出这样的决策？因为他担心社会公众对有毒物质的广泛认识和对他的公司有毒污染排放的关注会对为其公司的声誉和公众形象造成潜在危害，特别是当他的客户、投资者、政客以及新闻界等对揭露出来的真相做出消极反应时，会给公司业务造成严重损失。在此之前，企业还从未被如此大规模的要求披露他们有毒污染排放的数据。

在孟山都公司的带领下，其他大型公司和制造企业也给出了大幅减少污染排放的承诺。1992年，孟山都公司对外宣布他们实现了将污染排放降低至1988年水平的10%的承诺。这个案例说明了环境信息披露确实能够有效改变污染企业的行为。

其后几年中，TRI系统收集的信息越来越容易获取，包括国内的公益组织在内的民间组织以及积极草根环保人士、科研人员及其他人士开始充分利用这个资源。举例来说，地方环保组织开始使用TRI信息来向当地的社区居民普及工业企业生产排放有毒物质的相关知识和对大众健康的潜在危害。他们开始展开对如何减少这些危害的热烈讨论。一些组织甚至用TRI信息来告诉立法机关成员及其他政府官员这些污染对社会的危害，并开始用这个来改进现行环保法规系统以更好地为社区和民众服务。除此之外，有了TRI信息还可以更好地为环境健康以及有毒物质对环境的影响做出分析，并降低与私人企业及环境监管部门对话过程中社区环保人士对环境健康的担忧。

这些年来，TRI数据的使用以及它对企业污染行为的改变影响不断深化。TRI信息促进了环保管理行动的开展。它在不断向公众普及相关知识的同时，鼓励公众参与到环保问题上来。国家和地方环保组织、立法机关成员以及管理者有效地利用这一点解决了许多污染问题。

注意事项

然而，污染物排放与转移登记制度系统还有许多需要注意的事项。建立一个有效的污染减排计划，仅仅要求披露环境信息的立法是不够的。一个成功的信息披露制度通常需要三个条件：有能力正确实施该制度的政府、遵从制度的企业以及公众社会的参与。

能力：要充分利用这个工具的前提是监管体系必须有足够的能力去接收、存储、验证和管理信息，继而能以便捷的方式将信息提供给公众。

遵从：提交的数据越完整准确，报告的数字更容易在年份、领域和企业之间进行横向和纵向比较，这个披露制度才会更有利用价值。

公众参与：一个披露制度依赖于公众对这类信息的需求，公众的利益诉求以及社会对污染者和管理者的影响能力。前两个因素导致了最后一个因素的产生。通常向公众发布环保信息能够使公众以积极的角色参与其中，并会增强政府的努力。此外，获取信息的渠道和意识能够在社会、企业和政府之间鼓励社区的参与及环境伦理的发展。换句话说，它有助于公众参与的循环持续。因为披露制度允许公众参与，同时有效的公众参与也反过来推动披露制度的施行。

在很多方面，这些注意事项说明了污染物排放与转移登记制度和信息披露制度是镶嵌在一个大的管理系统中。在这个管理系统中，管理者、企业和社会有用以对这些通过强制或是其他责任机制收集来的信息进行反馈的工具，有鼓励使用披露信息的更为广泛的规章制度和法律法规，以及对这些要求的自愿遵守。

最后一个现象也是一个具有广泛意义的问题，它展示了信息披露制度如何嵌入一个更大的环境治理的系统中以及其他机构如何影响作为环保工具的信息披露制度的有效性。环境信息披露制度等工具的作用和重要性很大程度上取决于民间环保组织和学术界能否帮助公众和社会理解有关生态系统高技术的重要性，污染排放的风险危害及标准，以及关于污染和环境健康的复杂因果关系。换言之，在环境信息披露的制度下要求公众参与会有更大的效果，但也仍然需要其他机构特别是民间环保组织和学术研究人员的支持。在努力建立包含污染物排放与转移登记制度系统在内的法律改革时，应该牢记民间社会机构也是促使这种系统成功的重要主体之一。

结论

毫无疑问，环境信息披露制度在增强环境决策和促进公众在环保问题上的参与上具有很大的价值。此外，它使得民间社会能够教育公众，并更有效地向环境监管部门和企业传达他们对于环境问题的担忧。反过来说，它让企业和环保官员做出在保护环境和提高环境质量上更符合公众利益的反应。最终，类似美国的有毒物质排放清单（TRI）和污染物排放与转移登记制度（PRTR）等制度的有效的环境信息披露制度在其他地方有效实施，可以提高环境治理水平，最终达到改善环境的效果。

The US System of Environmental Information Disclosure and its Role in Encouraging Public Participation

Tseming Yang[*]

Introduction

This coming December will mark the 30th anniversary of the horrific gas leak disaster in Bhopal, one of India's largest cities. That fateful night of December 2, 1984, when many of the city's residents were asleep, a catastrophic industrial accident occurred at the Union Carbide Corporation's local pesticide manufacturing plant which released more than 40 tons of methyl isocyanate gas. Methyl isocyanate was a highly poisonous and deadly gas that was created as an intermediate substance during the production process at the plant.

A toxic gas cloud formed that spread to the surrounding areas, killing 2 000～3 000 people immediately, mostly residents of a poor slum settlement adjacent to the plant. Some estimate that another 15 000～30 000 more died in the aftermath from causes related to the toxic gas and hundreds of thousands more were injured or sickened by the industrial accident. Those who were lucky enough to survive suffered injuries and symptoms ranging from chemical burns, vomiting, and respiratory problems to decreased lung function and blindness. While the accident had been very serious in itself, it was exacerbated by the lack of an emergency plan or advice on what the population or emergency response personnel should do in the event of a release of

---

[*] Tseming Yang, Professor, Santa Clara University Law School, formerly Deputy General Counsel, US Environmental Protection Agency (2010-2012).

toxic gas such as methyl isocyanate. Furthermore, the hospitals that treated the victims brought to them also lacked critical information on the gas and were unprepared to provide the most effective treatments.

In the aftermath, Union Carbide attempted to blame a disgruntled worker for sabotage. However, in 2010, Indian courts found 7 management and technical staff members, then already in their 70s, guilty for the disaster and imposed 2-year prison sentences and small fines; however, most of them remained free on bail pending appeal. Tort litigation surrounding the gas leak lasted for decades and ultimately ended in a settlement of $470 million paid to the hundreds of thousands of victims. Even by the standards of that time, the settlement amount was a modest one. A separate earlier case brought by the gas leak victims in American courts did not succeed on procedural grounds and was sent back to India for resolution.

The impact of the Bhopal gas leak disaster and the shocking TV images of the dead and injured victims were tremendous in the United States. The American public's awareness about the risks of pollution and the dangers of chemicals used and released into communities increased greatly and eventually resulted in the enactment of the 1986 Emergency Planning and Community Right-to-Know Act (EPCRA) (42 U.S.C. §§ 11001-11050) .EPCRA, as expanded by the 1990 Pollution Prevention Act (PPA), had two specific goals: 1) promoting contingency planning for chemical emergencies, and 2) providing the public with information about the toxic and hazardous chemicals used in their communities.

To achieve these goals, EPCRA and the PPA created the Toxic Release Inventory, a database of toxic chemicals and pollutants that are used and released by industrial facilities into the environment. TRI is one version of what is more commonly known as a Pollution Release and Transfer Registry (PRTR), also referred to as Pollution Discharge Databases by some.

In recent years, PRTR systems have emerged as important tools for protecting the environment as well as engaging the public. PRTRs are essential tools for facilitating disclosure of complete and accurate information and for promoting wider participation in environmental decision-making when the information becomes available to the public, oftentimes through the internet. TRI is the primary PRTR used at the national level in the US and has become a key mechanism for making information about toxic chemicals releases and disposals available to the American public. It is administered by US EPA and its practical manifestation at present is as an EPA online database that is accessible for free.

In concept, TRI/PRTR requirements are relatively simple. They require industrial facilities to collect information about what kind of toxic chemicals and wastes are released into the environment and make such information available to the public. Its utility and effectiveness of PRTRs has led to their adoption into many other countries.

PRTR in the US at the National Level: The Toxic Release Inventory (TRI)

EPCRA section 313 specifically states that the fundamental purpose of releasing information is to "inform persons about releases of toxic chemicals to the environment, to assist government agencies, researchers, and other persons in the conduct of research and data gatherings; to aid in the development of appropriate regulations, guidelines, and standards; and for other similar purposes."

Under the TRI requirements, EPA is required on an annual basis to collect data about releases certain toxic chemicals from industrial facilities as well as their waste management and make the data available to the public through the Toxics Release Inventory (TRI). Currently, the TRI database contains data on disposal or other releases of over 600 toxic chemicals from 20 000 facilities in the US as well as information about how such facilities manage those chemicals through recycling energy recovery, and treatment (42 U.S.C. §11047 (8)).

What entities are subject to TRI requirements? It turns out that even though there are approximately 290 000 manufacturing facilities in the United States in total, TRI only covers less than 10 percent of the total facilities. That is primarily the result of limit on what kind of entities has to report releases as well as the threshold level of activities that trigger the reporting obligations.

Section 313 identifies a number of covered industrial sectors that are subject to the reporting requirements, primarily manufacturing industries. However, EPA was also given the power to add industrial sectors (42 U.S.C. §11023(b)(1)(B)). Since 1986, EPA has used that authority to add several other industrial sectors, including metal mining and electric utilities. The responsible units for TRI reporting purposes are individual industrial facilities rather than corporations (42 U.S.C. § 11049 (4)). The rationale for this can be found in TRI's objective of providing toxic chemical release information at a community level. Accomplishing this purpose requires knowledge of the specific releases from a specific facility rather than corporate-wide emission, as a corporation could own many individual facilities. Finally, TRI requirements apply only to facilities with 10 or more full-time employees, thereby exempting smaller facilities from the reporting burdens. The requirement also ensures that reporting is focused on those facilities that are substantial in size and thus likely have substantial releases.

In addition to limitations on the size of facilities that become subject to the TRI requirements, there are also activity thresholds that must be triggered. Covered facilities must manufacture or process more than 25 000 lbs of a toxic chemical or otherwise use more than 10 000 lbs of a toxic chemical in a calendar year (42 U.S.C. §11023 (f) (1)). For certain chemicals, EPA has lowered that reporting threshold for special toxic substances so as to capture a lower level of manufacturing, processing or use of substances that may pose special concern (42 U.S.C. §11023 (f) (2)). For example, the reporting threshold for mercury is 10 lbs per year.

Information collected by facilities for the TRI must be reported to EPA by July 1 of each year for the previous calendar year. That information is then the subject of data quality checks by EPA, though not a comprehensive and detailed review of the information, before it is made available to the public. Usually that means that TRI information does not become available to the public until a number of months after the statutory deadline. It is then made publicly available via a website maintained by EPA.

Trade secret protection have only limited applicability to TRI information, primarily because the very purpose of the TRI database is make toxic chemical release information available to the general public. The statute does allow withholding of information about specific chemical identity involved. If such information is withheld, facility must instead provide information about the "generic class or category" of the chemical involved (42 U.S.C. §1104).

Finally, the TRI system incorporates a set of enforcement provisions. EPA may impose penalties of up to $25 000 per day for violations of the reporting requirements (42 U.S.C.§11045). However, an inherent challenge to enforcement is the identification of facilities that should have reported TRI information but failed to do so. EPA may simply not know whether statutory thresholds have been crossed and thus whether reporting obligations were triggered.Furthermore, in order to alleviate the information collection obligations created by TRI, the reporting obligations allow facilities to use "readily available data (including monitoring data) collected pursuant to other provisions of law, or, where such data are not readily available, reasonable estimates of the amounts involved." The statute does not require "the monitoring or measurement of the quantities, concentration or frequency of any toxic chemical released into the environment" specifically for TRI reporting (42 U.S.C. §11023 (g) (2)).

Finally, like other US environmental laws, there is a citizen suit provision which allows private individuals to bring enforcement against facilities in violation of the TRI reporting requirements or against EPA to take legally required actions (42 U.S.C. §11046). As with other federal environmental citizen suit provisions, citizen plaintiffs have to provide a 60-day notice to defendants before filing a citizen suit. However, a substantive impediment to the success of such citizen actions with respect to reporting violations has been case law that allows industry to come into compliance by submitting the relevant forms and information before a suit is filed. As a result, an industry defendant can avoid a law suit once it becomes aware of an environmental group's intention to file a citizen enforcement action (Steel Company v. Citizens for a Better Environment, 523 U.S. 83 (1998)).

The Function of TRI (and PRTR)

Concretely, TRI/PRTRs provide communities toxic chemical releases and waste management activities as well as assist government officials, industry, civil society organizations, and individuals with environmental decision making. One of TRI's motivating purposes was to inform communities and federal, state, and local governments about toxic chemical releases to the environment. It is thus based on the premise that informed environmental decision-making is crucial for effective environmental governance, and that disclosure of environmental information is at the core of achieving environmental protection.

More specifically, PRTR systems can serve a number of functions that improve both environmental management and the role of the public in that process.

1. PRTRs can provide vital information to local communities on pollution that may directly affect them. The information may be used by communities to educate themselves about environmental hazards, and thereby allow them to manage or minimize their exposures, as well as allow them to better organize around issue of common concern.

2. The information can be used by industry/polluters themselves for evaluating the effectiveness of practices intended to reduce pollution or to identify opportunities for reduction. Because this information is disclosed to the public, it gets more attention from high-level company managers than routine emissions reports. These company managers often have the authority to implement internal changes to reduce pollution, such as adjustments towards more efficient materials management or the identification of alternative "greener" production methods.

3. Public access to environmental release data can inform public debate and consumer behavior, and leverage competitive pressure and the market as motivators for companies to reduce pollution. And while information about a facility's toxic chemical releases may be actively used to pressure and shame a company into taking appropriate environmental actions, just disclosure in itself allows public opinion and pressure to work on a polluter even without active intervention of others.

4. PRTRs can inform government decision making with respect to setting priorities and thus improve environmental governance. Naturally, PRTR information can supplement information that government officials already possess through other sources. However, it is important to note that this mission is closely inter-related with effective public participation. Good environmental governance and official decision-making takes into account the values and interests of the surrounding community affected by pollution and that are expressed through the active participation by communities.

Outcomes

How has TRI information been used in practice and what has it accomplished. A 2003 EPA study examined both of these questions and found the following:

• Communities use TRI data to begin dialogues with local facilities and to encourage them to reduce their emissions, develop pollution prevention (P2) plans, and improve safety measures.

•Public interest groups, government, academicians, and others use TRI data to educate the public about toxic chemical emissions and potential risk.

• Industry uses TRI data to identify P2 opportunities, set goals for toxic chemical release reductions, and demonstrate its commitment to and progress in reducing emissions.

• Federal, state, and local governments use TRI data to set priorities and allocate environmental protection resources to the most pressing problems.

• Regulators use TRI data to set permit limits, measure compliance with those limits, and target facilities for enforcement activities.

• Public interest groups use TRI data to demonstrate the need for new environmental regulations or improved implementation and enforcement of existing regulations.

• Investment analysts use TRI data to provide recommendations to clients seeking to make environmentally sound investments.

• Insurance companies use TRI data as one indication of potential environmental liabilities.

• Governments use TRI data to assess or modify taxes and fees based on toxic emissions or overall environmental performance.

• Consultants and others use TRI data to identify business opportunities, such as marketing P2 and control technologies to TRI reporting facilities.

Thus, TRI data is widely used within the federal and state governments, including for emergency planning, by businesses, including the reporting industrial facilities themselves, to improve their operations, by civil society, communities and environmental groups to help protect their communities, and by academic researchers to understand and suggest improvements to regulatory systems.

Beyond its utility to government officials, civil society, and businesses, TRI has also shown concrete environmental successes. In fact, the power of public disclosure of pollution releases harnessed by TRI made itself felt even before it became legally effective. In one of most widely reported stories about the early impact of the TRI, Richard Mahoney, the head of what was then still one of the world's largest chemical companies, Monsanto Corporation, announced to his managers that his company would reduce the release of toxic pollution by 90 percent within less than five years. And while the memo was released on June 30, 1988, the day before the first TRI reports were due to be submitted to EPA under the then-new law, the announcement had been prompted a few days earlier by Mahoney's own astonishment by information from his own staff that in 1987, Monsanto had "released 374 million pounds of toxic chemicals to the air, water, and land." Why such a response, especially because at the time, Mahoney actually did not know yet how his company would accomplish such reductions? Mahoney was concerned about the potential harm to its reputation and public image that widespread knowledge and attention to the toxic pollution from his company might have, especially harm to its business if its customers, investors, politicians, the press and others might react negatively to these revelations. Until then, companies had never before been required to disclose information about their toxic pollution in such a comprehensive fashion.

Other large companies and manufacturers followed Monsanto's lead and also promised large pollution reductions. In 1992, Monsanto was able to announce that it had been able to meet its pledge to reduce emissions to 10% of 1988 levels. The instance showed that environmental information disclosure could actually change behavior by polluting businesses.

In later years, when information collected through the TRI became widely available, civil society organization, including national environmental groups as well as small grass-roots/local community activists, researchers, and others began to make use of it. For example, local environmental organization started using TRI information to educate their local communities about the toxic chemical releases from local industrial facilities and potential public health effects. They then engaged in discussions with industrial facilities about those impacts and how they could be reduced. Some organizations even collected TRI information to educate legislators and other government officials about pollution releases of concern to communities and used this information as a starting point for examining how the environmental regulatory system could be improved to serve communities and people better. Finally, TRI information has also allowed for better analysis of health and environmental impacts of toxic chemical releases and enhanced the credibility of health concerns raised by community activists in their discussion with private companies and environmental regulators.

Over the years, the uses of TRI data as well as its effects on changing behavior have continued to evolve. TRI information has prompted and improved environmental regulatory actions. It has continued to engage communities on environmental issues and educated the public generally. And it continues to be used effectively by national and local environmental organizations, legislators, and regulators to address pollution issues.

Caveats

There are some caveats that must be kept in mind with PRTR systems, however. To create an effective PRTR program, simply enacting laws requiring disclosure of information is usually not enough. A successful disclosure system also requires: Capacity of the government to properly implement it, compliance by reporting facilities, and engagement by civil society through use of the information.

Capacity: To fully leverage this tool, a regulatory system must have the capacity to receive, store, verify, and manage the information and then must provide the information to the public in a timely and accessible manner.

Compliance: The more complete and accurate reporting is and the more easily the reported numbers can be compared from year to year, from sector to sector and from facility to facility, the more useful the disclosure system can be.

Public Engagement: And finally, a disclosure system relies on public demand for the information and the public's interest and ability to influence the behavior of the polluters and their regulators. This third factor can also grow from the first two. Routinely making environmental information available to the public enables society to take an active role, reinforcing and expanding upon government efforts. In addition, access to this information and the awareness of it fosters community engagement and development of an environmental ethic throughout civil society, industry, and government. In other words, it can contribute to self-perpetuating cycle of public involvement. The disclosure system both allows for and is improved by effective public participation.

In many respects, these caveats suggests that a PRTR and information disclosure system is embedded in a larger system of governance, where there should be tools for regulators, industry, and civil society to act in response to the information, whether it is through enforcement and other accountability mechanisms, and where the broader system of norms and the rule of law encourage use of the information that is disclosed and voluntary compliance with its requirements.

This last observation, how information disclosure systems are embedded in a larger system of environmental governance and how other institutions affect the effectiveness of information disclosure as a tool for environmental protection, is an issue that also has broader significance. The role and significance of tools such as environmental information disclosure depends in significant part on the role of environmental NGOs and academics to help the public and civil society understand the significance of highly technical information about ecological systems, the dangers or risks of pollution releases or standards, and the complicated cause-effect relationship with respect to pollution and environmental health. In other words, public participation is made more effective by environmental information disclosure requirements but also needs support from other institutions, especially environmental NGOs and academic researchers. Law-reform efforts, including the establishment of PRTR systems, should not forget that civil society institutions are an important ingredient to the success of such systems.

Conclusion

There can be little question that environmental information disclosure systems can be enormously valuable for enhancing environmental decision-making as well as promoting public participation on environmental issues. In addition, it enables civil society to educate the public and to make its concerns more effectively known to environmental regulators and industry. Conversely, it allows businesses and environmental regulators to be more responsive to interests of the public in protecting and enhancing environmental quality.Ultimately, effective environmental information disclosure systems such as the U.S. TRI and PRTRs elsewhere improve environmental governance and help protect and improve the environment more effectively.

Endnote

The paper is based on a presentation given at a EPA-MEP-NRDC Workshop in May 2012 and relies on research by Tim Epp and Erin Koch. Yorum Choe provided additional helpful research.

Reference

[1] Releases are defined to mean "any spilling, leaking, pumping, pouring, emitting, emptying,

discharging, injecting, escaping, leaching, dumping, or disposing into the environment (including the abandonment or discarding of barrels, containes, and other closed receptacles) of any hazardous chemical, extremely hazardous substance, or toxic chemical." 42 U.S.C. §11047 (8).

[2] A facility is defined as "all buildings, equipment, structures, and other stationary items which are located on a single site or on contiguous or adjacent sites and which are owned or operated by the same person (or by any person which controls, is controlled by, or under common control with, such person)." 42 U.S.C. § 11049 (4).

[3] EPA has published forms that help industry do their reporting of TRI data. http: //www2.epa.gov/toxics-release-inventory-tri-program.

[4] Toxic Release Inventory Program Division, Office of Environmental Information, US Environmental Protection Agency 1-2 (May, 2003), available at http: //www2.epa.gov/sites/production/files/documents/2003_TRI_Data_Uses_report.pdf.

[5] Mary Graham, Democracy by Disclosure: The Rise of Technopopulism 21 (2002).

环境民主——新的挑战

彼得·赫尔曼*

导言

或许我们会认为目前所存在的一个基本问题是，可持续性这个概念缺乏一个清晰的定义。然而实际情况却并不是缺乏一个定义本身，而是我们如何解释它。爱因斯坦曾说：环境是除了我之外的所有的东西（Einstein, 1931）。在此定义下，我们可以继而探讨环境问题的多个维度。以下是必不可少的当下语境，即有关界定环境民主的讨论：

它具有组织化的性质；

它也涉及那些人类改造的环境；

它可能与我们周围的人有关，同时它也是社会事务；

并且也包括我们作为人类主体与环境的关系。

在这些讨论中，一个重要的议题是如何区分环境自身和环境议题，把它们理解成为是一个环境概念或者是与这些议题相关的环境状况。

其最重要的是一个基本的解析和划分过程。就我们目前研究的议题来说，以下是具有决定性讨论的维度：

把建构人的存在作为中心，从而建立《圣经》所提供的规则。正如在《创世纪》中所说，充满这土地并征服它（基于基督教和开尔文教的原则）；

把人和人的存在从自然界中区分开来（基于基督教和开尔文教的原则）；

---

*彼得·赫尔曼，东芬兰大学社会科学学院教授，布达佩斯考文纽斯大学世界经济学院荣誉副教授。本文原文为英文，中译者为吕浩然（浙江大学公共管理学院）。

把个人和社会相关联（基于西方启蒙思想和部分地源于随后的利他主义）；

把经济和经济过程从社会和社会过程区分开来（基于卡尔·波兰尼的分析）。

在此过程中我们也要强调一些环境民主概念的基本缺点。这些批评要基于里约热内卢全球峰会和《奥胡斯公约》提供的关于理解这一概念的基本观点。在此背景中本文将讨论这些基本原则和概念，特别是强调从社会质量的视角来解决人们生产和再生产的基本问题。最后，通过以上讨论进一步提出政策建议，同时也促使我们进一步反思社会质量理论。由此，在研究的结尾我们将从社会责任的角度确定可持续性这一概念，而不仅仅是回答社会生态公民的民主问题。

环境民主是一个新词。它正在发展并已经进入讨论的议程。这一发展的原因由以下因素之间的联系构成：

环境和民主这两个核心概念都缺乏充分的界定；

这些因素的组合使得每个因素的定义更加困难；

然而，环境问题越来越严重且不可回避。实际上，这些问题已经存在了几个世纪，其严重性也在不断增强；

最后，任何环境问题讨论的背后都存在着合法性的挑战。

要回应这些问题，1998年6月25日在奥胡斯提出的《在环境问题上获取信息、公众参与决策和诉诸司法公约》，给出了一个高度简化的回答。其核心内容强调了三个支柱，即"获取环境信息"（第四条）、在特别的活动领域中，公众参与和决策过程（第六条）和"诉诸司法"（第九条）。这三个支柱有着两个决定性因素结构。一是环境的定义。1992年签署的《里约环境与发展宣言》，其核心原则是"人类处在关系持续发展的中心。他们有权和大自然协调一致从事健康的、创造财富的生活"（联合国，1992）。当然，这种观点是以人类为中心。第二个重要问题是，民主的原则十分有限。任何机会的缺乏带来的都是对相关权利的限制。换句话说，当人们强调过程的权利，它的后果的权力就常常被限制了。

批判性讨论

在前面的论述中我们已经提到了一些批判性观点，下面是对这些概念的进一步解释和说明：人类中心主义是定义环境民主问题的主要限制。它不允许从根本上质疑人类的所谓优越性，并且强调一种功用主义的视角。这意味着对环境的使用是以人类所用为原则主义；它并不能克服人类中心主义的缺点。这就意味着环境危机事实上是人类的危机。

以功用主义的视角来看待环境和民主问题，我们强调环境需要保护，但它本身并不被认为是人类的共同生产者。这就形成一个双元结构，即把环境问题看作是一方面，把人和人的行为问题看作是另一方面。在此我们可以引用一个小的例子：根据《地球正义》一书，在这个世界上的许多国家的宪法都确认清洁和保持健康环境或相关的权利。在全世界约 193 个国家和地区中有 117 个国家的宪法确立了保护环境或者自然资源的法律（地球正义，2005）。这种情况不仅仅解释了环境政策的工具性，也反映了在新千年的开始之际，人类活动已经造成严重的环境问题。反过来，严重的环境问题又会伤害人类自身。从积极的角度说，一个清洁、健康的环境是实现基本人权的必要条件（同上）。

然而，尽管分析仍有待深入，人的权利也只能够通过十分有限的个案反映出来（特别是在一些拉美国家）。但没有人会否认保护环境要采取的手段的重要性，不过必须消除其局限性。那种有害的控制在某种意义上是由其生产体制和不当的生活方式所造成的。由此，Jason W. More 批评说：从方法论上说社会和自然尽管是不同的事物它们是共生的，共生是一种将自然和社会连接起来的方式，从而把现代化与自然改变成在自然的中现代化。

以这一理解为基础，特别是国际相关的政治文件中，只有很少的定义是指更广泛的环境的意义，例如国际公正法庭也认为它认识到环境并不是抽象的，它代表了生活空间、生活质量和人类健康等问题，涉及未来的世代的利益（国际法院，1996）。

强化社会质量的思考

去批评现在的经济体系没有考虑环境成本而强调增长导向是相对容易的事情，虽然这样的解释可以从不同的视角展开。将责任归咎于个人的贪婪也是相对容易的事情，或者强调对于非物质价值的忽视在现有的经济体系综合导致了功利主义的行为导向。Brand und Wissen 评论说：我们常常假设在历史发展的某些阶段在生产和消费之间建立起平衡的理念。换句话说，人民广泛地接受和制度化了的追求生产模式已经十分深地扎根在了人们的日常生活行为和国家的保护中，并且得到导向的进步主义概念的支持。它与某些社会观念的进步现象相关：计算机会更强大和食物也会更便宜，无论在人们所处的社会和生态条件会如何（Brand and Wissen，2012）。

然而，这些讨论没有对这些进步的机制给予足够强调。为此，在本文中我们强调要把这两方面议题要放到一起，一是政治经济学问题，二是可持续行发展的定义。这两者都与社会质量情况相关。

I. 长周期理论

以此为起点，我们可以触及 Kondratieff 和 the bol'shie tsiklys，因为有关这个议题的最有代表性的讨论都与这本书相关。它力图解释导致这一现象的两方面的原因：首先，确定的主要问题是现有的生产方式与其对象的转化。它使我们能够反思社会质量。我们研究这一视角能够使我们探索社会结构的意义和政治经济体系的转化。同时，它也涉及更广泛的变化。由于这种视角在一定程度上是描述性的，它只能在严格意义上去界定社会性的定义。人作为行动者处在与之建构的自然环境之间的互动中。作为客观的研究，它涉及人文因素，生产性和再生产性的联系及进行交互作用（van der Maesen and Walker，2012）。

要理解生产和再生产关系，就必须从有说服力的解释方法开始，将下列方法和范式相结合：

• 延伸理解法国的监管理论，着重积累制度和监管模式，从具体的生活制度和生活方式——部分地讨论资本主义的种类，但这些辩论在两个方面简化：通常只关注微观层面和极其有限的变异程度。

• 世界体系理论使社会分工在全球范围内展开。

• 区别生产过程中的不同阶段，即严格意义上的生产、消费、分配和交换，这可以看作物质积累制度和生活制度。

• 波兰尼的方法区分了重要的主体者相互作用、分布和市场机制转向，讨论经济活动嵌入，讨论生活规律和模式的问题。

II. 结构和关联性——新的社会学方法

以社会质量的角度思考，当今社会和社会发展的两大挑战是个人主义和功利主义结构化的问题。这些都不是由"不受欢迎"规范性所造成的,其机制特点在于社会为它定义的条件以及宪法和法律规范在社区和机构之间的作用。

这意味着我们必须找一个符合两种标准的可持续发展定义:首先它具有超越常用参考环境的外部性;其次,我们必须确保社会人与环境的共同生产能够反映出可持续发展的定义。特别是第二点已经在前文介绍过,重要的是在发展中国家的具体政策。早期的可持续发展的定义将看它看作是"一个无生命的交互集成与生活实体之间的动态平衡状态,并作用在弹性系统的边界。这些生灵包括人类行为的复杂性。这些复杂性可导致可持续性或不可持续性的社会关系以及是否可持续条件下的弹性边界(IASQ,2012)。

因此,我们认为建立一个明确的和紧密的社会质量体系是可行的,它将着眼于环境民主问题并作为其参考。

权利不一定要以个人感觉作为物质的实体,它是一个社会的组成部分,这意味着权利被授予"人与人之间的相互作用或人与自然环境的相互作用"的意义,人与环境本身的承载权利主要涉及社区和机构之间的张力。

当然,权利和义务必须作为一个额外的维度来看,特别是处理规范性因素时。

权利(和义务)也需要在一定的条件因素的作用下再次讨论,要强调人类的规模和人文环境的关系。

最后要强调生产力和新生产方式的关系。一些理由可以看做是物质与非物质的增长。然而,缺乏辩护是由经济增长和缺乏后续监督所引起的。在很大程度上这是一个问题的构成要素。

环境民主是强调社会质量的语境,环境是社会的环境。这个定义已经显示出环境的属性,也就是说"人与人之间的相互作用和人类活动与自然环境的关系"。它需要承认环境的权利,这意味着要看环境相互作用的部分。这意味着经济增长不是物质和非物质的同时增长。因此,我们应该更加重视增长的质量。

促进城市可持续发展的途径

虽然对这一问题的讨论首先是理论性问题,但仍是政策制订所必须考虑到的具体事宜。在此,我们可以谈论 2007 年德国担任欧盟轮值主席国时制订的《关于可持续欧洲城市发展莱比锡宪章》开始,这一文件强调加强当地经济,并提出了需求。"创造更好的条件和手段加强地方经济和当地的劳动力市场,特别是促进社会经济及为公民提供友好服务"(莱比锡宪章,2007)。

这鼓励了生产方式的转变。这也重审了另一个被广泛讨论的问题:解析问题。主要方向通常是以技术的维度——以下是联合国环境规划署实例(UNEP,2011)。

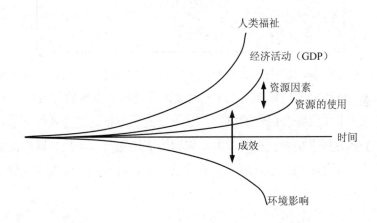

解析的两个方面

然而,这些讨论基本完全忽略了基本事实、经济活动、资源利用和被隔离的环境影响等主要问题。这些都是离开了人类福祉以外的思考。换句话说,它不能剥离经济本身,因为经济形态是社会发展的长期过程。

治理也是环境民主面临的主要挑战。它不仅仅是由于要改变城市和社会之间增加的距离的模式(见 Beguinot et al.,2011),更重要的是社会经济与政治之间分离的程度。需要进一步讨论的是,构建一个清晰且可操作的社会生态公民形态已被确认为环境民主的核心。我们可以从下面开始:

讨论中也必须提到一些必须解决的问题(Bell,2006;Dobson,2007)。首先关注的是常见的公民权受限,这是在参照国家中发现的。生态问题主要是全球性的,当然地方性的生态问题也存在。这就要求我们寻找加强局部生态建设的方法,然而同时牢固地连接这个全球性的维度。以国家作为参考点来定义好的治理,这可能会由社会生态可持续发展所取代。

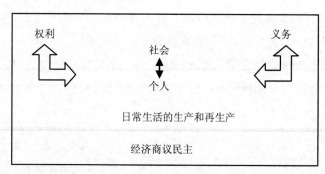

第二，系统联系的解决方法在处理公私问题上不是长久的。正如家庭经济出现超出家庭经济边界的现象表明（这一度被认为是私人事务的代表），它的私人性不能保持。这不仅意味着民营的概念，对于以盈利为目的的企业需要从根本上改变和适应社会的基本机制。在此，需要强调协商民主与"当地社区"的本质联系及监管模式。只有如此，才有可能克服利己主义取向和单一维度全球联盟霸权的出现。

当前的挑战

在本文的最后，笔者将提出四个可用于推动社会环境可持续发展相关辩论的工具/机制，它被视为主要的威胁和潜在的障碍。

公民教育。实际上是唯一可以长远有效同时作为工具的机制。它会改变大众态度，特别是通过"实践教学"，即让公民积极参与到政策制定的过程中并以此作为意识的连接。

公民社区替代服务导向。可以以不同方式应用于不同的地方——学校、民间社会组织和法定机构。重点是了解共同所有权并允许实时控制。

参与预算。考虑这是个重大的决策，不仅仅允许执行政治影响和经济层面建立直接连接，它也允许建立经济政策与日常生活之间的直接联系。

公共交通的使用空间。市民社会的发展与参与预算可以视为瞬间，但允许共享空间的发展则是长久的。连接公共交通和利用空间是一个非常具体的全球化的局部问题。

"谦虚"。持续的人类学的挑战是"认识"这一问题的边界。

改变态度和行为。客观力量–离开一旁的优势，我们面临的挑战是解决态度（和行为）。其实发展的公民作为参与审议上述方面可以作为工具，发展的态度从一开始就构成个体意识的社会的人。

自我限制（如工作条件）。需要注意的是，机制和态度是与自我限制和谦虚联系在一起的；但是，有两个其他角度。一是仍然存在并需要努力至最低的角度。一些情况下取决于可持续性资源，比如水。由私人使用过后遭到破坏，最终导致破坏可持续发展。在其他情况下，最小规定往往需要扩大生产，意味着增加能源使用的效率、合理开发原材料等。另一点是理性的局限，其中一个例子是在能源消耗量大的机器上使用机械工具，从而引起的污染排放等。另一个例子是流动性的问题，即我们应该如何流动？这应该意味着该如何利用？

平等、公正和强制性。虽然我们很容易地找到平等和正义等普世价值的有效性，但我们却很难具体界定它的程度。如果这样的普世价值能够被大多数人所同意，其应执行的所有——将会与 Rousseauean 问题有关。

这些都是非常现实并且严肃的问题。采用社会质量的方法可以得到简单答案，但它不以城市发展过程中的积累经验为基础，它是一种很有效的社会质量测量方法，也可以找到相关答案。它不仅看重生产过程中的系统性方法，也能进一步要求这一过程的自然和社会嵌入性识别属性。

Environmental Democracy – New Challenges

Peter Herrmann[*]

Introduction

One may assume that the ongoing problems about sustainability are the lack of a clear definition; however, we may say that the actual problem is not the lack of such a definition itself but the claim that we need such description. Einstein supposedly said "the environment is everything that isn't me" (Einstein, 1931). The debate over the definition invites us to imagine and accept the multiple dimensions of the environmental issues. At least the following are seen as essential for the present context, namely the discussion of environmental democracy:

- it is a matter of the organic nature;
- it is concerned with the inorganic, i.e. human-made surrounding;
- it is about the people around us, and with this a matter of the social itself;
- and not least it is about how we as human beings relate to this environment.

One important further moment at the outset is the necessity of distinguishing between the environment itself and environmental issues, understood as perception of and dealing with issues that concern the environment.

However, crucially important is the fundamental and widening process of disembedding and dichotomization. For the current subject, the following are the most decisive dimensions:

- the constitution of human existence as centre, thus establishing (and following)

---

[*]Peter Herrmann, professor at Department of Social Sciences, the University of Eastern Finland (Kuopio, Finland), and the honorary associate professor at Faculty of Economics, Corvinus University in Budapest.

the rule from the biblical order defined in Genesis 1/28 which states: "fill the earth and subdue it" (the roots being Christianity and capitalism);

• the subsequent separation of human beings and human existence from nature (the roots being Christian and Cartesian thinking);

• the juxtaposing of individuals and society (the roots being Western enlightenment and the [in part] subsequent utilitarianism);

• and finally the disembedding of the economy and economic processes from the social and societal processes (analyzed by Karl Polanyi).

Against this background we have to highlight some fundamental shortcomings of the concept of environmental democracy. Such critique will be based in a very short presentation of the main lines of the understanding as they are defined by the Summit in Rio and the Aarhus Declaration. On this background the understanding of some basic principles and concepts will be outlined, especially aiming on developing the social quality approach further by highlighting the aspect of production and reproduction of everyday's life. Finally, by way of conclusion this will allow some policy recommendations as challenge of mainstream thinking, but also in order to develop social quality thinking further. Sure, at the end we still face the question of defining sustainability in the light of social quality thinking, not least by asking for democracy in the era of socio-ecological citizenship.

Environmental democracy has the character of a neologism that has already been a torso since it entered the agenda. The reason is simple, consisting of the connection between the following factors:

• both core terms, environment and democracy, are ill-defined;

• their combination actually makes a definition of each of them even more difficult;

• nevertheless, the environmental problems and their increasing severity are undeniable, and also the perception of these problems, some of them existing in fact already for centuries, is increasing;

• and finally any solution is confronted with the challenge of legitimacy not least in the light of fundamental changes that are standing at the end of any attempt dealing with environmental issues.

A highly simplified answer is given by the Convention on Access to Information, Public Participation in Decision-Making and Access to Justice in Environmental Matters done at Aarhus, Denmark, on 25 June 1998 (Convention, 1998). At the core there are three pillars outlined in the title, namely "Access to Environmental Information" (article 4), "Public Participation in Decisions on Specific Activities' (article 6) and 'Access to Justice" (article 9). These three pillars are complemented by two decisive structural moments. The first is the underlying definition of environment. As the reference of the convention is the Rio Declaration from 1992, the core principle has to be seen in the claim that

Human beings are at the centre of concerns for sustainable development.
They are entitled to a healthy and productive life in harmony with nature.
(United Nations, 1992)

This, of course, maintains an anthropocentric worldview.

The second important issue is that the principle of democracy is highly limited; one point being the lack of opportunities to decide what is actually relevant. In other words, the substantive rights are widely limited whereas the procedural rights are emphasized.

A Critical Reproach

Some critical points had been already mentioned, immediately in connection with the presentation of the concept. The following have to be seen as fundamental:

Anthropocentrism defines a principal limitation to the question of environmental democracy. It does not allow to fundamentally questioning the supposed superiority of human beings, which means in turn that the utilitarian approach towards the environment is set in stone as long as this anthropocentrism is not overcome. This means the other way round that we have to emphasize again the fact: The environmental crisis is in actual fact an anthropological crisis.

However, we have to be careful, as this anthropological crisis is not simply about humankind as such – understood as a homogenous agency. Instead, anthropocentrism veils the fact that we are dealing with a historically specific relationship not only between humankind and "external nature". In other words, centre stage is the relationship between human beings and the historically given understanding of what "good life" and "good society" is about. In other words, we are dealing with an understanding that fosters an instrumental understanding of resources: understanding human grandeur as being limitless means that the subordinated environment has to serf this strive to unlimited growth.

Without developing this further, this leads in consequence to maintaining a utilitarian approach to both, environment and democracy: the environment needs to be protected but is itself not considered to be part of a co-productive setting. In other words, this approach maintains the dualist structure that sees relevant issues as environmental problems, not as problems of humans and their (mis-) behavior. Here it is only possible to give a short hint: according to Earthjustice "Numerous constitutions of the nations of the world guarantee a right to a clean and healthy environment or a related right. Of the approximately 193 countries of the world, there are now 117 whose national constitutions mention the protection of the environment or natural resources." (Earthjustice, 2005). – This stands not least in connection with an instrumentalist interpretation of the environment: At the beginning of this third millennium, there should be no doubt that human activities can cause serious environmental problems, or that those problems, in turn, often result in grave harm to human beings. Put positively, a clean and healthy environment is essential to the realization of fundamental human rights. (ibid.)

However, though a thorough analysis remains to be undertaken, it is well known that the right of nature remains by and large limited to a few cases, being especially some Latin America countries. Without denying the importance of protective measures, the limitation should be clear: the approach is geared to damage control, leaving by and large a more fundamental questioning of the mode of production and the related life regimes out of consideration. Consequently, Jason W. More criticizes rightly an ontological agreement: the "co-production of society with nature", as if these were two independent entities. While co-production is the right way to put it, its association with a Nature/Society vocabulary short-circuits the effort to move from modernity with nature towards modernity-in-nature. (More, 2014a)

With this consideration in mind, it is remarkable that – especially when it comes to internationally relevant political documents – only very few definitions refer to the environment in a wider sense, for instance the *International Court of Justice contending* that it also recognizes that the environment is not an abstraction but represents the living space, the quality of life and the very health of human beings, including generations unborn (International Court of Justice, 1996).

Consolidating Social Quality Thinking

It is somewhat easy – and surely necessary – to criticize the current economic system for being negligibly considering the environmental costs of a one sided growth orientation – although such an interpretation may surely start from many different viewpoints. It is surely also easy – and perhaps even more so – to blame individuals' greediness. And finally it is of course relatively easy, necessary and still also misleading to emphasize the neglect of immaterial values by the current economic growth dynamic and the subsequent utilitarian orientation. Brand and Wissen contend

> *the important point for our argument is the assumption that in certain historical phases, and building on a coherence between norms of production and of consumption, a hegemonic – or in other words – broadly accepted and institutionally secured mode of living can emerge that is deeply rooted in the everyday practices of people and safeguarded by the state, and it is further associated with certain concepts of progress: computers must be ever more powerful and food ever cheaper – regardless of the social or ecological conditions under which they are produced. (Brand and Wissen, 2012)*

However, this remains so far problematic as it does not sufficiently address the underlying mechanisms. In the following two strands will be drawn together, the first is concerned with political economy, the second a definition of sustainability in the light of social quality will be approached.

I. Theory of Long Waves

An initial point of reference can be seen in the work of Kondratieff and the *bol'shie tsiklys*, i.e. major cycles and the respective discussion that followed his proposals. The aim of the specific interpretation I want to suggest is twofold. Form here we can determine major shifts of the current modes of production and their objective dimension. As such it also allows pushing the social quality thinking further, working on the meaning of the structure, structure and change of political economy (as reality of the political economy and the analysis alike). Instead of simply referring to the broad changes – actually Kondratieff himself claimed that his approach would be by and large descriptive, not aiming on providing an analysis in the strict sense – it is necessary to draw on other paradigms – the reason for this can be found in the definition of the social itself, which refers to the social as "an outcome of the interaction between people (constituted as actors) and their constructed and natural environment. Its subject matter refers to people's interrelated productive and reproductive relationships." (*van der Maesen and Walker*, 2012)

For developing a sound understanding of these productive and reproductive relationships it is fruitful to interpret the approach to major cycles with the help of a systematic combination of the following approaches and paradigms:

• an extended understanding of the French theory of regulation, looking at the accumulation regime and mode of regulation but exploring from here specific life regimes and modes of life – part of this has to take the debate on the varieties of capitalism into account, though those debates are reductionist in two ways: (a)they are commonly only concerned with the micro level and(b)the proposed degree of variation is in fact extremely limited.

• World Systems Theory, allowing to understand the social division of labor on a global level.

• the distinction and drawing together of the different moments of the process of production, namely production in the strict sense, consumption, distribution and exchange – this can be seen as matter of accumulation regime and life regime.

• and not least is Karl Polanyi's approach of importance, distinguishing importantly between reciprocity, distribution and market as mechanisms of steering and – not less important – discussing the embeddedness of economic activities – this can be seen as matter of mode of regulation and mode of living.

II. Structuration and Relationality – New Sociological Approaches

In the perspective of social quality thinking two major challenges of today's societies and societal developments are individualism and utilitarianism as principles and driving forces of structuration. These are not "unwelcome" normative strives, but mechanism that characterize the social as it is defined as interplay of conditional, constitutional and normative factors along the two tensional lines between communities and institutions on the one hand and biographical and societal development on the other hand.

This means that we have to look for a definition of sustainability that fulfils two criteria: first it has to go beyond the commonly used reference to environment as by and large undefined externality that needs to be protected; second, we have to make sure that the aspect of co-production between social human beings and the environment is reflected in the definition of sustainability. The second point has already been presented in the previous section – it is important to keep this in mind when it comes to developing concrete policies. An earlier definition of sustainability had been given, seeing it as

> *"a state of dynamic equilibrium between the entire interactive ensemble of non-living and living entities, functioning within the boundaries of a resilient system". These living entities include the complexities of human actions. These complexities can cause either sustainable or unsustainable societal relationships as well as sustainable or unsustainable conditions concerning the resilience boundaries.* (IASQ, 2012)

Accordingly, we consider it be possible to establish an explicit and thus closer link to the social quality architecture, having in mind environmental democracy as an additional point of reference.

- Rights are not bound to agency in the sense of individual actors but seen as matter of entities that are a constitutive part of the social – this means that rights are granted to "the interaction between people (constituted as actors) and their constructed and natural environment": both interaction, people and the environment itself are bearers of rights.–This concerns mainly the tension between communities and institutions.

- "Identifiable interest", legitimately constitutive for defining and granting rights, are not self-referential; instead, they are defined by the relationality which reflects the definition of the social, speaking of "people's interrelated productive and reproductive relationships" – This mainly concerns the tension between biographical and societal development.

- Rights and obligations have to be discussed in this light and responsibility as an additional dimension has to be looked at, especially dealing with the normative factors of the architecture.

- Rights (and obligations) also need to be discussed again in the light of the conditional factors, which themselves have to be defined (or redefined) by way of the relational dimension of (re-) production. This allows emphasizing the productive dimension of human beings, humanity and the environment.

- Finally it is required to look at the productive forces and a new take by way emphasizing relationality. With some justification this can be seen as matter of material and non-material growth. However, the limit of the justification is given by (a) maintaining the centrality of growth and (b) the subsequent lack of questioning the character of production. This is largely a matter of the constitutional factors.

In short, environmental democracy in the context of social quality thinking first requires highlighting the fact that the environment is co-producer of the social – the definition of the social already points into this direction, speaking of "the interaction between people (constituted as actors) and their constructed and natural environment". However, secondly it requires acknowledging the right of the environment, which implies to see the environment also as active part of the interaction. Third, it means that the question of growth is not about juxtaposing material and non-material growth. Instead, the question is not even concerned with the quality of growth. Instead it has to be centered on the quality of production. Finally this leads to an orientation on the most fundamental question: the re-embedding of production into the entirety of social and societal relationships.

Furthering the Way to Sustainable Urban Development

Though this is first and foremost highly theoretical, it is relevant when it comes to concrete matters of policy making. We can start for instance with the reference to the *Leipzig Charter on Sustainable European Cities* which was brought forward in 2007 under the German EU-presidency. It speaks of strengthening local economies and importantly it also mentions the need.

> *to create better conditions and instruments to strengthen the local economy and thus the local labour markets, in particular by promoting the social economy and providing citizen-friendly services.* (Leipzig Charter, 2007)

In other words, it also encourages the change of the way of production. It should be added that the concept of commons requires to be revisited and utilized for further development. This highlights another problem that is widely discussed: the issue of decoupling. The main orientation is usually seeing a technical dimension of it – in the following for instance presented by the UNEP. (UNEP, 2011)

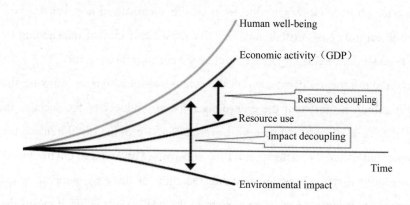

Two Aspects of Decoupling

However, this completely neglects the main problematique, namely the underlying fact that economic activities, resource use and environmental impact are seen in isolation, leaving the human well-being outside the consideration. In other words, it does not look at the disembedded economy itself – the economy, we may term provocatively, is seen as a-social issue.

Governance – and this is also the major challenge in terms of environmental democracy – is not simply about changing the pattern of the increasing distance between city and society (*see Beguinot et al., 2011*). More important is the distance between the social and the economy and the separation of politics from both. Further discussions are needed, striving for a clear and operational understanding of socio-ecological citizenship which has to be developed as core of environmental democracy. We may start from the following:

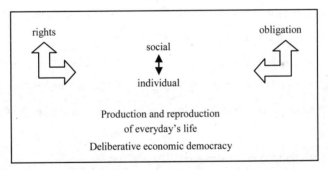

In this discussion, some general points have to be addressed – here only mentioned as questions (Bell, 2006; Dobson, 2007). The first is concerned with the well-known common limitation of citizenship, founded in the maintained reference to the (nation) state. Ecological problems are mainly global and also local. This requires looking for ways of strengthening the local dimension, however at the same time firmly linking this to the global dimension. As much as the nation-state served as point of reference for defining a common good, this may have to be replaced by socio-ecological sustainability.

A second and systematically linked moment deals with the public-private which as such cannot be maintained. As much as *oikonomia* emerged beyond its original border of a household economy (as such it had been considered as "private affair"), its private character cannot be maintained. This means not least that also the concept of private, for profit business needs to be fundamentally transformed and adapted to some fundamental mechanisms of socialisation (*see in this context for instance Sukhdev*). Furthermore this requires emphasizing the importance of deliberative democracy that intrinsically links aspects of "local communities" with the dimension of the mode of regulation. With this link, it can be possible to overcome NIMBY-orientations and also the hegemony of one-dimensional global alliances.

Immediate Challenges Ahead

In the final step of this presentation, I want to pose four major tools or mechanisms that may be applied for pushing the relevant debates on socio-environmental sustainability forward and four major challenges that have to be seen as major potential threats or hindrances.

- *Citizenship Education* – it is surely only a mechanism that can work in a very long term perspective, at the same time it can be used as instrument that possibly leads to a long term change of attitudes, especially if teaching takes place by way of "practice teaching", i.e. developing the awareness in connection with active involvement of pupils into processes of policy making.

- *"Citizenry of Communities" instead of Service Orientation* – it can be developed and applied in different ways and different settings – reaching from schools, civil society organizations, and statutory organizations. The crucial point is to develop an understanding and reality of co-ownership that goes beyond lip services, allowing real control.

- *Participative Budgeting* – considering this can be a major step when it comes to decision making, not least allowing establishing an immediate link between executing political influence and part of the economic dimension, it also allows establishing an immediate link between economic policies and everyday life.

• *Public Transport and Use of Space as Part of Commons* – both, the development of Citizenry and (in connection with) Participative Budgeting can be considered as moments that allow the development of commons. Linking public transport and use of space in this context may enforce linking local and global issues in very concrete terms.

• *"Modesty"* – an ongoing and anthropological[①] challenge is that of "recognizing sufficiency" which a matter of accepting borders and limits is also.

• *Changing Attitudes and Behavior* – leaving the dominance of the objective forces aside, the challenge remains to address attitudes (and behavior). Actually the aspect of developing citizenry as matter of participatory deliberation mentioned above can serve as tool, developing attitudes from the outset as constitutive for individuals acting consciously as social beings.

• *"Limits to Self-Limitation"* (e.g. working conditions, "luxury") – one point should never be forgotten: considering mechanisms and attitudes that are concerned with self-limitation and modesty while living on a level of high standards, is one thing; however, there are two other perspectives. The one is about a still existing need to strive even for the minimum of provisions. In several cases this is actually depends on sustainability – we may think about the right to (access to) water for all, being undermined by private control, the latter undermining sustainable development. In other cases this however is not the case: provision of the minimum often requires expansion of production, meaning increased energy use, exploitation of raw materials etc. The other point is about the reasonableness of limitations – a very crude example is the use of mechanical tools for heavy labor versus applying easy-to-use machines which consume energy, cause emission etc..Another example is concerned with mobility, i.e. the questions *How mobile should we be?* and *Which means should we use?*

• *Equality, Justice and Enforceability* – though we will easily find a general

---

① To be sure, anthropological here has always to been seen in its historical context, not being matter of humankind in general.

agreement on the validity of equality and justice as universal values, we will surely find it difficult to define such values in concrete terms. An additional moment has to look at the question if such general values, if agreed by the majority, should be enforced – the old Rousseauean question of relating the will of all with the general will.

These are very practical and very serious questions at the same time. It would be presumptuous to say that the social quality approach has simple answers to them. But not in the least based on the experience gathered during the projects on urban development, it is promising to consider social quality offering a useful methodology that allows finding relevant answers. It allows a systemic approach that does not solely and primarily look at the processes of production. It goes further by also asking for recognition of the social and natural embeddedness of this process.

References

[1] Beguinot, C. et al. 2011. The City Crises: For a UN World Conference and for a UN Resolution. The Priority of the XXI Century. Napoli. Giannini, 323 Eighth Tome Series of Urban Studies.

[2] Brand, U. and Wissen, M. 2012. Global Environmental Politics and the Imperial Mode of Living: Articulations of State–Capital Relations in the Multiple Crisis; in: Globalizations, 9: 4, 547-560; http: //dx.doi.org/10.1080/14747731.2012.699928). Accessed on?

[3] Convention on Access to Information, Public Participation in Decision-Making and Access to Justice in Environmental Matters done at Aarhus, Denmark, on 25 June 1998（http: //www.unece.org/env/pp/treatytext.html; English version at http: //www.unece.org/fileadmin/DAM/env/pp/documents/cep43e.pdf; accessed 15/06/2014.

[4] Dobson, A. 2007. Environmental Citizenship: Towards Sustainable Development; in: Sustainable Development. Sust. Dev. 15, 276–285（2007）; Published online in Wiley InterScience（www.interscience.wiley.com）DOI: 10.1002/sd.344.

[5] Dobson, A. and Bell, D. . 2006.: Introduction; in: Dobson, Andrew/Bell, Derek（eds.）: Environmental Citizenship; Cambridge, Massachusetts/London, England: The MIT Press.

[6] Earthjustice. 2005. Environmental Rights Report. Human Rights and the Environment;

Materials for the 61st Session of the United Nations Commission on Human Rights. Geneva. March 14 – April 22. Oakland. .

[7] Einstein, A. 1931. Cosmic Religion: With Other Opinions and Aphorisms p. 97; also in Transformation: Arts, Communication, Environment. 1950 by Holtzman, H. p. 138; http://en.wikiquote.org/wiki/Albert_Einstein. accessed on 15/06/2014.

[8] IASQ. 2012. Final draft: Development toward sustainability. The need for a Comprehensive Conceptual and Methodological. Framework for new politics and policies: A social quality perspective. A contribution to the Rio+20 Conference on the sustainability of human existence on Earth. A Report presented by the International Association on Social Quality (i.s.n.): 3; http://www.socialquality.org/site/ima/FRStudyGroupSustainability19June2012.pdf accessed on 14/06/2014.

[9] International Court of Justice. 1996. Reports of Judgments, Advisory Opinions and Orders: Legality of the Threat or Use of Nuclear Weapons. Advisory Opinion of 8 July 1996; http://www.icj-cij.org/docket/files/95/7495.pdf. Accessed on 15/06/2014.

[10] Leipzig Charter on Sustainable European Cities: 6 - http://ec.europa.eu/regional_policy/archive/themes/urban/leipzig_charter.pdf.Accessed on16/06/2014

[11] More, J. and W. March. 2014（a）: The Capitalocene. Part I: On the Nature and Origins of Our Ecological Crisis; March 2014. http://www.jasonwmoore.com/uploads/The_Capitalocene__Part_I__June_2014.pdf accessed on 15/06/2014.

[12] More, Jason W., March 2014（b）: The Capitalocene. Part II: Abstract Social Nature and the Limits to Capital; March 2014. Minor revisions June 2014. http://www.jasonwmoore.com/uploads/The_Capitalocene___Part_II__June_2014.pdf accessed on15/06/2014.

[13] More, J. and W. March. 2014: The Capitalocene. Part I: On the Nature and Origins of Our Ecological Crisis.; March 2014: 6. http://www.jasonwmoore.com/uploads/ The_Capitalocene__Part_I__June_2014.accessed on - 15/06/2014.

[14] Sukhdev, P. Corporation 2020. Transforming Business for Tomorrows World; Washington: Island Press.

[15] UNEP. 2011. Decoupling natural resource use and environmental impacts from economic growth, A Report of the Working. Group on Decoupling to the International Resource Panel.

Fischer-Kowalski, M., Swilling, M., von Weizsäcker, E.U., Ren, Y., Moriguchi, Y., Crane, W., Krausmann, F., Eisenmenger, N., Giljum, S., Hennicke, P., Romero Lankao, P., Siriban Manalang, A., Sewerin, S.; http://www.unep.org/resourcepanel/decoupling/files/pdf/decoupling_report_english.pdf - 15/06/2014

[16] United Nations, General Assembly. August 12th, 1992. Rio Declaration on Environment and Development; Principle 1; http://www.un.org/documents/ga/conf151/aconf15126-1annex1.htm. Accessed on 15/06/2014.

[17] van der Maesen, L. .J.G., assisted by Herrmann, P. 2012. Welfare Arrangements, Sustainable Urban Development, and New forms of Governance: the current 'demonstration project' of the City of The Hague as example Plans for the start of comparative urban studies between The Hague, Sheffield and Hangzhou as point of departure for the European GOSUD-project; Draft Working-paper nr.8, The Hague, EFSQ.

[18] van der Maesen, L. J.G., November 20th. 2012: Working-paper no 9: Elaboration of a Lecture on the Orientation, Strategies and Model [or Experiences] of the City of Hangzhou [Zhejiang province of mainland China], from a comparative perspective; working papers at http://socialquality.eu/. accessed on 20/09/2013.

[19] van der Maesen, L.J.G. and Walker, A.(eds.). 2012. Social Quality. From Theory to Indicators: Basingstoke. Palgrave Macmillan/.

[20] van der Maesen, L.J.G. and Walker, A. . 2012. Social Quality and Sustainability; in: van der Maesen, Laurent J.G./Walker, Alan (eds.): Social Quality. From Theory to Indicator. Basingstoke. Palgrave Macmillan. pp. 250-274.

后 记

本书汇集了有关环境保护公众参与的9篇论文,着重论述这方面的国际经验。这些论文是为期30个月的中欧环境治理项目浙江公众参与地方伙伴项目——嘉兴模式中的公众参与环境治理及其在浙江的可推广性的研究成果之一。在项目执行期间中外学者进行交流和相互学习,共同探讨与环境公共参与相关的重要问题,并举行了三次国际学术会议进行研讨,在相关问题上达成了一定的共识。本书所包含的论文大多已在这些会议上发表。其中,沃斯的论文已在《国际社会质量期刊》上发表。这些研究成果可以为中国学者和相关机构了解欧盟在环境治理方面的经验提供帮助,也加深了对一些理论问题的讨论。

本书用中英文对照的形式出版,英文为原文,中文译者来自浙江大学公共管理学院、浙江大学外国语学院以及香港教育学院。在文章翻译成文后我们对译稿进行了一些编辑工作,在一些地方采取意译的方式,因而译文与原文未必一一对应。在论文的选择中我们以欧洲经验为主,辅以个别非欧洲国家的相关经验作为补充。对于本书的出版,我们要感谢欧盟对本研究项目的支持以及介入该项目研究的中外学者。我们衷心期待这一论著能够为中国读者了解环境保护公众参与的国际经验有所帮助。

林卡

2015年4月3日

致　谢

中欧环境治理项目浙江公众参与地方伙伴项目——嘉兴模式中的公众参与环境治理及其在浙江的可推广性，从 2012 年 9 月实施以来，历经 30 余月，成果终于付诸出版。本项目由浙江省环境宣传教育中心牵头协调，浙江大学、浙江工商大学、英国格拉斯哥大学、英国利兹大学、荷兰国际社会质量协会作为合作伙伴，浙江省公共政策研究院、嘉兴市环保联合会、荷兰阿姆斯特丹自由大学、芬兰图尔库大学、瑞典斯德哥尔摩大学、英国谢菲尔德大学等作为协作单位，在环保部环境与经济政策研究中心指导下共同完成。感谢合作伙伴浙江大学公共管理学院林卡教授、浙江工商大学朱狄敏博士以及荷兰国际社会质量协会 Dr. Laurent J.G. van der Maesen、英国格拉斯哥大学 Dr. Neil Munro、英国利兹大学 Dr. Hinrich Voss 等在项目前期论证申请过程的合作，使我们的申请得以成功通过。在项目执行过程中，非常感谢项目团队其他中外伙伴们的精诚合作，他（她）们是荷兰国际社会质量协会 Dr.Kai Wang、英国格拉斯哥大学 Dr. Nai Rui Chng、瑞典乌普萨拉大学 Dr.Mattias Burell、李心悦博士、浙江大学光华法学院钱水苗教授、巩固副教授、嘉兴学院朱海伦副教授、浙江省社会科学院王釜灿副研究员、唐明良副研究员，以及北京大学阳平坚博士、杭州电子科技大学方建中副教授、浙江农林大学陈海嵩博士、浙江财经大学冯涛博士等都在与嘉兴模式相关的问题上进行了研究或作出贡献。同时也感谢荷兰国际社会质量协会 Ms.Helma Verklej，英国格拉斯哥大学 Dr.Bettina Bluemling，瑞典斯德哥尔摩国际水研究院 Mr.Frank Zhang 对项目国际交流以及论文编撰等的支持。感谢浙江大学公共管理学院朱浩、易龙飞、侯百谦、吕浩然、张育琴、付志宇、黄立婉等同学以及浙江大学北欧班的同学对项目调研和成果研究作出的贡献。

项目实施以来，我们也非常感谢浙江省有关领导、各级环保和相关部门人士的大力支持。他（她）们是十一届浙江省政协副主席、时任浙江省政府副省长陈加元、浙江省环保厅厅长方敏、浙江省环保厅副厅长卢春中，浙江省环保厅生态处处长马青骏、法规处处长叶俊、副处长陈云娟、时任法规处副处长徐妙芳（挂职）以及戴任重，建设项目管理处处长周碧河，科技与合作处处长李晓伟、副调研员邱中云，浙江省环境宣传教育中心主任潘林平、时任主任黄渭、时任副主任兼中国环境报浙江记者站站长赵晓，杭州市环保局总工程师沈海峰、邵煜琦、孟祥胜、林燕、余中平、裴斐，杭州市发展研究中心政治文明研究处处长孙颖，宁波市环保局巡视员吴建伟、谢小诚、陈晓众、王璐，温州市环保局副局长胡正武、副局长林曙、刘卓谞，湖州市环保局纪检组长黄一平、副局长余加伟、吴婧、邵波，嘉兴市环保局纪检组长邱再青、副局长朱伟强、杨建强、蔡华晨、王黎，海盐县人民政府县长章剑，绍兴市环保局副局长胡剑、沈慧惠，金华市环保局副局长李荣军、李期辉、金丹华，衢州市环保局副局长胡耀龙、王峰，台州市环保局党组成员马银来、李展明、丁华慧，舟山市环保局副局长於敏峰、邬国桢、黄最惠，丽水市环保局副局长胡晓红、陈锋。

此外，对于此项目的执行，我们要特别感谢中欧环境治理项目主管方对本项目的支持。他（她）们是欧盟驻华代表团项目主管黄雪菊女士、欧盟驻坦桑尼亚代表团 Ms.Maria Chiara Femiano（时任欧盟驻华代表团中欧环境治理项目官员），环保部政策法规司副司长别涛先生、环保部政策法规司法规处副处长李静云博士、商务部国际经贸关系司一等商务秘书罗煜女士、一等商务秘书陈红英女士，环保部环境与经济政策研究中心副主任、中欧环境治理项目主任原庆丹先生，环保部环境与经济政策研究中心战略室主任、中欧环境治理项目执行主任俞海博士，中欧环境治理项目欧方主任 Mr.Dimitri de Boer、项目专家 Mr.Richard Hardiman 以及项目办张会君、尚宏博、潘泓、李华蕾、刘梦星、Merav Cohen，他们的支持为本研究项目的顺利进行提供了基本的保障。另外，我们也感谢环保部宣教中心主任贾峰先生，环保部宣教司综合处赵莹处长，全国人大法工委行政法室刘海涛处长，环保部环境与经济政策研究中心首席专家王华博士，浙江省法制办立法二处童剑峰处长，浙江省公共政策研究院副院长蓝蔚青研究员以及河北省人大常委会城环工委白刚副主任和武晓雷处长等对项目成果推广交流的支持。

在项目的执行过程中，我们也要感谢各类社会组织及其相关人士参与到项目各类活动。这些社会组织的人士包括,嘉兴市环保联合会副秘书长万加华、阿里巴巴公益基金会杨方义、杭州市上城区艾绿环保科普服务中心郑元英、温州绿色水网公益中心白洪鲍,浙江工业大学生物与环境工程学院党委副书记马骏、分团委书记张烽,以及浙江外国语学院张英、李玲玲、祁晓茵、孙福汝、陈瑶、王怡文、苏洁和浙江工业大学潘娇娇、浙江大学宁波理工学院黄慧珍等同学的翻译志愿服务。同时,也感谢中欧环境治理项目地方伙伴关系项目其他兄弟机构的支持,他们是中华环保联合会、天津泰达低碳经济促进中心、北京市朝阳区公众环境研究中心、布莱克·史密斯环境研究所、德国国际合作机构、瑞典环境科学院等。

同时,我们也要感谢有关媒体伙伴对项目宣传的支持,他(她)们是中新社赵小燕,中国网焦梦,《中国环境报》陈媛媛、晏利扬,《世界环境》刘茜,《新环境》丁瑶瑶,《浙江日报》吴妙丽、陈文文,浙江在线潘杰、陈铖、孙璐、仲瑶卿,浙江之声余昌伟、吴轶颖,浙江交通之声王桔,浙江卫视周菁,浙江经视金昆,《青年时报》朱敏,《都市快报》王中亮,《浙江人大》林龙、《今日环境》许佐民等(以发稿时所在单位为准),还有其他很多媒体伙伴要感谢,在这不一一列出。

最后,还要特别感谢浙江省环境宣传教育中心项目助理王雯,以及寿颖慧、邵甜、杨贡江、吴涓、俞桂英、陆俊超、王希莉、沈焕壮、梁婧婧、任依依、金元森、蒋和平和实习生李猛等同事在项目推进过程中的努力、配合和支持。

项目执行主任　虞伟

2015 年 11 月